Space Technology Library

Volume 40

Editor-in-Chief
James R. Wertz, Microcosm, Inc., El Segundo, CA, USA

The *Space Technology Library* is a series of high-level research books treating a variety of important issues related to space missions. A wide range of space-related topics is covered starting from mission analysis and design, through a description of spacecraft structure to spacecraft attitude determination and control. A number of excellent volumes in the *Space Technology Library* were provided through the US Air Force Academy's Space Technology Series. The quality of the book series is guaranteed through the efforts of its managing editor and well-respected editorial board. Books in the *Space Technology Library* are sponsored by ESA, NASA and the United States Department of Defense.

More information about this series at https://link.springer.com/bookseries/6575

James M. Longuski • Felix R. Hoots
George E. Pollock IV

Introduction to Orbital Perturbations

 Springer

James M. Longuski
School of Aeronautics and Astronautics
Purdue University
Lafayette, IN, USA

Felix R. Hoots
The Aerospace Corporation
Colorado Springs, CO, USA

George E. Pollock IV
The Aerospace Corporation
Colorado Springs, CO, USA

Space Technology Library
ISBN 978-3-030-89760-4   ISBN 978-3-030-89758-1   (eBook)
https://doi.org/10.1007/978-3-030-89758-1

This Springer imprint is published by the registered company Springer Nature Switzerland AG
The registered company address is: Gewerbestrasse 11, 6330 Cham, Switzerland

Preface

The student or engineer who looks beyond a first course in Keplerian orbital mechanics soon encounters the field of orbital perturbations and Lagrange's planetary equations. In modern texts, these equations (for the time-varying orbital elements) are often presented without proof and with a pronouncement that the derivation is provided elsewhere—often in sources that are out of print. To make Lagrange's equations even more mysterious and perhaps discouraging, the student is warned that the planetary equations are found after "some labor." (And yes, this turns out to be an understatement!)

So how is the student to proceed? In my graduate education, one of my dear professors told his class, "You cannot use an equation that you have not—at least once in your life—derived." (I believe he excepted $F = ma$.) I went ahead and derived, in my doctoral thesis, an analytical theory of orbit decay (due to the perturbative effect of atmospheric drag), which made heavy use of Lagrange's planetary equations. But I did not derive the planetary equations!

It was not until I taught a graduate course in general orbital perturbations that I took on the task considered "de rigueur" by my professor.

The book you are reading provides details of the derivation—details not available elsewhere in the current print—of Lagrange's planetary equations and of the closely related and more general Gauss's variational equations. This work fills the gaps that have crept into the literature. Following closely the work of Fitzpatrick and McCuskey (and others listed in the bibliography), we take on the massive problem step by step—in which each step is elementary and rigorous. We encourage the reader to verify these simple steps to ensure their understanding. (The level of detail is left to the reader's conscience.)

In this text, we concentrate our effort on analytical work in which closed-form (usually approximate) solutions can be found for an orbit that is perturbed from its Keplerian trajectory. The method is called "variation of parameters" and it was first presented by (who else?) Leonard Euler. From this concept, Lagrange developed his famous planetary equations. Gauss followed Lagrange with his more general variational equations.

The field we are discussing is called "general perturbations" as opposed to "special perturbations." Paraphrasing Vallado, general perturbations allow the analyst to replace the exact equations of motion with an approximation that captures the essential behavior of the original problem—and, at the same time, permits analytical integration. Typically the method involves series expansion of the perturbing accelerations. The result is that a slightly degraded solution is produced that reveals the character of the "osculating orbit" that is faster (but less accurate) than the exact solution found by precise numerical integration of the original equations. And, of course, the numerical solution does not provide any insights into the broad behavior of the perturbed orbit.

The importance of analytical solutions in areas of astrodynamics, astronomy, dynamics, physics, planetary science, spacecraft missions, and many others cannot be overemphasized. However, due to the pervasive use of computers to perform highly accurate numerical integration of the governing equations, there is a tendency among engineers to "throw the equations on the computer and grind out the solution."

Such an attitude ignores the power and value of analytical solutions. Analytical solutions permit the analyst to see the explicit terms that drive the motion of the system under study. Such solutions can help verify the results of numerical work, giving the engineer confidence that his or her analysis is correct.

Typical insights gained by analytical theory are to identify when the behavior of a system has secular terms (that drive the answer to infinity as time goes to infinity) or if the solution is bounded by periodic or semi-periodic behavior. Analytical solutions can be used to study the effect of changing behavior due to changing numerical values of the system parameters.

Say, for example, we have an analytical solution of the state of a spacecraft at a given time in the future. In that case, we can write a subroutine that gives a closed-form solution at that given time. Next suppose that the spacecraft mass is changed slightly, then the analytical solution provides an instant final state via a single subroutine call. On the other hand, if we do not have an analytical solution, then we must numerically integrate the differential equations of motion with the new mass and propagate the solution to the given time.

What we just described highlights the difference between "general perturbations" and "special perturbations." The general perturbation approach gives solutions for all possible changes to initial conditions and parameters (provided the perturbed solution is not too perturbed). Special perturbations (are not so special and) give only one numerically generated solution at a time, so that for all new values of parameters and initial conditions, we must re-create the solution from scratch.

Once the variational equations are derived, we apply them to many interesting problems including the Earth–Moon system, the effect of an oblate planet, the perturbation of Mercury's orbit due to general relativity, and the perturbation due to atmospheric drag. In these applications, we introduce several useful techniques such as the method of averaging and the Lindstedt–Poincaré method of small parameters.

In the end, we hope to encourage students, practicing engineers, and those in related fields to be on the lookout for the potential of finding analytical solutions in their endeavors.

They are the ones who should read this book.

West Lafayette, IN, USA   James M. Longuski
July 2021

Acknowledgments

I would like to thank all those who contributed to this book by their positive support, their helpful suggestions, as well as their sharp eyes for typos. Among them are Jackson Kulik and my doctoral students, Rachana Agrawal, Archit Arora, Weston Buchanan, Athul P. Girija, James W. Moore, Alec Mudek, Jeffrey Pekosh, Robert Potter, and Paul Witsberger. By the time this book is published, many of them will have completed their dissertations.

I thank Hannah Kaufman, Associate Editor of Astronomy, Astrophysics, Astronautics, and Space Studies, at Springer for her enthusiastic and professional support (for answering her phone whenever I called).

I also thank my friend and colleague, Dr. James R. Wertz of Microcosm Press, for co-publishing our book with Springer.

Thanks to all those students who took my course, AAE 690 Orbital Perturbations, upon which this book is based—who helped in many significant ways to make the final manuscript as clear, accurate, and useful as possible and for their kind words along the way to publication. In particular, I thank Dr. Dan Grebow, Prof. Kaela Martin (née Rasmussen), James W. Moore, Joshua Barnett, Michael Barton, Prof. Rohan Sood, Mark Mendiola, Alex Gonring, Megan Tadge, Dr. Sarag Saikia, David Simpson, Dr. Emily Spreen (née Zimovan), Dr. Kshitij Mall, Cassie Alberding, Dr. Alex Friedman, Dr. Akihisa Aikawa, Dr. Mike Sparapany, Nick Frey, Jim Less, Kevin Koch, Denon Wang, Joseph Whaley, Wes McVay, Juan Gutierrez, Emily Leong, Kevin Gosselin, Nicola Baumann, Andrew Piwowarek, Jake Covington, Jessica Wedell, Michele Ziegler, Nathan Fergot, Cynthia Rose, Luca Ferretti, Enrique Babio, Krista Farrell, Shaid Rajani, Alex Burton, Dr. Siwei Fan, Allen Qin, Dr. Ted Wahl, Dr. Bryan Little, and Kevin Vicencio.

Finally my most grateful thanks to my wife, Holly, for her unwavering support, her enthusiastic encouragement, and especially for her love.

James M. Longuski

I am deeply indebted to a few individuals whose influence put me squarely on the path to the fascinating field of astrodynamics. Professor Philip M. Fitzpatrick of Auburn University first introduced me to this subject that eventually led to my dissertation research. More importantly, he taught me a unique way of approaching problems that has served me well for nearly 50 years.

My first astrodynamics job was working for Max H. Lane, who had been a colleague of Fitzpatrick, and who had assembled a group of engineers and mathematicians to provide Government astrodynamics expertise. The transition from theoretical problems at a university to real-world problems on the job is facilitated by knowledgeable and caring mentors. I had two of the best—Paul Major and Bob Morris. Basic orbit determination was first explained to me on the back of a napkin by Paul, and Bob is still teaching me new things even today.

Felix R. Hoots

I thank Professor J. M. Longuski for introducing me to the subject of General Perturbations and for mentoring me in both technical analysis and technical writing. Thanks to Professor K. C. Howell for her rigorous instruction in the fundamentals of orbit mechanics.

I am profoundly grateful to my parents and grandparents who each fanned the flames of wonder about space that developed into a rewarding career in astrodynamics. Thanks to my four wonderful children for their love, for their encouragement during this project, and for sharing with me a fascination about space.

Sincerest thanks to my wife, Amy, for her enthusiastic support of this project, including designing most of the figure artwork in this book, and above all, for her unconditional love.

George E. Pollock IV

Contents

Advance Praise for *Introduction to Orbital Perturbations*

"An extensive, detailed, yet still easy-to-follow presentation of the field of orbital perturbations."

Prof. Hanspeter Schaub, Smead Aerospace Engineering Sciences Department, University of Colorado, Boulder

Author of *Analytical Mechanics of Space Systems*, Fourth Edition (with Prof. John Junkins)

===

"This book, based on decades of teaching experience, is an invaluable resource for aerospace engineering students and practitioners alike who need an in-depth understanding of the equations they use."

Dr. Jean Albert Kéchichian, The Aerospace Corporation, Retired

Author of *Applied Nonsingular Astrodynamics: Optimal Low-Thrust Orbit Transfer*

===

"Today we look at perturbations through the lens of the modern computer. But knowing the why and the how is equally important. In this well organized and thorough compendium of equations and derivations, the authors bring some of the relevant gems from the past back into the contemporary literature."

Dr. David A Vallado, Senior Research Astrodynamicist, COMSPOC

Author of *Fundamentals of Astrodynamics and Applications*, Second Edition

===

"The book presentation is with the thoroughness that one always sees with these authors. Their theoretical development is followed with a set of Earth orbiting and Solar System examples demonstrating the application of Lagrange's planetary equations for systems with both conservative and nonconservative forces, some of which are not seen in orbital mechanics books."

Prof. Kyle T. Alfriend, Professor and Holder of the Jack E. & Francis Brown Chair II, University Distinguished Professor, Texas A&M University

===

About the Authors

Professor James M. Longuski (long - gŭś - skē) has authored or coauthored over 250 conference and journal papers in the area of astrodynamics on topics that include designing spacecraft trajectories to explore the Solar System and a new idea to test Einstein's General Theory of Relativity. He has also coauthored several papers with Dr. Buzz Aldrin on a human Earth-to-Mars transportation system known as the "Aldrin Cycler."

He has published three books: *Advice to Rocket Scientists: A Career Survival Guide for Scientists and Engineers* (2004, AIAA), *The Seven Secrets of How to Think Like a Rocket Scientist* (2007, Springer), and *Optimal Control with Aerospace Applications* with José J. Guzmán and John E. Prussing (2014, Springer). In 2008, Dr. Longuski was inducted into Purdue University's Book of Great Teachers.

Dr. Felix R. Hoots, a Fellow at The Aerospace Corporation, received his Ph.D. in mathematics with an emphasis in astrodynamics from the Auburn University in 1976. His nearly 50-year professional career began as a civil servant for the 14th AF and NORAD/ADCOM and later as a defense contractor for GRC International supporting the Air Force and Navy as well as the intelligence community. During this time, he worked with the Naval Research Lab on the first demonstration of an all Special Perturbations satellite catalog. His current work at The Aerospace Corporation includes the development of an improved accuracy General Perturbations model for production of the widely used Two-Line Element sets. Hoots is a Fellow of the American Astronautical Society and a Fellow of the American Institute of Aeronautics and Astronautics. He is a coauthor (with George Chao) of *Applied Orbit Perturbation and Maintenance*. He has been an invited speaker at NATO and International Astronomical Union conferences and was a technical organizer and speaker at a series of US/Russian Space Surveillance Workshops beginning in 1994 and continuing until 2012.

Dr. George E. Pollock IV, Director of the Astrodynamics Department at The Aerospace Corporation, received his Ph.D. in aeronautical and astronautical engineering with a focus in astrodynamics from Purdue University in 2010. He leads a team of analysts providing space domain awareness and advanced space mission analyses for the U.S. Space Force, NASA, and other customers. In over a decade of professional practice, he has directly contributed to national security space through innovative mission design, system concept analysis, and architecture studies. He received Purdue University's highest award for graduate student educators in 2010 and was a member of a team recognized with The Aerospace Corporation's Innovation Award in 2019.

Nomenclature

In this section, we summarize the primary notation and symbols used in the book. Where a symbol is reused, the particular meaning will be apparent from the context.

Coordinate Systems

r, θ	Polar coordinates for motion in the orbital plane
$\mathbf{b}_1$, $\mathbf{b}_2$, $\mathbf{b}_3$	Body-fixed unit vectors for a coordinate system affixed to a rigid body
$\mathbf{e}_1$, $\mathbf{e}_2$, $\mathbf{e}_3$	Unit vectors for an arbitrary coordinate system
$\mathbf{i}$, $\mathbf{j}$, $\mathbf{k}$	Unit vectors for inertial frame, Cartesian coordinate system
$\mathbf{u}_r$, $\mathbf{u}_\theta$, $\mathbf{u}_A$	Unit vectors for polar coordinate system
$\mathbf{u}_P$, $\mathbf{u}_Q$, $\mathbf{u}_W$	Unit vectors for perifocal PQW coordinate system
$\mathbf{u}_N$, $\mathbf{u}_T$, $\mathbf{u}_W$	Unit vectors for NTW coordinate system
$\mathbf{u}_R$, $\mathbf{u}_S$, $\mathbf{u}_W$	Unit vectors for RSW coordinate system
D_{ij}	Elements of rotation matrix transforming from $\mathbf{i}$, $\mathbf{j}$, $\mathbf{k}$ to $\mathbf{u}_R$, $\mathbf{u}_S$, $\mathbf{u}_W$
P_i, Q_i, W_i	Components of PQW unit vectors
ξ, η, ζ	Components in PQW system

Latin Characters

$\mathbf{A}$	Vector representing area
A	Area
A_{ij}	Partial derivative relating residuals to solution vector
AU	Astronomical unit
a	Semimajor axis
B_i	Measurement residual

b	Semiminor axis, or 1/2 times the satellite ballistic coefficient [i.e., $C_D A/(2m)$]
C_D	Dimensionless drag coefficient
c	Speed of light
E	Eccentric anomaly, or total energy of the system of particles
$\mathbf{e}$	Eccentricity vector
e	Eccentricity
$\mathcal{E}$	Total specific energy
$\mathbf{F}$	External force
$\mathcal{F}$	Linear impulse
$F_{lmp}(i)$	Inclination functions
$\mathbf{f}_{12}$	Internal force on body 1 by body 2
f	True anomaly
G	Universal gravitational constant
$G_{lpq}(e)$	Eccentricity functions
$\mathbf{H}^O$	Angular momentum about O
H	Scale height for atmospheric density model
$\mathbf{h}$	Specific angular momentum
$I_1,\ I_2,\ I_3$	Moments of inertia
I_n	Modified Bessel function of the first kind
i	Inclination
J_n	Coefficients of the zonal harmonics
j	$\sqrt{-1}$
l	Rotation matrix
M	Mean anomaly
$\mathbf{M}^O$	Moment about point O
$\mathcal{M}^O$	Angular impulse about point O
$m_1,\ m_2$	Mass of body 1, mass of body 2
n	Mean motion
osc	Subscript denoting osculating orbit elements
P	Orbit period
$P_1,\ P_2$	Points in space
$P_i \cos\phi$	Legendre polynomials of the first kind
$P_{lm}(\sin\phi)$	Legendre associated function
p	Semi-latus rectum (also called the parameter)
p_0	Reference altitude for atmospheric density model
$\mathcal{R}$	Disturbing potential function
$\mathbf{r}$	Position vector
$\mathbf{r}^{OC}$	Position vector from O to center of mass, C
$\mathbf{r}^{OP_i}$	Position vector from O to P_i
r	Magnitude of position vector
r_a	Radius of apoapsis
r_e	Equatorial radius of the Earth

r_p	Radius of periapsis
r_{p0}	Radial distance at reference altitude p_0
T	Time of periapsis passage
T_i	Kinetic energy of the ith body, or time span of the ith track
$\hat{T}_i$	Midpoint time of the ith track
t	Time
U	Potential function due to gravity
U_0	Two-body gravitational potential function
U_i	Zonal harmonics of the gravitational field
u	True argument of latitude
V_i	Potential energy of the ith body
V_{lm}	Geopotential term of degree l and order m
v	Velocity magnitude
v_∞	Hyperbolic excess speed
W	Work done
$\mathbf{X}$	Least squares solution vector
$x,\ y,\ z$	Components of a vector in inertial frame, Cartesian coordinate system
y_n	Ratio of modified Bessel $= I_n / I_1$
Z_0	Dimensionless constant used in drag modeling

Greek Characters

α_i	Shape and position orbital elements
β_i	Orientation orbital elements
β	$= \sqrt{1 - e^2}$
β^*	Inverse scale height, H^{-1}
Δ	Incremental change in a variable
$\delta_{\rho i}$	Range noise for the ith measurement
$\delta_{\theta i}$	Angle noise for the ith measurement
ε	Unitless small parameter denoting size, as in $O\left(\varepsilon^2\right)$
λ	Longitude angle in spherical coordinates
μ	Gravitational parameter, Gm for a central body of mass m
ϕ	Latitude angle in spherical coordinates
$\boldsymbol{\rho}$	Relative position vector, e.g., from satellite to perturbing body
ρ	Relative position magnitude, or atmospheric density
ρ_{p0}	Atmospheric density at reference altitude, p_0
σ_ρ	Range noise characterization for a tracker
σ_θ	Angle noise characterization for a tracker
$\sigma_{\rho\rho}^2$	Range variance
$\sigma_{\theta\theta}^2$	Angle variance
$\sigma_{\rho\theta}^2$	Range–angle covariance
τ	Dimensionless time
ψ	Turn angle of a gravity-assist flyby

Ω Ascending node angle; right ascension (or longitude) of the ascending node
ω Argument of periapsis

Mathematical Symbols and Operators

∇ Gradient operator
∇^2 Laplacian operator
$[c_i, c_j]$ Lagrange bracket
J^x Jacobian
$\Re$ Real part of complex number
$'$ and $''$ Rotated coordinate systems
∂ Partial derivative
$'$ Denotes a perturbing body or the derivative of an alternate variable
$\langle \text{function} \rangle$ Average value of the function
$\langle \langle \text{function} \rangle \rangle$ Doubly averaged function

Boldface denotes vector quantities, a single dot ($\dot{\ }$) over a variable indicates the first time derivative, two dots ($\ddot{\ }$) is the second time derivative, and a subscript 0 denotes an initial condition. We use the standard convention for matrices, where superscript T denotes the transpose operation and superscript -1 denotes the inverse.

Method of Averaging Symbols

$M_i^{(j)}$ jth-order secular part of the ith slow variable
$\eta_i^{(j)}$ jth-order periodic part of the ith slow variable
$\Omega_\alpha^{(j)}$ jth-order secular part of the αth fast variable
$\phi_\alpha^{(j)}$ jth-order periodic part of the αth fast variable
ω_α 0th-order secular part of the αth fast variable
δx_{SP} Short-period portion of the variable x
δx_{LP} Long-period portion of the variable x

Chapter 1
The n-Body Problem

1.1 System of Particles

Let us start by considering a system consisting of only two particles: P_1 and P_2 as shown in Fig. 1.1. Later we generalize to a system of n particles. In this analysis, we closely follow the text by Greenwood (1988) and the paper by Tragesser and Longuski (1999). Figure 1.2 gives the free-body diagram for the particle P_1.

The positions of the particles are given with respect to the inertial frame (with an arbitrary coordinate system defined by the orthogonal unit vectors $\mathbf{e}_1, \mathbf{e}_2, \mathbf{e}_3$), and we assume that the only internal forces are due to the gravitational interaction between the particles. We denote the internal force on particle P_1 by particle P_2 as $\mathbf{f}_{12}$ and the external force on P_1 as $\mathbf{F}_1$. Let $\mathbf{u}_{12}$ be defined as the unit vector directed from P_1 toward P_2. The gravitational force on P_1 due to P_2 is given by Newton's law of gravitation:

$$\mathbf{f}_{12} = \frac{Gm_1m_2}{r_{12}^2}\mathbf{u}_{12}, \tag{1.1}$$

where m_1 and m_2 are the masses of the particles, G is Newton's universal gravitational constant, and r_{12} is the distance between P_1 and P_2:

$$r_{12} = \left|\mathbf{r}^{OP_1} - \mathbf{r}^{OP_2}\right| = \left|-\mathbf{r}^{P_1P_2}\right| = \left|\mathbf{r}^{P_2P_1}\right|. \tag{1.2}$$

Let the position vectors of the two particles be given by

$$\mathbf{r}^{OP_1} = x_1\mathbf{e}_1 + y_1\mathbf{e}_2 + z_1\mathbf{e}_3$$

$$\mathbf{r}^{OP_2} = x_2\mathbf{e}_1 + y_2\mathbf{e}_2 + z_2\mathbf{e}_3, \tag{1.3}$$

© The Author(s), under exclusive license to Springer Nature Switzerland AG 2022
J. M. Longuski et al., *Introduction to Orbital Perturbations*, Space Technology
Library 40, https://doi.org/10.1007/978-3-030-89758-1_1

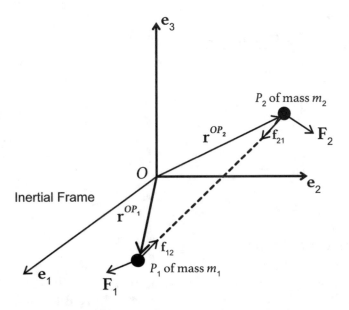

Fig. 1.1 A system of particles in the inertial *e* frame in which *O* is an inertially fixed point

Fig. 1.2 Free-body diagram for particle P_1 where $\mathbf{f}_{12}$ is an internal gravitational force and $\mathbf{F}_1$ is an arbitrary external force

where the superscripts indicate the origin and endpoint of each position vector (e.g., $\mathbf{r}^{OP_1}$ is the position vector from O to P_1). We may write r_{12} as

$$
\begin{aligned}
r_{12} &= \left| \mathbf{r}^{OP_1} - \mathbf{r}^{OP_2} \right| \\
&= \left| (x_1 - x_2)\,\mathbf{e}_1 + (y_1 - y_2)\,\mathbf{e}_2 + (z_1 - z_2)\,\mathbf{e}_3 \right| \\
&= \left[(x_1 - x_2)^2 + (y_1 - y_2)^2 + (z_1 - z_2)^2 \right]^{1/2}.
\end{aligned}
\tag{1.4}
$$

To write $\mathbf{u}_{12}$ in Eq. 1.1, we put

$$
\begin{aligned}
\mathbf{u}_{12} &= \frac{\mathbf{r}^{P_1 P_2}}{\left| \mathbf{r}^{P_1 P_2} \right|} \\
&= \frac{\mathbf{r}^{P_1 O} + \mathbf{r}^{O P_2}}{\left| \mathbf{r}^{P_1 O} + \mathbf{r}^{O P_2} \right|}
\end{aligned}
$$

$$= \frac{\mathbf{r}^{OP_2} - \mathbf{r}^{OP_1}}{\left|\mathbf{r}^{OP_2} - \mathbf{r}^{OP_1}\right|}$$

$$= \frac{[(x_2 - x_1)\,\mathbf{e}_1 + (y_2 - y_1)\,\mathbf{e}_2 + (z_2 - z_1)\,\mathbf{e}_3]}{\left[(x_2 - x_1)^2 + (y_2 - y_1)^2 + (z_2 - z_1)^2\right]^{1/2}}. \tag{1.5}$$

Writing out the internal force $\mathbf{f}_{12}$, we have (from Eqs. 1.4 and 1.5)

$$\mathbf{f}_{12} = \frac{Gm_1 m_2}{r_{12}^2}\mathbf{u}_{12}$$

$$= \frac{Gm_1 m_2 \left[(x_2 - x_1)\,\mathbf{e}_1 + (y_2 - y_1)\,\mathbf{e}_2 + (z_2 - z_1)\,\mathbf{e}_3\right]}{\left[(x_2 - x_1)^2 + (y_2 - y_1)^2 + (z_2 - z_1)^2\right]^{3/2}}, \tag{1.6}$$

or in more compact notation:

$$\mathbf{f}_{12} = \frac{Gm_1 m_2}{\left|\mathbf{r}^{P_1 P_2}\right|^2} \frac{\mathbf{r}^{P_1 P_2}}{\left|\mathbf{r}^{P_1 P_2}\right|} = \frac{Gm_1 m_2 \mathbf{r}^{P_1 P_2}}{\left|\mathbf{r}^{P_1 P_2}\right|^3}. \tag{1.7}$$

This expression may be readily *generalized* to give all $\mathbf{f}_{ij}$ in an arbitrary system of n massive particles

$$\mathbf{f}_{ij} = \frac{Gm_i m_j \mathbf{r}^{P_i P_j}}{\left|\mathbf{r}^{P_i P_j}\right|^3}.$$

Newton's first law gives us the equation of motion for P_1

$$m_1 \ddot{\mathbf{r}}^{OP_1} = \sum_{j=1}^{n} \mathbf{f}_{1j} + \mathbf{F}_1, \tag{1.8}$$

where $\mathbf{f}_{11} \equiv \mathbf{0}$, as a particle cannot exert a force upon itself. We use a single dot above a quantity to denote the first time derivative and two dots to indicate the second time derivative. Since the frame is inertial, we can directly write the acceleration of P_1 with respect to O, as

$$\ddot{\mathbf{r}}^{OP_1} = \ddot{x}_1 \mathbf{e}_1 + \ddot{y}_1 \mathbf{e}_2 + \ddot{z}_1 \mathbf{e}_3, \tag{1.9}$$

and we have

$$m_1 \left(\ddot{x}_1 \mathbf{e}_1 + \ddot{y}_1 \mathbf{e}_2 + \ddot{z}_1 \mathbf{e}_3 \right) = \sum_{j=2}^{n} \frac{Gm_1 m_j}{\left|\mathbf{r}^{P_1 P_j}\right|^3} \mathbf{r}^{P_1 P_j} + \mathbf{F}_1, \tag{1.10}$$

where we start the summation at $j = 2$ because $\mathbf{f}_{11} \equiv \mathbf{0}$. Thus, for the ith particle, the equation of motion is

$$m_i \left(\ddot{x}_i \mathbf{e}_1 + \ddot{y}_i \mathbf{e}_2 + \ddot{z}_i \mathbf{e}_3 \right) = \sum_{j=1}^{n} \frac{Gm_i m_j}{\left| \mathbf{r}^{P_i P_j} \right|^3} \mathbf{r}^{P_i P_j} + \mathbf{F}_i, \qquad j \neq i \qquad (1.11)$$

$$m_i \ddot{\mathbf{r}}^{O P_i} = \sum_{j=1}^{n} \frac{Gm_i m_j}{\left| \mathbf{r}^{P_i P_j} \right|^3} \mathbf{r}^{P_i P_j} + \mathbf{F}_i, \qquad j \neq i. \qquad (1.12)$$

We note that the gravitational forces are internal to the system of particles and obey the relation:

$$\mathbf{f}_{ij} = -\mathbf{f}_{ji} \qquad (1.13)$$

according to Newton's third law (of action and reaction). Furthermore, we have

$$\sum_{i=1}^{n} \sum_{j=1}^{n} \mathbf{f}_{ij} = \mathbf{f}_{11} + \mathbf{f}_{12} + \mathbf{f}_{13} + \cdots + \mathbf{f}_{1n}$$

$$+ \mathbf{f}_{21} + \mathbf{f}_{22} + \mathbf{f}_{23} + \cdots + \mathbf{f}_{2n}$$

$$\vdots \qquad \ddots$$

$$+ \mathbf{f}_{n1} + \mathbf{f}_{n2} + \mathbf{f}_{n3} + \cdots + \mathbf{f}_{nn}$$

$$= \mathbf{0} \qquad (1.14)$$

because

$$\mathbf{f}_{ii} \equiv \mathbf{0}, \qquad (1.15)$$

so the diagonal terms in Eq. 1.14 vanish and the off-diagonal terms cancel in pairs, e.g.,

$$\mathbf{f}_{12} + \mathbf{f}_{21} = \mathbf{0}. \qquad (1.16)$$

This last fact is due to Newton's third law, Eq. 1.13.

The center of mass, C, or barycenter of the system of particles is located (with respect to O) by the weighted average of the mass locations in the system:

$$\mathbf{r}^{OC} \equiv \frac{1}{m} \sum_{i=1}^{n} m_i \mathbf{r}^{O P_i}, \qquad (1.17)$$

where

$$m \equiv \sum_{i=1}^{n} m_i$$

is the total system mass. Differentiating Eq. 1.17 twice (and multiplying through by the mass of the system), we arrive at the following expressions:

$$m\mathbf{r}^{OC} = \sum_{i=1}^{n} m_i \mathbf{r}^{OP_i} \tag{1.18}$$

$$m\dot{\mathbf{r}}^{OC} = \sum_{i=1}^{n} m_i \dot{\mathbf{r}}^{OP_i} \tag{1.19}$$

$$m\ddot{\mathbf{r}}^{OC} = \sum_{i=1}^{n} m_i \ddot{\mathbf{r}}^{OP_i}. \tag{1.20}$$

Summing over i in Eq. 1.12, we obtain

$$\sum_{i=1}^{n} m_i \ddot{\mathbf{r}}^{OP_i} = \sum_{i=1}^{n} \sum_{j=1}^{n} \frac{G m_i m_j}{\left|\mathbf{r}^{P_i P_j}\right|^3} \mathbf{r}^{P_i P_j} + \sum_{i=1}^{n} \mathbf{F}_i, \qquad j \neq i. \tag{1.21}$$

We now use Eq. 1.20 to rewrite the left-hand side of Eq. 1.21, and from Eq. 1.14, the first term on the right-hand side of Eq. 1.21 is the zero vector. The simplified result may be written as

$$m\ddot{\mathbf{r}}^{OC} = \mathbf{F}, \tag{1.22}$$

where

$$\mathbf{F} = \sum_{i=1}^{n} \mathbf{F}_i$$

is the resultant of all external forces acting on the system. Equation 1.22 is Newton's second law for a system of particles. It states that the motion of the center of mass is the same as if the entire mass of the system were concentrated there and acted upon by the resultant of the external forces. (See Greenwood, 1988.)

Next, we consider changing the reference point in Eq. 1.12 from the origin, O, to the center of mass, C, using the following relationships:

$$\begin{aligned} \mathbf{r}^{OP_i} &= \mathbf{r}^{OC} + \mathbf{r}^{CP_i} \\ \dot{\mathbf{r}}^{OP_i} &= \dot{\mathbf{r}}^{OC} + \dot{\mathbf{r}}^{CP_i} \\ \ddot{\mathbf{r}}^{OP_i} &= \ddot{\mathbf{r}}^{OC} + \ddot{\mathbf{r}}^{CP_i}. \end{aligned} \tag{1.23}$$

Substituting Eqs. 1.23 into Eq. 1.12 provides

$$m_i\ddot{\mathbf{r}}^{OC} + m_i\ddot{\mathbf{r}}^{CP_i} = \sum_{j=1}^{n} \frac{Gm_im_j}{\left|\mathbf{r}^{P_iC} + \mathbf{r}^{CP_j}\right|^3} \left(\mathbf{r}^{P_iC} + \mathbf{r}^{CP_j}\right) + \mathbf{F}_i, \qquad j \neq i. \quad (1.24)$$

From Eq. 1.22, we note that

$$m_i\ddot{\mathbf{r}}^{OC} = m_i\frac{\mathbf{F}}{m}, \qquad (1.25)$$

which we substitute for the first term on the left-hand side of Eq. 1.24 to write

$$m_i\ddot{\mathbf{r}}^{CP_i} = \sum_{j=1}^{n} \frac{Gm_im_j}{\left|\mathbf{r}^{P_iC} + \mathbf{r}^{CP_j}\right|^3} \left(\mathbf{r}^{P_iC} + \mathbf{r}^{CP_j}\right) + \mathbf{F}_i - m_i\frac{\mathbf{F}}{m}, \qquad j \neq i. \quad (1.26)$$

Equation 1.26 gives the accelerations of the particles, P_i, relative to the center of mass (or barycenter) of the system. We can use Eq. 1.22 (which gives the motion of the center of mass) in combination with Eq. 1.26 to find the motion of the particles with respect to the origin, O.

1.2   Equations of Relative Motion

Another possibility is to use P_1 as the reference point. In which case, by analogy with Eqs. 1.23, we have

$$\mathbf{r}^{OP_i} = \mathbf{r}^{OP_1} + \mathbf{r}^{P_1P_i}$$
$$\dot{\mathbf{r}}^{OP_i} = \dot{\mathbf{r}}^{OP_1} + \dot{\mathbf{r}}^{P_1P_i} \qquad (1.27)$$
$$\ddot{\mathbf{r}}^{OP_i} = \ddot{\mathbf{r}}^{OP_1} + \ddot{\mathbf{r}}^{P_1P_i}.$$

Substituting Eqs. 1.27 into Eq. 1.12 gives (for the ith particle)

$$m_i\left(\ddot{\mathbf{r}}^{OP_1} + \ddot{\mathbf{r}}^{P_1P_i}\right) = \sum_{j=1}^{n} \frac{Gm_im_j}{\left|\mathbf{r}^{P_iP_j}\right|^3}\mathbf{r}^{P_iP_j} + \mathbf{F}_i, \qquad (1.28)$$

where it is understood that

$$\mathbf{r}^{P_iP_j} = \mathbf{r}^{P_iP_1} + \mathbf{r}^{P_1P_j} \qquad (1.29)$$

wherever $\mathbf{r}^{P_iP_j}$ appears in Eq. 1.28. Setting $i = 1$ in Eq. 1.28 provides

$$m_1 \ddot{\mathbf{r}}^{OP_1} = \sum_{j=2}^{n} \frac{Gm_1 m_j}{\left| \mathbf{r}^{P_1 P_j} \right|^3} \mathbf{r}^{P_1 P_j} + \mathbf{F}_1, \tag{1.30}$$

where we have used $\ddot{\mathbf{r}}^{P_1 P_1} \equiv \mathbf{0}$ and have changed the lower summation index ($j = 2$) since it is always assumed that $j \neq i$ in all sums having $\left| \mathbf{r}^{P_i P_j} \right|^{-3}$ terms. Multiplying Eq. 1.30 by m_i / m_1 and subtracting the result from Eq. 1.28, we can eliminate the leading term in parentheses on the left-hand side of Eq. 1.28 to obtain

$$m_i \ddot{\mathbf{r}}^{P_1 P_i} = \sum_{j=1}^{n} \frac{Gm_i m_j}{\left| \mathbf{r}^{P_i P_j} \right|^3} \mathbf{r}^{P_i P_j} + \mathbf{F}_i - \sum_{j=2}^{n} \frac{Gm_i m_j}{\left| \mathbf{r}^{P_1 P_j} \right|^3} \mathbf{r}^{P_1 P_j} - \frac{m_i}{m_1} \mathbf{F}_1, \qquad i \neq 1. \tag{1.31}$$

Equation 1.31 consists of $n - 1$ vector equations that provide the motion of the particles *relative to* the reference point P_1. The motion of P_1 relative to O is given by Eq. 1.30.

We now wish to make all the P_1 terms explicit in our equations of relative motion. In Eq. 1.31, we can extract the term corresponding to $i = 1$ from the summation and rewrite the first summation term on the right-hand side as

$$\sum_{j=1}^{n} \frac{Gm_i m_j}{\left| \mathbf{r}^{P_i P_j} \right|^3} \mathbf{r}^{P_i P_j} = \frac{Gm_i m_1}{\left| \mathbf{r}^{P_i P_1} \right|^3} \mathbf{r}^{P_i P_1} + \sum_{j=2}^{n} \frac{Gm_i m_j}{\left| \mathbf{r}^{P_i P_j} \right|^3} \mathbf{r}^{P_i P_j}, \tag{1.32}$$

where the left-hand side summation is over terms where $j \neq i$, but the summation on the right-hand side includes the terms where $j = i$. Substituting Eq. 1.32 into Eq. 1.31 and dividing by m_i provide

$$\ddot{\mathbf{r}}^{P_1 P_i} = \sum_{j=2}^{n} \frac{Gm_j}{\left| \mathbf{r}^{P_i P_j} \right|^3} \mathbf{r}^{P_i P_j} - \frac{Gm_1}{\left| \mathbf{r}^{P_i P_1} \right|^3} \mathbf{r}^{P_1 P_i} + \frac{\mathbf{F}_i}{m_i} - \sum_{j=2}^{n} \frac{Gm_j}{\left| \mathbf{r}^{P_1 P_j} \right|^3} \mathbf{r}^{P_1 P_j} - \frac{\mathbf{F}_1}{m_1}$$

$$j \neq i, \qquad i = 2, 3, \dots, n. \tag{1.33}$$

We have reversed the order of the superscripts in the first term on the right-hand side in Eq. 1.32 and switched the sign; this term now appears as the second term on the right-hand side of Eq. 1.33. Dividing Eq. 1.30 by m_1 provides

$$\ddot{\mathbf{r}}^{OP_1} = \sum_{j=2}^{n} \frac{Gm_j}{\left| \mathbf{r}^{P_1 P_j} \right|^3} \mathbf{r}^{P_1 P_j} + \frac{\mathbf{F}_1}{m_1}. \tag{1.34}$$

Equations 1.33 and 1.34 are identical in meaning (with slight differences in nomenclature) to Eqs. 5 and 4 of Tragesser and Longuski (1999). In their paper, "Modeling Issues Concerning Motion of the Saturnian Satellites," Tragesser and Longuski investigate various perturbing accelerations, including spherical gravity harmonics, the Saturnian tides and rings, satellite attitude coupling, other Solar System bodies, and general relativity.

In Eq. 1.33, by solving for the position relative to P_1, we have reduced the number of degrees of freedom from $3n$ to $3(n-1)$. If we are interested in inertial positions of the particles in the system or in the motion of P_1, Eq. 1.34 must be used.

1.3   The Ten Known Integrals

Now we assume that our system of particles is not subjected to any external forces so that

$$\mathbf{F}_i = \mathbf{0}. \tag{1.35}$$

Using Eq. 1.35 in Eq. 1.22 gives

$$m\ddot{\mathbf{r}}^{OC} = \mathbf{0}. \tag{1.36}$$

Dividing Eq. 1.36 by m and integrating, we obtain

$$\dot{\mathbf{r}}^{OC} = \mathbf{a}, \tag{1.37}$$

where the vector $\mathbf{a}$ is an integration constant that gives the velocity of the barycenter. Integrating Eq. 1.37 provides

$$\mathbf{r}^{OC} = \mathbf{a}t + \mathbf{b}. \tag{1.38}$$

Equations 1.37 and 1.38 state that the center of mass of the system moves through space with constant velocity. The constant vectors $\mathbf{a}$ and $\mathbf{b}$ provide six constants of integration. Equation 1.38 is a direct consequence of the Principle of Linear Impulse and Linear Momentum for a system of particles.

By performing a time integral of Eq. 1.22, we have

$$\mathcal{F} = \int_{t_1}^{t_2} \mathbf{F}dt = m\dot{\mathbf{r}}^{OC}\Big|_{t_1}^{t_2}, \tag{1.39}$$

where the left-hand side is the linear impulse and the right-hand side is the change in linear momentum. Since $\mathbf{F} = \mathbf{0}$, we have

$$m\dot{\mathbf{r}}^{OC}\Big|_{t_1}^{t_2} = \mathbf{0}, \tag{1.40}$$

which is the Conservation of Linear Momentum for a system of particles, where the linear momentum can be computed from Eq. 1.19:

$$m\dot{\mathbf{r}}^{OC} = \sum_{i=1}^{n} m_i \dot{\mathbf{r}}^{OP_i} = \text{constant.} \tag{1.41}$$

To find more constants of the motion, we take the vector product of $\mathbf{r}^{OP_i}$ and $\ddot{\mathbf{r}}^{OP_i}$ for each of the particles, using Eq. 1.12 (after setting $\mathbf{F}_i = \mathbf{0}$) and summing:

$$\sum_{i=1}^{n} m_i \mathbf{r}^{OP_i} \times \ddot{\mathbf{r}}^{OP_i} = \sum_{i=1}^{n} \sum_{j=1}^{n} \frac{Gm_i m_j}{\left|\mathbf{r}^{P_i P_j}\right|^3} \mathbf{r}^{OP_i} \times \mathbf{r}^{P_i P_j}, \qquad j \neq i. \tag{1.42}$$

We note that the cross product on the right-hand side of Eq. 1.42 can be written as

$$\mathbf{r}^{OP_i} \times \mathbf{r}^{P_i P_j} = \mathbf{r}^{OP_i} \times \left(\mathbf{r}^{OP_j} - \mathbf{r}^{OP_i}\right) = \mathbf{r}^{OP_i} \times \mathbf{r}^{OP_j}, \tag{1.43}$$

and we also have (by analogy with Eq. 1.43)

$$\mathbf{r}^{OP_j} \times \mathbf{r}^{P_j P_i} = \mathbf{r}^{OP_j} \times \left(\mathbf{r}^{OP_i} - \mathbf{r}^{OP_j}\right) = \mathbf{r}^{OP_j} \times \mathbf{r}^{OP_i} = -\mathbf{r}^{OP_i} \times \mathbf{r}^{OP_j}. \tag{1.44}$$

Since the right-hand side of Eq. 1.42 reduces in pairs (by Eqs. 1.43 and 1.44) to zero, we obtain

$$\sum_{i=1}^{n} m_i \mathbf{r}^{OP_i} \times \ddot{\mathbf{r}}^{OP_i} = \mathbf{0}. \tag{1.45}$$

Integrating Eq. 1.45, we have

$$\sum_{i=1}^{n} m_i \mathbf{r}^{OP_i} \times \dot{\mathbf{r}}^{OP_i} = \mathbf{c}. \tag{1.46}$$

We may verify this integral by differentiating Eq. 1.46 as follows:

$$\frac{d}{dt}\left(m_i \mathbf{r}^{OP_i} \times \dot{\mathbf{r}}^{OP_i}\right) = m_i \left(\dot{\mathbf{r}}^{OP_i} \times \dot{\mathbf{r}}^{OP_i} + \mathbf{r}^{OP_i} \times \ddot{\mathbf{r}}^{OP_i}\right)$$

$$= m_i \mathbf{r}^{OP_i} \times \ddot{\mathbf{r}}^{OP_i}.$$

Equation 1.46 states that the sum of the angular momenta of the masses in the system is a constant vector.

The constant vector, $\mathbf{c}$, defines a plane known as the invariable plane of Laplace. For the Solar System, this plane is inclined about $1.5°$ to the ecliptic plane and lies between the orbital planes of Jupiter and Saturn. The constant vector, $\mathbf{c}$, provides three constants of integration, bringing our total to nine.

Equation 1.46 is a direct consequence of the Principle of Angular Impulse and Angular Momentum for a system of particles:

$$\mathcal{M}^O = \int_{t_1}^{t_2} \mathbf{M}^O dt = \mathbf{H}^O \Big|_{t_1}^{t_2}, \tag{1.47}$$

where $\mathbf{M}^O$ is the moment about O of the resultant external force, $\mathbf{F}$. We note that the moment of the internal forces sums to zero

$$\sum_{i=1}^{n} \sum_{j=1}^{n} \mathbf{r}^{OP_i} \times \mathbf{f}_{ij} = \mathbf{0} \tag{1.48}$$

because $\mathbf{f}_{ij} = -\mathbf{f}_{ji}$ (Newton's third law) and because $\mathbf{f}_{ij}$ and $\mathbf{f}_{ji}$ are collinear for mutually gravitating bodies. Since $\mathbf{F} = \mathbf{0}$, then $\mathbf{M}^O$ vanishes in Eq. 1.47, and we have Conservation of Angular Momentum for a system of particles:

$$\mathbf{H}^O = \sum_{i=1}^{n} \mathbf{r}^{OP_i} \times m_i \dot{\mathbf{r}}^{OP_i} = \text{constant}, \tag{1.49}$$

which is identical to Eq. 1.46.

So far, we have found nine integrals of the motion for our system of particles (the vectors $\mathbf{a}$, $\mathbf{b}$, and $\mathbf{c}$ of Eqs. 1.38 and 1.46). To find the tenth integral, we use the Principle of Work and Kinetic Energy for a system of particles. Taking the dot (or scalar) product of $\dot{\mathbf{r}}^{OP_i}$ and $\ddot{\mathbf{r}}^{OP_i}$ for each of the particles, using Eq. 1.12 (after setting $\mathbf{F}_i = \mathbf{0}$), integrating over time, and summing, we obtain

$$\sum_{i=1}^{n} \int_{t_1}^{t_2} m_i \ddot{\mathbf{r}}^{OP_i} \cdot \dot{\mathbf{r}}^{OP_i} dt = \sum_{i=1}^{n} \int_{t_1}^{t_2} \sum_{j=1}^{n} \frac{G m_i m_j}{\left| \mathbf{r}^{P_i P_j} \right|^3} \mathbf{r}^{P_i P_j} \cdot \dot{\mathbf{r}}^{OP_i} dt, \qquad j \neq i. \tag{1.50}$$

The term on the right-hand side of Eq. 1.50 represents the work done by the internal gravitational forces. The term on the left-hand side can be written as

$$\sum_{i=1}^{n} \int_{t_1}^{t_2} m_i \ddot{\mathbf{r}}^{OP_i} \cdot \dot{\mathbf{r}}^{OP_i} dt = \sum_{i=1}^{n} \frac{1}{2} m_i \int_{t_1}^{t_2} \frac{d}{dt} \left[\dot{\mathbf{r}}^{OP_i} \cdot \dot{\mathbf{r}}^{OP_i} \right] dt$$

$$= \sum_{i=1}^{n} \frac{1}{2} m_i \dot{\mathbf{r}}^{OP_i} \cdot \dot{\mathbf{r}}^{OP_i} \Big|_{t_1}^{t_2}$$

$$= T(t_2) - T(t_1), \tag{1.51}$$

which is the change in the kinetic energy of the system. Thus, Eq. 1.50 can be written succinctly as

$$W = \Delta T. \tag{1.52}$$

Now we consider the term on the right-hand side of Eq. 1.50. We seek to demonstrate that the right-hand side of Eq. 1.50 can be expressed as an exact differential of a potential function, U. We have the relations:

$$\dot{\mathbf{r}}^{OP_i} \cdot \mathbf{r}^{P_iP_j} = \dot{\mathbf{r}}^{OP_i} \cdot \left(\mathbf{r}^{OP_j} - \mathbf{r}^{OP_i} \right) \tag{1.53}$$

and

$$\dot{\mathbf{r}}^{OP_j} \cdot \mathbf{r}^{P_jP_i} = \dot{\mathbf{r}}^{OP_j} \cdot \left(\mathbf{r}^{OP_i} - \mathbf{r}^{OP_j} \right). \tag{1.54}$$

Adding Eqs. 1.53 and 1.54 and collecting terms on the right-hand side give

$$\dot{\mathbf{r}}^{OP_i} \cdot \mathbf{r}^{P_iP_j} + \dot{\mathbf{r}}^{OP_j} \cdot \mathbf{r}^{P_jP_i} = \left(\dot{\mathbf{r}}^{OP_i} - \dot{\mathbf{r}}^{OP_j} \right) \cdot \left(\mathbf{r}^{OP_j} - \mathbf{r}^{OP_i} \right)$$

$$= -\dot{\mathbf{r}}^{P_iP_j} \cdot \mathbf{r}^{P_iP_j}. \tag{1.55}$$

The terms on the right-hand side of Eq. 1.55 (after summing over i and j) give twice the value of

$$\sum_{i=1}^{n} \sum_{j=1}^{n} \dot{\mathbf{r}}^{OP_i} \cdot \mathbf{r}^{P_iP_j},$$

so we can write

$$\sum_{i=1}^{n} \sum_{j=1}^{n} \dot{\mathbf{r}}^{OP_i} \cdot \mathbf{r}^{P_iP_j} = -\frac{1}{2} \sum_{i=1}^{n} \sum_{j=1}^{n} \dot{\mathbf{r}}^{P_iP_j} \cdot \mathbf{r}^{P_iP_j}. \tag{1.56}$$

Here we note that the sums in Eq. 1.56 are over all i and j, while sums in Eq. 1.50 exclude $j = i$. Substituting Eq. 1.56 into the right-hand side of Eq. 1.50 removes the $\dot{\mathbf{r}}^{OP_i}$ terms to yield

$$-\sum_{i=1}^{n} \sum_{j=1}^{n} \frac{1}{2} \int_{t_1}^{t_2} \frac{Gm_i m_j}{\left| \mathbf{r}^{P_iP_j} \right|^3} \mathbf{r}^{P_iP_j} \cdot \dot{\mathbf{r}}^{P_iP_j} \, dt = \frac{1}{2} \sum_{i=1}^{n} \sum_{j=1}^{n} \frac{Gm_i m_j}{\left| \mathbf{r}^{P_iP_j} \right|} \Big|_{t_1}^{t_2}. \tag{1.57}$$

Here we have used the following steps to compute the integral in Eq. 1.57

$$\int -\frac{\dot{\mathbf{r}} \cdot \mathbf{r}}{|\mathbf{r}|^3} dt = \int -\frac{\dot{\mathbf{r}} \cdot \mathbf{r}}{(\mathbf{r} \cdot \mathbf{r})^{3/2}} dt$$

$$= -\int \frac{1}{2} (\mathbf{r} \cdot \mathbf{r})^{-3/2} 2 (\dot{\mathbf{r}} \cdot \mathbf{r}) \, dt$$

$$= -\int \frac{1}{2} (\mathbf{r} \cdot \mathbf{r})^{-3/2} (\dot{\mathbf{r}} \cdot \mathbf{r} + \mathbf{r} \cdot \dot{\mathbf{r}}) \, dt$$

$$= \int \frac{d}{dt} (\mathbf{r} \cdot \mathbf{r})^{-1/2} \, dt$$

$$= \int d \left(\frac{1}{|\mathbf{r}|} \right)$$

$$= \frac{1}{|\mathbf{r}|}. \tag{1.58}$$

Thus we have demonstrated that both sides of Eq. 1.50 can be integrated to provide

$$\sum_{i=1}^{n} \frac{1}{2} m_i \dot{\mathbf{r}}^{O P_i} \cdot \dot{\mathbf{r}}^{O P_i} \Big|_{t_1}^{t_2} = \frac{1}{2} \sum_{i=1}^{n} \sum_{j=1}^{n} \frac{G m_i m_j}{|\mathbf{r}^{P_i P_j}|} \Big|_{t_1}^{t_2} \tag{1.59}$$

or, more compactly:

$$T_2 - T_1 = -V_2 + V_1, \tag{1.60}$$

so that we have

$$T_2 + V_2 = T_1 + V_1 = E, \tag{1.61}$$

where V is the potential energy. Equation 1.61 states that the total energy of the system of particles is a constant, E, which is the tenth constant of integration. No further integrals have been discovered. (We note that the potential function, $U = -V/m$, is often used throughout the text.)

1.3.1   The Question of New Integrals

Moulton (1914) discusses the question of whether additional integrals may exist. He points out that Bruns demonstrated that no new algebraic integrals exist when the rectangular coordinates are selected as the dependent variables. Moulton comments that this does not preclude the existence of further algebraic integrals when other variables are used. Moulton reports that Poincaré has shown that the three-body problem admits no new uniform transcendental integrals, even in the case where two of the masses are negligible in comparison to the mass of the third body. Poincaré used the orbital elements of the bodies as the dependent variables. Moulton remarks that the non-existence of additional integrals of the type considered by Poincaré does not follow from Poincaré's theorem when other dependent variables are chosen. Moulton concludes by stating that "the practical importance of the theorems of Bruns and Poincaré have often been overrated by those who have forgotten the conditions under which they have been proved to hold true."

1.4   Equations of Relative Motion for the n-Body Problem

Later, in the development of General Perturbations, we see that in the case of a planet in the Solar System and in the case of a close satellite about a non-spherical planet, a potential function, U, can be formed such that

$$U = U_0 + \mathcal{R}, \tag{1.62}$$

where U_0 is the potential function due to the point-mass two-body problem and $\mathcal{R}$ is a potential function due to any other forces, which could include other attracting masses in the system or the oblateness of the planet about which the body revolves.

The term $\mathcal{R}$ is called the "Disturbing Function" and is usually an order of magnitude less than the two-body point mass potential term U_0. Under these conditions, General or Special Perturbation methods can be used.

Let us consider the Equations of Relative Motion for the n-body problem, Eq. 1.33, with no external forces (so we set $\mathbf{F}_1 = \mathbf{F}_i = \mathbf{0}$):

$$\ddot{\mathbf{r}}^{P_1 P_i} = \sum_{j=2}^{n} \frac{Gm_j}{\left|\mathbf{r}^{P_i P_j}\right|^3} \mathbf{r}^{P_i P_j} - \frac{Gm_1}{\left|\mathbf{r}^{P_1 P_i}\right|^3} \mathbf{r}^{P_1 P_i} - \sum_{j=2}^{n} \frac{Gm_j}{\left|\mathbf{r}^{P_1 P_j}\right|^3} \mathbf{r}^{P_1 P_j}, \quad i = 2, 3, \ldots, n. \tag{1.63}$$

We note that the first summation in Eq. 1.63 necessarily has $j \neq i$ since the particle cannot exert a force on itself (i.e., $\mathbf{f}_{ii} = \mathbf{0}$), but the second summation may have $j = i$. On the other hand, we can extract the m_i term from the last summation and specify $j \neq i$; then, the m_i term is combined with the $Gm_1 \mathbf{r}^{P_1 P_i}$ term.

To see how the m_i term is removed from the summation, we consider the case of $i = 3$. Then Eq. 1.63 becomes

$$\ddot{\mathbf{r}}^{P_1 P_3} = \sum_{j=2}^{n} \frac{Gm_j}{\left|\mathbf{r}^{P_3 P_j}\right|^3} \mathbf{r}^{P_3 P_j} - \frac{Gm_1}{\left|\mathbf{r}^{P_1 P_3}\right|^3} \mathbf{r}^{P_1 P_3}$$

$$- \frac{Gm_2}{\left|\mathbf{r}^{P_1 P_2}\right|^3} \mathbf{r}^{P_1 P_2} - \frac{Gm_3}{\left|\mathbf{r}^{P_1 P_3}\right|^3} \mathbf{r}^{P_1 P_3} - \frac{Gm_4}{\left|\mathbf{r}^{P_1 P_4}\right|^3} \mathbf{r}^{P_1 P_4} - \cdots$$

$$= \sum_{j=2}^{n} \frac{Gm_j}{\left|\mathbf{r}^{P_3 P_j}\right|^3} \mathbf{r}^{P_3 P_j} - \frac{G(m_1 + m_3)}{\left|\mathbf{r}^{P_1 P_3}\right|^3} \mathbf{r}^{P_1 P_3} - \frac{Gm_j}{\left|\mathbf{r}^{P_1 P_j}\right|^3} \mathbf{r}^{P_1 P_j}, \qquad (1.64)$$

where we emphasize the critical distinction that in Eq. 1.64 $j \neq i$ in either summation (in contrast to Eq. 1.63 in which the second summation may have $j = i$). Thus, having extracted the m_i from the second summation in Eq. 1.63, we can write

$$\ddot{\mathbf{r}}^{P_1 P_i} = \sum_{j=2}^{n} \frac{Gm_j}{\left|\mathbf{r}^{P_i P_j}\right|^3} \mathbf{r}^{P_i P_j} - \frac{G(m_1 + m_i)}{\left|\mathbf{r}^{P_1 P_i}\right|^3} \mathbf{r}^{P_1 P_i} - \frac{Gm_j}{\left|\mathbf{r}^{P_1 P_j}\right|^3} \mathbf{r}^{P_1 P_j},$$

$$j \neq i, \quad i = 2, 3, \ldots, n. \qquad (1.65)$$

Here we again note the important difference between Eqs. 1.63 and 1.65: Eq. 1.63 includes $i = j$ in the last summation term, while Eq. 1.65 does not. Rearranging Eq. 1.65, we obtain

$$\ddot{\mathbf{r}}^{P_1 P_i} + \frac{G(m_1 + m_i)}{\left|\mathbf{r}^{P_1 P_i}\right|^3} \mathbf{r}^{P_1 P_i} = G \sum_{j=2}^{n} m_j \left(\frac{\mathbf{r}^{P_i P_j}}{\left|\mathbf{r}^{P_i P_j}\right|^3} - \frac{\mathbf{r}^{P_1 P_j}}{\left|\mathbf{r}^{P_1 P_j}\right|^3} \right),$$

$$j \neq i, \quad i = 2, 3, \ldots, n. \qquad (1.66)$$

Equation 1.66 is identical in meaning (and similar in nomenclature) to the result found in Roy (2005).

We make the following notes on Eq. 1.66:

1. It provides the motion of the particles P_i having masses m_i with respect to the particle P_1, which has mass m_1.
2. If other particles, P_j ($j \neq i$), do not exist (or are vanishingly small), then the right-hand side is zero and Eq. 1.66 reduces to the two-body motion of P_i about P_1.
3. The right-hand side consists of perturbations from the P_j ($j \neq i$) on the orbit of P_i about P_1. As an example in our Solar System, m_1 is the mass of the Sun and we have m_j/m_1 no larger than 10^{-3}—even for Jupiter—so the right-hand side effects are small.
4. For artificial Earth-orbiting satellites, the primary perturbing effects are due to the non-spherical Earth, atmospheric drag, lunar gravity, and solar gravity.

1.5   The Disturbing Function

Let a scalar function, $\mathcal{R}$, be defined by

$$\mathcal{R} \equiv G \sum_{j=2}^{n} m_j \left(\frac{1}{\left| \mathbf{r}^{P_i P_j} \right|} - \frac{\mathbf{r}^{P_1 P_i} \cdot \mathbf{r}^{P_1 P_j}}{\left| \mathbf{r}^{P_1 P_j} \right|^3} \right), \qquad j \neq i, \quad i = 2, 3, \dots, n, \qquad (1.67)$$

where

$$\mathbf{r}^{P_1 P_i} = x_i \mathbf{e}_1 + y_i \mathbf{e}_2 + z_i \mathbf{e}_3$$

$$\mathbf{r}^{P_1 P_j} = x_j \mathbf{e}_1 + y_j \mathbf{e}_2 + z_j \mathbf{e}_3$$

$$\left| \mathbf{r}^{P_i P_j} \right| = \left[(x_j - x_i)^2 + (y_j - y_i)^2 + (z_j - z_i)^2 \right]^{1/2} \qquad (1.68)$$

$$\left| \mathbf{r}^{P_1 P_j} \right|^3 = \left[x_j^2 + y_j^2 + z_j^2 \right]^{3/2}.$$

We use $\mathcal{R}_i$ to denote the potential function due to the ith body. Next, we show that $\nabla \mathcal{R}_i$ equals the right-hand side of Eq. 1.66 where the symbol ∇ (pronounced "nabla" or "del") denotes the gradient operator, which produces a vector:

$$\nabla \equiv \mathrm{grad} = \mathbf{e}_1 \frac{\partial}{\partial x_i} + \mathbf{e}_2 \frac{\partial}{\partial y_i} + \mathbf{e}_3 \frac{\partial}{\partial z_i}. \qquad (1.69)$$

We note that for the ith body

$$\nabla \frac{1}{\left| \mathbf{r}^{P_i P_j} \right|} = \nabla \left[(x_j - x_i)^2 + (y_j - y_i)^2 + (z_j - z_i)^2 \right]^{-1/2}$$

$$= \mathbf{e}_1 \frac{\partial}{\partial x_i} \left[(x_j - x_i)^2 + (y_j - y_i)^2 + (z_j - z_i)^2 \right]^{-1/2}$$

$$+ \mathbf{e}_2 \frac{\partial}{\partial y_i} \left[(x_j - x_i)^2 + (y_j - y_i)^2 + (z_j - z_i)^2 \right]^{-1/2}$$

$$+ \mathbf{e}_3 \frac{\partial}{\partial z_i} \left[(x_j - x_i)^2 + (y_j - y_i)^2 + (z_j - z_i)^2 \right]^{-1/2}$$

$$= \mathbf{e}_1 \left(-\frac{1}{2} \right) \left[(x_j - x_i)^2 + (y_j - y_i)^2 + (z_j - z_i)^2 \right]^{-3/2} 2 (x_j - x_i) (-1)$$

$$+ \mathbf{e}_2 \left(-\frac{1}{2} \right) \left[(x_j - x_i)^2 + (y_j - y_i)^2 + (z_j - z_i)^2 \right]^{-3/2} 2 (y_j - y_i) (-1)$$

$$+ \mathbf{e}_3 \left(-\frac{1}{2} \right) \left[(x_j - x_i)^2 + (y_j - y_i)^2 + (z_j - z_i)^2 \right]^{-3/2} 2 (z_j - z_i) (-1)$$

$$= \left| \mathbf{r}^{P_i P_j} \right|^{-3} \left[(x_j - x_i) \mathbf{e}_1 + (y_j - y_i) \mathbf{e}_2 + (z_j - z_i) \mathbf{e}_3 \right], \qquad (1.70)$$

or, written more compactly:

$$\nabla \frac{1}{\left|\mathbf{r}^{P_i P_j}\right|} = \frac{\mathbf{r}^{P_i P_j}}{\left|\mathbf{r}^{P_i P_j}\right|^3}. \tag{1.71}$$

We also note that

$$\mathbf{e}_1 \frac{\partial}{\partial x_i}\left[\frac{-\mathbf{r}^{P_1 P_i} \cdot \mathbf{r}^{P_1 P_j}}{\left|\mathbf{r}^{P_1 P_j}\right|^3}\right] = \mathbf{e}_1 \frac{\partial}{\partial x_i}\left[\frac{-(x_i\mathbf{e}_1 + y_i\mathbf{e}_2 + z_i\mathbf{e}_3) \cdot (x_j\mathbf{e}_1 + y_j\mathbf{e}_2 + z_j\mathbf{e}_3)}{\left(x_j^2 + y_j^2 + z_j^2\right)^{3/2}}\right]$$

$$= \mathbf{e}_1 \frac{\partial}{\partial x_i}\left[\frac{-x_i x_j - y_i y_j - z_i z_j}{\left(x_j^2 + y_j^2 + z_j^2\right)^{3/2}}\right]$$

$$= \mathbf{e}_1 \left(\frac{-x_j}{\left|\mathbf{r}^{P_1 P_j}\right|^3}\right), \tag{1.72}$$

where the denominator is unaffected by the derivative operation because it only contains terms in j and not in i. In a similar manner to Eq. 1.72, we have

$$\mathbf{e}_2 \frac{\partial}{\partial y_i}\left[\frac{-\mathbf{r}^{P_1 P_i} \cdot \mathbf{r}^{P_1 P_j}}{\left|\mathbf{r}^{P_1 P_j}\right|^3}\right] = \mathbf{e}_2 \left(\frac{-y_j}{\left|\mathbf{r}^{P_1 P_j}\right|^3}\right) \tag{1.73}$$

$$\mathbf{e}_3 \frac{\partial}{\partial z_i}\left[\frac{-\mathbf{r}^{P_1 P_i} \cdot \mathbf{r}^{P_1 P_j}}{\left|\mathbf{r}^{P_1 P_j}\right|^3}\right] = \mathbf{e}_3 \left(\frac{-z_j}{\left|\mathbf{r}^{P_1 P_j}\right|^3}\right). \tag{1.74}$$

Summing Eqs. 1.72–1.74, we obtain

$$\nabla \left[\frac{-\mathbf{r}^{P_1 P_i} \cdot \mathbf{r}^{P_1 P_j}}{\left|\mathbf{r}^{P_1 P_j}\right|^3}\right] = -\frac{\mathbf{r}^{P_1 P_j}}{\left|\mathbf{r}^{P_1 P_j}\right|^3}. \tag{1.75}$$

Thus, from Eqs. 1.71 and 1.75, we may write the gradient of the disturbing function for the ith body as

$$\nabla \mathcal{R}_i = G \sum_{j=2}^{n} m_j \left(\frac{\mathbf{r}^{P_i P_j}}{\left|\mathbf{r}^{P_i P_j}\right|^3} - \frac{\mathbf{r}^{P_1 P_j}}{\left|\mathbf{r}^{P_1 P_j}\right|^3}\right), \qquad j \neq i. \tag{1.76}$$

We observe that the right-hand side of Eq. 1.76 is identical to the right-hand side of Eq. 1.66, so we can write the equations of motion of P_i with respect to P_1 as

$$\ddot{\mathbf{r}}^{P_1 P_i} = \nabla \left(U_i + \mathcal{R}_i\right), \tag{1.77}$$

where

$$U_i = \frac{G\,(m_1 + m_i)}{\left|\mathbf{r}^{P_1 P_i}\right|}. \tag{1.78}$$

From Eq. 1.71, we can show that the term ∇U_i in Eq. 1.77 gives the second term on the left-hand side of Eq. 1.66:

$$\nabla U_i = G\,(m_1 + m_i)\,\nabla \frac{1}{\left|\mathbf{r}^{P_1 P_i}\right|} = \frac{G\,(m_1 + m_i)\,\mathbf{r}^{P_1 P_i}}{\left|\mathbf{r}^{P_1 P_i}\right|^3}, \tag{1.79}$$

where we can pull the numerator of Eq. 1.78 in front of the ∇ operator since it does not depend on the position coordinates x_i, y_i, or z_i. The treatment of the Disturbing Function, $\mathcal{R}$, is the major problem in General Perturbations and is considered in detail in the following chapters. Of course, each particle P_i of mass m_i has a different disturbing function $\mathcal{R}_i$, which is defined by Eq. 1.67.

References

Greenwood, D. T. (1988). *Principles of dynamics* (2nd ed.). Englewood Cliffs: Prentice Hall.

Moulton, F. R. (1914). *An introduction to celestial mechanics* (2nd ed.). New York: The Macmillan Company.

Roy, A. E. (2005). *Orbital motion* (4th ed.). New York: Taylor & Francis Group.

Tragesser, S. G., & Longuski, J. M. (1999). Modeling issues concerning motion of the Saturnian satellites. *Journal of the Astronautical Sciences, 47*(3 and 4), 275–294.

Chapter 2
The Two-Body Problem

2.1 The Special Case of Two Particles

We now consider the two-body problem. This special, unperturbed case provides a point of departure for describing perturbed motion throughout the remainder of the book.

Returning to the equations of relative motion for the n-body problem with no external forces, Eq. 1.63, setting $n = 2$, and using A and B to denote the masses at points P_1 and P_2, we find

$$\ddot{\mathbf{r}}^{AB} = -\frac{Gm_1}{\left|\mathbf{r}^{AB}\right|^3}\mathbf{r}^{AB} - \frac{Gm_2}{\left|\mathbf{r}^{AB}\right|^3}\mathbf{r}^{AB}, \tag{2.1}$$

which simplifies to

$$\ddot{\mathbf{r}}^{AB} = -\frac{G(m_1 + m_2)}{\left|\mathbf{r}^{AB}\right|^3}\mathbf{r}^{AB}. \tag{2.2}$$

Equation 2.2 describes the motion of particle B referenced to an origin at A. The general solution of this vector second-order differential equation requires six independent constants of integration.

2.1.1 Angular Momentum: The First Three Constants of Integration

Taking the cross product of $\mathbf{r}$ with Eq. 2.2 (and dropping the superscripts A and B), we note that the new quantity can be written as the time derivative of the specific angular momentum (i.e., the angular momentum per unit mass)

© The Author(s), under exclusive license to Springer Nature Switzerland AG 2022
J. M. Longuski et al., *Introduction to Orbital Perturbations*, Space Technology
Library 40, https://doi.org/10.1007/978-3-030-89758-1_2

$$\mathbf{r} \times \ddot{\mathbf{r}} = \frac{d}{dt}(\mathbf{r} \times \dot{\mathbf{r}}) = \dot{\mathbf{h}} = \mathbf{0}. \tag{2.3}$$

Integrating Eq. 2.3 yields a constant vector, the specific angular momentum vector of B in its motion relative to A, which can be written as

$$\mathbf{h} = c_1\mathbf{e}_1 + c_2\mathbf{e}_2 + c_3\mathbf{e}_3 \tag{2.4}$$

in some convenient nonrotating rectangular reference frame, where c_1, c_2, and c_3 are the first three constants of the motion for the two-body problem. The relation

$$\mathbf{r} \times \dot{\mathbf{r}} = \mathbf{h} \tag{2.5}$$

means that the position vector $\mathbf{r}$ is orthogonal to $\mathbf{h}$ at all times. Since $\mathbf{h}$ is a constant vector, the motion of B occurs in a fixed plane in space that contains the origin at A and has $\mathbf{h}$ as its normal vector. This fixed plane is known as the orbital plane.

2.1.2   Spatial Orientation of the Orbital Plane

We define the orientation of the orbital plane to facilitate interpretation of the motion in Eq. 2.2 in three-dimensional space. The orbital plane is given with respect to a rectangular reference frame with origin at A. Let $\mathbf{i}$, $\mathbf{j}$, and $\mathbf{k}$ denote an orthogonal basis for the starting inertial reference frame, having axes x, y, and z. In this development, the orientation of this starting reference frame with respect to space is not specified; therefore, one is free to define its orientation as is convenient for the problem at hand.

A common choice for problems of Earth-orbiting satellites takes the fundamental $x–y$ plane to be Earth's equatorial plane. This system is referred to as an equatorial system. In contrast, if one were dealing with the orbit of Earth around the Sun, the Earth's orbital plane is taken as the fundamental plane. This second case is referred to as an ecliptic system. The line of intersection between the ecliptic plane and the equatorial plane defines the vernal equinox as the direction at which the Sun crosses the equatorial plane from South to North (at the Spring equinox in the northern hemisphere). Returning to the equatorial system, the x axis is directed toward the vernal equinox at a given epoch, the z axis is aligned with Earth's axis of rotation and pointing North, and the y axis completes the orthogonal right-handed set.

When the orbital plane is not parallel to the $x–y$ plane, the intersection of the orbital plane with the $x–y$ plane is defined and is referred to as the line of nodes (Fig. 2.1).

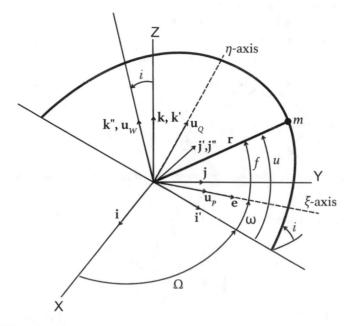

Fig. 2.1 Orientation of the orbital plane in inertial space

Beginning with the inertial unit vectors **i**, **j**, **k**, we consider the following sequence of three rotations (which comprise a 3–1–3 Euler angle sequence) as illustrated in Fig. 2.2:

1. Rotate about **k** by the angle Ω to get to the **i′**, **j′**, **k′** frame. Let Ω have the domain $[0, 2\pi]$ and be defined positive in the sense of a counterclockwise rotation about the z axis (or **k** unit vector).
2. Next, rotate about **i′** by the angle i to get to the **i″**, **j″**, **k″** frame. Let i have the domain $[0, \pi]$ and be defined positive in the sense of a counterclockwise rotation about the **i′** direction.
3. Finally, rotate about **k″** by the angle ω to get to the **u**$_P$, **u**$_Q$, **u**$_W$ frame. Let ω have the domain $[0, 2\pi]$ and be defined positive in the sense of a counterclockwise rotation about the **k″** direction. (We note that counterclockwise is also referred to as the right-hand rule.)

If the x, y, and z components in the original frame are represented, respectively, by ξ, η, and ζ components in the new coordinate frame, then the coordinate transformation is

$$
\begin{Bmatrix} \xi \\ \eta \\ \zeta \end{Bmatrix} = \begin{bmatrix} \cos\omega & \sin\omega & 0 \\ -\sin\omega & \cos\omega & 0 \\ 0 & 0 & 1 \end{bmatrix} \begin{bmatrix} 1 & 0 & 0 \\ 0 & \cos i & \sin i \\ 0 & -\sin i & \cos i \end{bmatrix} \begin{bmatrix} \cos\Omega & \sin\Omega & 0 \\ -\sin\Omega & \cos\Omega & 0 \\ 0 & 0 & 1 \end{bmatrix} \begin{Bmatrix} x \\ y \\ z \end{Bmatrix}.
$$

$$(2.6)$$

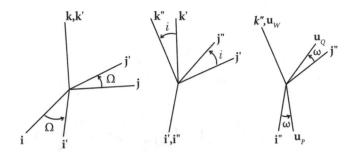

Fig. 2.2 The sequence of three rotations starting from the inertial unit vectors **i, j, k**

Equation 2.6 can be written as

$$
\left\{ \begin{array}{c} \xi \\ \eta \\ \zeta \end{array} \right\} = [l] \left\{ \begin{array}{c} x \\ y \\ z \end{array} \right\},
\tag{2.7}
$$

where, having multiplied the three direction cosine matrices in Eq. 2.6, we obtain

$$
[l] = \begin{bmatrix}
C_\omega C_\Omega - C_i S_\omega S_\Omega & C_\omega S_\Omega + S_\omega C_i C_\Omega & S_\omega S_i \\
-S_\omega C_\Omega - C_\omega C_i S_\Omega & -S_\omega S_\Omega + C_\omega C_i C_\Omega & C_\omega S_i \\
S_i S_\Omega & -S_i C_\Omega & C_i
\end{bmatrix},
\tag{2.8}
$$

where we use the shorthand notation C for cos and S for sin.

From Eqs. 2.4 and 2.8, we have

$$
h\mathbf{u}_W = c_1\mathbf{i} + c_2\mathbf{j} + c_3\mathbf{k} = h\sin i \sin \Omega \mathbf{i} - h\sin i \cos \Omega \mathbf{j} + h\cos i\mathbf{k},
\tag{2.9}
$$

so that the three components may be isolated as

$$
\sin i \sin \Omega = \frac{c_1}{h}
\tag{2.10}
$$

$$
\sin i \cos \Omega = -\frac{c_2}{h}
\tag{2.11}
$$

$$
\cos i = \frac{c_3}{h},
\tag{2.12}
$$

where

$$
h = \sqrt{c_1^2 + c_2^2 + c_3^2}.
$$

The angle i may be found uniquely from Eq. 2.12 since we adopt the convention where $0 \le i \le \pi$ radians, so the principal value of the inverse cosine is used.

Having computed the value of i, one may substitute it into Eqs. 2.10 and 2.11. Then, the signs of $\sin \Omega$ and $\cos \Omega$ may be used to uniquely determine the quadrant of Ω, which can take values from $0 \le \Omega \le 2\pi$ radians. We refer to i as the inclination of the orbit with respect to the x–y plane and Ω as the right ascension of the ascending node, both of which are depicted in Fig. 2.1.

2.1.3 Kepler's Law of Areas

Following Fitzpatrick (1970), we note the geometrical significance of the vector **h**. The rate at which the radius vector from an origin O to a moving point P sweeps out a two-dimensional surface is known as the areal velocity with respect to the origin O.

Let $\Delta \mathbf{A}$ be a vector representing the area of the small triangle OPP' swept out by the radius vector in time Δt, as depicted in Fig. 2.3. The sense of $\Delta \mathbf{A}$ is given by the equation:

$$\Delta \mathbf{A} = \frac{1}{2}[\mathbf{r} \times (\mathbf{r} + \Delta \mathbf{r})] = \frac{1}{2}\mathbf{r} \times \Delta \mathbf{r}. \tag{2.13}$$

From the definition of the areal velocity $\dot{\mathbf{A}}$ as $\lim\limits_{\Delta t \to 0} \dfrac{\Delta \mathbf{A}}{\Delta t}$, we obtain

$$\dot{\mathbf{A}} = \frac{1}{2}\mathbf{r} \times \dot{\mathbf{r}} = \frac{1}{2}\mathbf{h}. \tag{2.14}$$

Fig. 2.3 Vector diagram of area, $\Delta \mathbf{A}$, swept out in time Δt

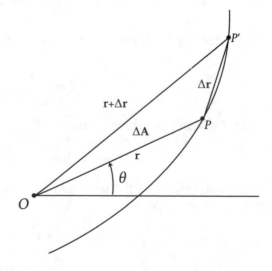

Therefore, the areal velocity (the rate at which area is being swept out by the radius vector **r**, as viewed at O) is a constant equal to half the specific angular momentum as viewed at O. This result verifies Kepler's second law, which states that a radius vector joining any planet to the Sun sweeps out equal areas in equal lengths of time.

We may express the position, velocity, and acceleration in polar coordinates as

$$\mathbf{r} = r\mathbf{u}_r$$

$$\dot{\mathbf{r}} = \dot{r}\mathbf{u}_r + r\dot{\theta}\mathbf{u}_\theta \qquad (2.15)$$

$$\ddot{\mathbf{r}} = \left(\ddot{r} - r\dot{\theta}^2\right)\mathbf{u}_r + \left(r\ddot{\theta} + 2\dot{r}\dot{\theta}\right)\mathbf{u}_\theta,$$

where $\mathbf{u}_r = \mathbf{r}/r$ and $\mathbf{u}_\theta$ is a unit vector in the orbital plane, orthogonal to $\mathbf{u}_r$, in the direction of increasing θ. The coordinate system is completed with $\mathbf{u}_A = \mathbf{u}_r \times \mathbf{u}_\theta$, which is in the direction of **h** and normal to the orbital plane. Then the angular momentum per unit mass may be written as

$$\mathbf{h} = \mathbf{r} \times \dot{\mathbf{r}} = r\mathbf{u}_r \times \left(\dot{r}\mathbf{u}_r + r\dot{\theta}\mathbf{u}_\theta\right) = r^2\dot{\theta}\mathbf{u}_A = h\mathbf{u}_A \qquad (2.16)$$

and

$$r^2\dot{\theta} = h, \qquad (2.17)$$

where $h = |\mathbf{h}|$ is a positive constant and θ must be measured such that $\dot{\theta}$ is positive.

2.1.4   The Remaining Constants of Integration

Returning to Eq. 2.2, again dropping the superscripts, and defining $\mu \equiv G(m_1+m_2)$, we have

$$\ddot{\mathbf{r}} = -\frac{\mu}{|\mathbf{r}|^3}\mathbf{r}, \qquad (2.18)$$

and we also have the constant relation

$$\mathbf{h} = \mathbf{r} \times \dot{\mathbf{r}}. \qquad (2.19)$$

To rearrange Eq. 2.18 into a form that can be integrated, we take the cross product with the constant vector **h**:

$$\ddot{\mathbf{r}} \times \mathbf{h} = -\frac{\mu}{r^3}\mathbf{r} \times \mathbf{h}$$

$$= -\frac{\mu}{r^3}\mathbf{r} \times (\mathbf{r} \times \dot{\mathbf{r}})$$

$$= \frac{\mu}{r^3} \left[\dot{\mathbf{r}} (\mathbf{r} \cdot \mathbf{r}) - \mathbf{r}(\mathbf{r} \cdot \dot{\mathbf{r}}) \right]$$

$$= \mu \left(\frac{\dot{\mathbf{r}}}{r} - \frac{\mathbf{r}\dot{r}}{r^2} \right), \tag{2.20}$$

where we use the vector identity: $\mathbf{A} \times (\mathbf{B} \times \mathbf{C}) = (\mathbf{A} \cdot \mathbf{C})\mathbf{B} - (\mathbf{A} \cdot \mathbf{B})\mathbf{C}$ and the fact (which is apparent with the convenient use of polar coordinates in Eq. 2.15 for the planar motion) that $\mathbf{r} \cdot \dot{\mathbf{r}} = (r\mathbf{u}_r) \cdot (\dot{r}\mathbf{u}_r + r\dot{\theta}\mathbf{u}_\theta) = r\dot{r}$. Then we note that the inertial time derivative of the unit vector along $\mathbf{r}$ may be written as

$$\frac{d}{dt} \left(\frac{\mathbf{r}}{|\mathbf{r}|} \right) = \frac{\dot{\mathbf{r}}}{r} - \frac{\mathbf{r}\dot{r}}{r^2}. \tag{2.21}$$

Thus Eq. 2.20 may be written as

$$\ddot{\mathbf{r}} \times \mathbf{h} = \mu \frac{d}{dt} \left(\frac{\mathbf{r}}{r} \right), \tag{2.22}$$

which can be integrated to give

$$\dot{\mathbf{r}} \times \mathbf{h} = \mu \left(\frac{\mathbf{r}}{r} + \mathbf{e} \right), \tag{2.23}$$

where $\mathbf{e}$ is a dimensionless vector constant of integration that lies in the orbital plane. This vector is referred to as the eccentricity vector, shown in Fig. 2.1. The eccentricity vector defines the third rotation angle ω, to be the argument of periapsis. This new constant is the angle between the line of nodes and the eccentricity vector, measured in the orbital plane and defined as positive with a right-handed rotation about the orbit normal vector, $\mathbf{h}$.

Taking the dot product of $\mathbf{r}$ with Eq. 2.23 produces the scalar equation:

$$\mathbf{r} \cdot (\dot{\mathbf{r}} \times \mathbf{h}) = \mathbf{r} \cdot \mu \left(\frac{\mathbf{r}}{r} + \mathbf{e} \right), \tag{2.24}$$

and, using the identity $\mathbf{A} \cdot \mathbf{B} \times \mathbf{C} = \mathbf{C} \cdot \mathbf{A} \times \mathbf{B}$, we can write

$$\mathbf{h} \cdot (\mathbf{r} \times \dot{\mathbf{r}}) = h^2 = \mathbf{r} \cdot \mu \left(\frac{\mathbf{r}}{r} + \mathbf{e} \right)$$

$$= \mu (r + \mathbf{r} \cdot \mathbf{e})$$

$$= \mu (r + re \cos f), \tag{2.25}$$

where the angle, f, is known as the true anomaly. The true anomaly is the angle between the vectors $\mathbf{e}$ and $\mathbf{r}$, taken to have the positive sense in the direction of travel in the orbital plane (or by the right-hand rule relative to the angular momentum vector, $\mathbf{h}$). Solving for r gives

$$r = \frac{h^2/\mu}{1 + e \cos f},$$
(2.26)

which is the equation of a conic section in polar coordinates with the origin at the focus of the conic section and the eccentricity, e. Equation 2.26 is also known as the conic equation.

Recall that we set out to find six independent constants of integration to specify the motion in Eq. 2.2, and it might at first appear that we have succeeded by finding $\mathbf{h}$ and $\mathbf{e}$. However, since $\mathbf{h} \cdot \mathbf{e} = 0$, we presently only have five independent constants. While we found a closed-form solution to the nonlinear equation of motion in Eq. 2.2, the independent variable in Eq. 2.26 is not time but rather the angle, true anomaly, f. We therefore have in Eq. 2.26 a geometric description of the orbit in which we can determine r for all values of f if we know μ, h, and e. The position of the orbiting mass at a given time is absent in this geometric description.

Before obtaining the missing sixth constant, let us first discuss the distinct cases of conic sections. Examining Eq. 2.26, we can make several observations based on the values the nonnegative eccentricity e may take. The radius r is constant, and the orbit is circular when $e = 0$. Elliptic orbits have $0 \le e < 1$ because the denominator must not vanish for any value $0 \le f < 2\pi$ if the value of r is to have a finite maximum radius. Circular orbits are merely a special case of elliptical orbits.

If $e = 1$, then Eq. 2.26 indicates that the value of r approaches infinity as f approaches $\pm\pi$, and a parabolic orbit is the result. If $e > 1$, the orbit is a hyperbola wherein the value of r approaches infinity along asymptotes defined by values of f obtained from $e \cos f_\infty = -1$, where the asymptotic true anomaly is $f_\infty < \pi$. Thus, orbital motion in the two-body problem follows the form of the conic sections: ellipses, parabolas, and hyperbolas.

2.1.5 Elliptic Orbits

Kepler's first law asserts that the orbit of each planet is an ellipse with the Sun at one focus, which has been verified with Eq. 2.26 and the fact that elliptic orbits provide one solution for the relative motion of a planet with respect to the Sun. The polar equation of an ellipse is

$$r = \frac{a\left(1 - e^2\right)}{1 + e \cos f},$$
(2.27)

where a is the semimajor axis of the ellipse.

Comparing Eqs. 2.26 and 2.27, we can express the specific angular momentum as a function of the two masses, the semimajor axis, and eccentricity of the orbit:

$$h = \sqrt{\mu a \left(1 - e^2\right)}.$$
(2.28)

We now introduce three particular points of interest in the orbit.

Setting $f = 0$ results in the minimum value of r,

$$r_p = a\,(1-e)\,. \tag{2.29}$$

At this point of closest approach between a satellite and the attracting body at one focus of the ellipse, the satellite is at periapsis. For an Earth-orbiting object, this point is called the perigee; for an object orbiting the Sun, it is known as the perihelion.

Setting $f = \pi/2$ results in $r = a\left(1-e^2\right)$. This separation distance is termed the semi-latus rectum, or the parameter, p, of the ellipse, as shown in Fig. 2.4. Thus, we may write Eq. 2.27 as

$$r = \frac{p}{1 + e\cos f}\,. \tag{2.30}$$

Comparing Eqs. 2.26 and 2.30, the specific angular momentum is related to the parameter by

$$p = \frac{h^2}{\mu} = a\left(1-e^2\right)\,. \tag{2.31}$$

Setting $f = \pi$ results in the maximum value of r

$$r_a = a\,(1+e)\,. \tag{2.32}$$

At this farthest point of separation, the satellite is at apoapsis (or apogee, aphelion, etc., depending on which attracting body it is orbiting).

Integrating the scalar equality obtained from the vector relationship in Eq. 2.14

$$\frac{d\mathbf{A}}{dt} = \frac{1}{2}\mathbf{h}$$

yields

Fig. 2.4 The geometry of the ellipse, where $r = p/(1 + e \cos f)$

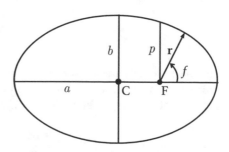

$$A = \frac{1}{2}ht + c, \tag{2.33}$$

where c is a constant of integration. Setting $A = 0$ at some initial time, t_0, we obtain

$$c = -\frac{1}{2}ht_0. \tag{2.34}$$

The area swept out in the elapsed time $t - t_0$ is

$$A(t) = \frac{1}{2}h(t - t_0). \tag{2.35}$$

The area of the ellipse is πab, and therefore we may express the orbit period P, the time required to complete one revolution of the elliptical path, as

$$P = \frac{\pi ab}{\dot{A}} = \frac{2\pi ab}{h} = \frac{2\pi a \cdot a\sqrt{(1 - e^2)}}{\sqrt{(\mu a)}\sqrt{(1 - e^2)}} = 2\pi\sqrt{\frac{a^3}{\mu}}, \tag{2.36}$$

where $b = a\sqrt{1 - e^2}$ is the semiminor axis of the ellipse, and we have used Eqs. 2.14 and 2.28. Equation 2.36 verifies Kepler's third law (the squares of the orbital periods of the planets are proportional to the cubes of their mean distances from the Sun) to a good approximation wherein the masses of the planets are much less than the mass of the Sun. For objects of negligibly small mass orbiting a massive attracting body (i.e., $m_2 << m_1$), μ is well-approximated by $\mu \approx Gm_1$, and the mass, m_2, of the satellite has no effect on the orbit period, P, which is determined solely by μ and the size of the orbit given by its semimajor axis.

2.1.6   Conservation of Energy

The total mechanical energy of the satellite is the combination of the kinetic and potential energies

$$T + V = \frac{mv^2}{2} - \frac{m\mu}{r}, \tag{2.37}$$

and we have only the conservative force of mutual gravitational attraction in the present two-body problem. Therefore this sum is a constant. Dividing through by the mass of the satellite gives the specific energy

$$\mathcal{E} = \frac{v^2}{2} - \frac{\mu}{r} = \text{constant}. \tag{2.38}$$

To find the value of this constant, we consider the satellite at periapsis, where (for any of the types of conics) the radius and velocity vectors are orthogonal and

$$h = |\mathbf{r} \times \mathbf{v}| = r_p v_p = \sqrt{\mu p}. \tag{2.39}$$

Rearranging to isolate v_p and using Eqs. 2.29 and 2.31 give

$$v_p = \frac{\sqrt{\mu p}}{r_p} = \frac{\sqrt{\mu p}}{a\,(1-e)} = \sqrt{\mu p}\frac{(1+e)}{p} = (1+e)\sqrt{\frac{\mu}{p}}. \tag{2.40}$$

Thus the constant $\mathcal{E}$ in Eq. 2.38 must equal

$$\mathcal{E} = \frac{v^2}{2} - \frac{\mu}{r} = \frac{(1+e)^2\,\mu}{2p} - \frac{(1+e)\,\mu}{p}, \tag{2.41}$$

and we note that the energy is $\mathcal{E} = 0$ for a parabolic orbit ($e = 1$, $p \neq 0$). For elliptic and hyperbolic orbits, we substitute the relation $p = a(1-e^2)$ from Eq. 2.31 to obtain

$$
\begin{aligned}
\mathcal{E} &= \frac{(1+e)^2\,\mu}{2a\left(1-e^2\right)} - \frac{(1+e)\,\mu}{a\left(1-e^2\right)} \\
&= \frac{\left(1+2e+e^2\right)\mu}{2a\left(1-e^2\right)} - \frac{2\,(1+e)\,\mu}{2a\left(1-e^2\right)} \\
&= -\frac{\mu}{2a}.
\end{aligned} \tag{2.42}
$$

This total specific energy depends only on the semimajor axis of the orbit and the gravitational parameter, μ, for all conic orbits. We use the convention where the semimajor axis of a hyperbolic orbit is considered to be negative ($a < 0$), so that the form of the conic equations is unchanged between elliptical and hyperbolic orbits. Combining Eqs. 2.38 and 2.42 and rearranging give the *vis-viva* equation for the velocity at a given radius

$$v^2 = \mu \left(\frac{2}{r} - \frac{1}{a}\right). \tag{2.43}$$

In the case of hyperbolic orbits, it is common to refer to the total energy not by the semimajor axis but instead by the speed at infinite radius, or the hyperbolic excess speed, v_∞. From Eq. 2.43, setting $r = \infty$, we obtain

$$v_\infty = \sqrt{-\frac{\mu}{a}}, \tag{2.44}$$

Table 2.1 Types of conics

Orbit type	Eccentricity	Energy	Semimajor axis
Ellipse	$0 \leq e < 1$	$\mathcal{E} < 0$	$a > 0$
Parabola	$e = 1$	$\mathcal{E} = 0$	$a = \infty$
Hyperbola	$e > 1$	$\mathcal{E} > 0$	$a < 0$

which produces a real value because we take a to be negative for hyperbolic orbits. Table 2.1 summarizes the properties of the three types of conic orbits.

To develop an expression for the eccentricity of a general conic orbit, we pursue an alternate derivation for Eq. 2.26. We recall Eqs. 2.17 and 2.38

$$h = r^2 \dot{\theta}$$

$$\mathcal{E} = \frac{v^2}{2} - \frac{\mu}{r},$$

where from Eq. 2.15

$$|\dot{\mathbf{r}}|^2 = v^2 = \dot{r}^2 + r^2 \dot{\theta}^2,$$

so that we have two first-order differential equations in r and θ:

$$h = r^2 \dot{\theta}$$

$$\mathcal{E} = \frac{1}{2} \left(\dot{r}^2 + r^2 \dot{\theta}^2 \right) - \frac{\mu}{r}.$$

Solving the first of these equations for dt yields

$$dt = \frac{r^2}{h} d\theta,$$

and substituting into the second differential equation gives

$$\mathcal{E} = \frac{1}{2} \left[\left(\frac{h}{r^2} \frac{dr}{d\theta} \right)^2 + r^2 \left(\frac{h}{r^2} \right)^2 \right] - \frac{\mu}{r}$$

$$= \frac{1}{2} \left[\left(\frac{h}{r^2} \frac{dr}{d\theta} \right)^2 + \frac{h^2}{r^2} \right] - \frac{\mu}{r}.$$

Rearranging gives the differential equation:

$$\frac{2}{h^2}\left(\mathcal{E} + \frac{\mu}{r}\right) = \frac{1}{r^2} + \frac{1}{r^4}\left(\frac{dr}{d\theta}\right)^2.$$

Next, we introduce a change of variable to facilitate integration, defining

$$\rho = \frac{1}{r} = r^{-1}$$

$$\frac{d\rho}{d\theta} = -r^{-2}\frac{dr}{d\theta}. \tag{2.45}$$

Substituting into the differential equation produces

$$\frac{2}{h^2}\left(\mathcal{E} + \mu\rho\right) = \rho^2 + \left(\frac{d\rho}{d\theta}\right)^2,$$

and we solve for $d\rho/d\theta$ to obtain

$$\frac{d\rho}{d\theta} = \pm\sqrt{\frac{2}{h^2}\left(\mathcal{E} + \mu\rho\right) - \rho^2}.$$

Isolating $d\theta$ gives

$$d\theta = \frac{d\rho}{\pm\sqrt{\frac{2}{h^2}\left(\mathcal{E} + \mu\rho\right) - \rho^2}},$$

which can be solved by consulting a table of integrals to obtain

$$\theta = \cos^{-1}\left(\frac{\rho - \frac{\mu}{h^2}}{\sqrt{\frac{\mu^2}{h^4} + \frac{2\mathcal{E}}{h^2}}}\right) + w,$$

where w is a constant of integration related to the direction of periapsis in the orbital plane.

Rearranging gives

$$\rho = \frac{1}{r} = \frac{\mu}{h^2} + \sqrt{\frac{\mu^2}{h^4} + \frac{2\mathcal{E}}{h^2}}\cos\left(\theta - w\right),$$

and we solve for r to obtain

Fig. 2.5 The perifocal
coordinate system in which
$\mathbf{u}_P$ is along the periapsis

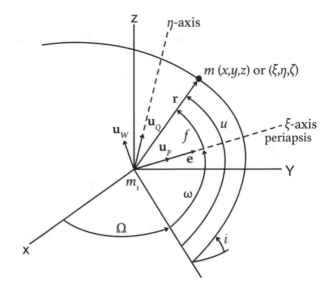

$$r = \frac{h^2/\mu}{1 + \sqrt{1 + \dfrac{2\mathcal{E}h^2}{\mu^2}} \cos(\theta - \omega)}, \qquad (2.46)$$

where we take the reference direction from which θ is measured to be the line of
nodes so that $\theta = u$ and $w = \omega$, the argument of periapsis, as depicted in Fig. 2.5.

Comparing Eqs. 2.26 and 2.46, we note that

$$f = u - \omega \qquad (2.47)$$

$$e = \sqrt{1 + \frac{2\mathcal{E}h^2}{\mu^2}}, \qquad (2.48)$$

with Eq. 2.48 providing a general expression of the eccentricity for any conic orbit.

We now summarize the physical interpretation of the five constants thus far
obtained. In the PQW system shown in Fig. 2.5, the fundamental plane is the orbital
plane of the satellite, and the origin is located at the center of the central body
(e.g., the planet or the Sun).

We recall that $\mathbf{u}_P$ is a unit vector in the direction of the periapsis, $\mathbf{u}_Q$ is a unit
vector in the direction $f = 90°$ (true anomaly in the orbital plane), and $\mathbf{u}_W$ is a unit
vector normal to the orbital plane such that

$$\mathbf{u}_W = \mathbf{u}_P \times \mathbf{u}_Q. \qquad (2.49)$$

Next we find the perifocal unit vectors $\mathbf{u}_P$, $\mathbf{u}_Q$, $\mathbf{u}_W$ in terms of the inertial unit vectors $\mathbf{i}$, $\mathbf{j}$, $\mathbf{k}$.

From Eq. 2.7, we can write the perifocal unit vectors in terms of the inertial unit vectors:

$$\mathbf{u}_P = P_1\mathbf{i} + P_2\mathbf{j} + P_3\mathbf{k}$$

$$\mathbf{u}_Q = Q_1\mathbf{i} + Q_2\mathbf{j} + Q_3\mathbf{k} \tag{2.50}$$

$$\mathbf{u}_W = W_1\mathbf{i} + W_2\mathbf{j} + W_3\mathbf{k},$$

where the P_i, Q_i, and W_i are the appropriate elements of the direction cosine matrix in Eq. 2.8:

$$P_1 = \cos\omega\cos\Omega - \cos i \sin\omega\sin\Omega$$

$$P_2 = \cos\omega\sin\Omega + \sin\omega\cos i \cos\Omega$$

$$P_3 = \sin\omega\sin i$$

$$Q_1 = -\sin\omega\cos\Omega - \cos\omega\cos i \sin\Omega$$

$$Q_2 = -\sin\omega\sin\Omega + \cos\omega\cos i \cos\Omega \tag{2.51}$$

$$Q_3 = \cos\omega\sin i$$

$$W_1 = \sin i \sin\Omega$$

$$W_2 = -\sin i \cos\Omega$$

$$W_3 = \cos i$$

The five constants we have thus far determined using the specific angular momentum vector, $\mathbf{h}$, and the eccentricity vector, $\mathbf{e}$, may be summarized as a, e, i, Ω, and ω. The semimajor axis is denoted by a and the eccentricity by e. The orbital inclination is denoted by i, the right ascension of the ascending node (RAAN) is represented by Ω, and the argument of periapsis is denoted by ω. When working in an equatorial system, we use RAAN, whereas in an ecliptic system Ω is the longitude of the ascending node because the angle is measured in a different fundamental plane. In our development, where the equations are general and could—depending on the application—refer to either RAAN or the longitude of the ascending node, we refer to Ω as the "ascending node angle."

For further discussion of hyperbolic and parabolic orbits, refer to Vallado (2013) or Prussing and Conway (2013). We focus primarily on the ellipse for the remainder of the book.

2.1.7   Position in Orbit as a Function of Time

We now return to the matter of the sixth constant of the motion. A convenient selection is T, the time of periapsis passage. This completes the set of six constants needed to completely define the orbit; however, we have not yet solved the problem of locating the position of the orbiting mass on the orbit at a given time.

2.1.7.1   Eccentric Anomaly, Mean Anomaly, and Kepler's Equation

Following McCuskey (1963), we consider an elliptical orbit with semimajor axis, a, and an auxiliary circle with radius equal to a.

The mass m_1 is at the focus F, and the orbiting mass m_2 travels the ellipse counterclockwise with an instantaneous position denoted by the point H. C is the common center for both the circle and ellipse, D is the periapsis point on the ellipse (nearest to the occupied focus, F). The location of H is given by the polar coordinates: the radial position r and the true anomaly f. The line segment BA is perpendicular to CD and passing through H, the position of m_2. We define the auxiliary angle, E, the eccentric anomaly, as shown in Fig. 2.6. As the true anomaly evolves over 2π radians, so does E.

From the properties of the ellipse and Fig. 2.6,

$$\xi = r \cos f = a \cos E - ae \qquad (2.52)$$

$$\eta = r \sin f = a \sin E - HA. \qquad (2.53)$$

Fig. 2.6 Elliptical orbit and auxiliary circle, with the auxiliary angle, E, the eccentric anomaly

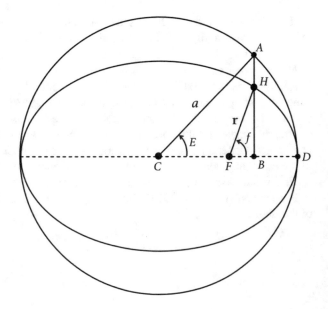

It can be shown (with the first equality demonstrable via elementary geometry) that

$$\frac{\text{area } BDH}{\text{area } BDA} = \frac{BH}{BA} = \frac{b}{a} = \sqrt{1 - e^2}, \tag{2.54}$$

where b is the semiminor axis of the ellipse. So we have

$$HA = BA - BH = BA \left(1 - \sqrt{1 - e^2}\right), \tag{2.55}$$

and Eq. 2.53 may be written as

$$\eta = r \sin f = a \sin E \sqrt{1 - e^2}. \tag{2.56}$$

Summing the squares of Eqs. 2.52 and 2.56 and simplifying yield

$$r = a \left(1 - e \cos E\right). \tag{2.57}$$

In Eq. 2.57, we now have an expression for r, provided that we can determine E as a function of the time, t. We use a geometric argument to find this expression and note that

$$\text{area } BDH = \text{area } DFH - \text{area } BFH \tag{2.58}$$

$$\text{area } BDA = \text{area } DCA - \text{area } BCA, \tag{2.59}$$

and from Kepler's second law, the area DFH swept out in the time $t - T$ (from integrating Eq. 2.14 and substituting Eq. 2.28) is

$$\text{area } DFH = \frac{1}{2} na^2 \sqrt{1 - e^2} \, (t - T), \tag{2.60}$$

where n is the mean motion (the average angular rate of the satellite) given by

$$n = \frac{2\pi}{P} = \left(\frac{\mu}{a^3}\right)^{1/2}. \tag{2.61}$$

Thus we can write Eq. 2.54 as

$$\sqrt{1 - e^2} = \frac{\frac{1}{2} na^2 \sqrt{1 - e^2} \, (t - T) - \frac{1}{2} r^2 \sin f \cos f}{\frac{1}{2} a^2 E - \frac{1}{2} a^2 \sin E \cos E}. \tag{2.62}$$

Using Eqs. 2.52 and 2.56, we obtain

$$\sqrt{1 - e^2} = \frac{\frac{1}{2} na^2 \sqrt{1 - e^2} \, (t - T) - \frac{1}{2} a \sin E \sqrt{1 - e^2} \, (a \cos E - ae)}{\frac{1}{2} a^2 E - \frac{1}{2} a^2 \sin E \cos E}, \tag{2.63}$$

which, by some algebraic manipulation, simplifies to

$$n\,(t - T) = E - e\sin E. \tag{2.64}$$

Then we can write

$$M = n\,(t - T) = M_0 + nt = E - e\sin E, \tag{2.65}$$

where we define M to be the mean anomaly. The mean anomaly is the angle through which the radius vector would have traveled if it were moving uniformly with the average angular rate of $2\pi/P$. M_0 is the value of the mean anomaly at some initial time, t_0, which need not correspond to the time of periapsis passage, T.

Equation 2.65 is Kepler's equation and relates time to position (as described by the eccentric anomaly) for an elliptic orbit. For a given elliptic orbit, it is straightforward to compute the time corresponding to the orbiting body being at a particular position (i.e., value of E) in the orbit. However, determining the position of the body at a specified time is not so easy, as Kepler's equation is transcendental in the variable E. Prussing and Conway (2013) show that the solution to Kepler's equation for E exists and is unique, and thus iterative methods may be reliably employed to numerically solve for the eccentric anomaly.

We have previously developed an expression for the radius vector in terms of the eccentric anomaly. Often, it is geometrically more intuitive to use the true anomaly, f, so we now develop an expression for f in terms of E.

From Eq. 2.52, we can write

$$a\cos E = ae + r\cos f, \tag{2.66}$$

and from the polar equation of the orbit (Eq. 2.27),

$$r = \frac{a\left(1 - e^2\right)}{1 + e\cos f}. \tag{2.67}$$

Combining these two expressions gives

$$\cos E = \frac{e + \cos f}{1 + e\cos f}. \tag{2.68}$$

Therefore, we have

$$1 - \cos E = 1 - \frac{e + \cos f}{1 + e\cos f}$$

$$= \frac{1 + e\cos f - e - \cos f}{1 + e\cos f}$$

$$= \frac{(1-e)(1-\cos f)}{1+e\cos f} \tag{2.69}$$

and

$$1 + \cos E = 1 + \frac{e + \cos f}{1 + e \cos f}$$

$$= \frac{1 + e \cos f + e + \cos f}{1 + e \cos f}$$

$$= \frac{(1+e)(1+\cos f)}{1+e\cos f}, \tag{2.70}$$

such that dividing gives

$$\frac{1 - \cos E}{1 + \cos E} = \frac{(1-e)(1-\cos f)}{(1+e)(1+\cos f)}. \tag{2.71}$$

Using the trigonometric identity

$$\tan^2 \theta = \frac{1 - \cos \theta}{1 + \cos \theta} \tag{2.72}$$

yields

$$\tan^2 \frac{E}{2} = \frac{1-e}{1+e} \tan^2 \frac{f}{2}, \tag{2.73}$$

and we have

$$\tan \frac{E}{2} = \left(\frac{1-e}{1+e}\right)^{1/2} \tan \frac{f}{2}. \tag{2.74}$$

Since $\tan(f/2)$ and $\tan(E/2)$ always have the same sign, there is no ambiguity in the sign of the square root. Equation 2.74 enables conversion between true anomaly and eccentric anomaly.

Later, we require expressions for $\sin E$ as a function of the true anomaly and for $\sin f$ and $\cos f$ as functions of the eccentric anomaly, so we develop them here. From Eq. 2.56, we write

$$\sin E = \frac{r \sin f}{a\sqrt{1-e^2}},$$

and substituting Eqs. 2.30 and 2.31, we obtain

$$\sin E = \frac{\sin f \sqrt{1 - e^2}}{1 + e \cos f}.$$ (2.75)

From Fig. 2.6 and Eqs. 2.54, 2.56, and 2.57, we may write

$$\cos f = \frac{BF}{r} = \frac{\cos E - e}{1 - e \cos E}$$ (2.76)

$$\sin f = \frac{BH}{r} = \frac{\sin E \sqrt{1 - e^2}}{1 - e \cos E}.$$ (2.77)

2.2   Summary of the Two-Body Problem

We now summarize several useful formulas for the elliptical orbital elements, including some from Fitzpatrick (1970) (the derivation of which is left as an exercise for the reader). These equations provide a methodical, step-by-step approach one can use to convert from position and velocity vectors to the orbital elements.

$$\mathbf{h} = \mathbf{r} \times \dot{\mathbf{r}} = c_1 \mathbf{i} + c_2 \mathbf{j} + c_3 \mathbf{k}$$ (2.78)

$$\sin i \sin \Omega = \frac{c_1}{h}$$ (2.79)

$$\sin i \cos \Omega = -\frac{c_2}{h}$$ (2.80)

$$\cos i = \frac{c_3}{h}$$ (2.81)

$$a = \frac{r}{2 - \dfrac{r\dot{r}^2}{\mu}}$$ (2.82)

$$e = \left[1 - \frac{r\dot{r}^2}{\mu} \left(2 - \frac{r\dot{r}^2}{\mu} \right) \sin^2 \gamma \right]^{1/2}$$ (2.83)

$$\cos E = \frac{a - r}{ae}$$ (2.84)

$$\mathbf{e} = \frac{\dot{\mathbf{r}} \times \mathbf{h}}{\mu} - \frac{\dot{\mathbf{r}}}{r}$$ (2.85)

$$\mathbf{i}' = \cos \Omega \mathbf{i} + \sin \Omega \mathbf{j}$$ (2.86)

$$\cos \omega = \mathbf{i}' \cdot \mathbf{e}$$ (2.87)

$$M = E - e \sin E$$ (2.88)

$$T = t - M \left(\frac{a^3}{\mu} \right)^{1/2} \tag{2.89}$$

$$P = 2\pi \sqrt{\frac{a^3}{\mu}} \tag{2.90}$$

$$n = \frac{2\pi}{P} = \left(\frac{\mu}{a^3} \right)^{1/2} \tag{2.91}$$

$$v^2 = \mu \left(\frac{2}{r} - \frac{1}{a} \right), \tag{2.92}$$

where γ is the angle between $\mathbf{r}$ and $\dot{\mathbf{r}}$.

The six orbital elements $(a, e, i, \Omega, \omega,$ and T or $M_0)$ are numerically equivalent to the six parameters in the position and velocity vectors. The pair of vectors provides an instantaneous picture of the orbit state (but one cannot begin to imagine the size, shape, and orientation of the orbit), whereas the set of six orbital elements gives an intuitive geometric description. Both sets of six numbers are useful for different purposes. Indeed, there are other sets of orbital elements, such as equinoctial elements, that can be used to mitigate certain limitations of the classical Keplerian elements presented here.

Now we have found six constants of the motion (the orbital elements) that provide the exact solution of the two-body problem. We next turn our attention to perturbations of this two-body motion.

References

Fitzpatrick, P. M. (1970). *Principles of celestial mechanics*. New York: Academic Press.

McCuskey, S. W. (1963). *Introduction to celestial mechanics*. Reading: Addison-Wesley Publishing Company.

Prussing, J. E., & Conway, B. A. (2013). *Orbital mechanics* (2nd ed.). New York: Oxford University Press.

Vallado, D. A. (2013). *Fundamentals of astrodynamics and applications* (4th ed.). El Segundo: Microcosm Press.

Chapter 3
General Perturbations

The Solar System represents a classic example of the n-body problem. The Sun is the dominant force on the orbital motion of Earth, while the perturbations due to the gravity of the other planets give rise to deviations from the 2-body motion. Another example is the case of an artificial satellite in Earth orbit. The dominant force center is Earth, with perturbations due to the gravity of the Sun and Moon, the aspherical (oblate) shape of Earth, atmospheric drag, solar radiation pressure, and general relativity.

3.1 General Perturbations vs. Special Perturbations

General Perturbations make use of the fact that the 2-body orbit due to the dominant potential term U_0 only changes slowly due to the disturbing function $\mathcal{R}$ (see Eq. 1.62). The approach of General Perturbations is to (where possible) develop analytical expressions for the changes in the orbital elements due to $\mathcal{R}$. These expressions are valid within a certain time interval. The major limitations of General Perturbations are: (1) it is not always possible to find appropriate analytical expressions for the important perturbations and (2) the analytical expressions are frequently more difficult to ascertain than the force models necessary for Special Perturbations.

The method of *Special Perturbations* is to numerically integrate the equations of motion in some form, especially when it is not possible to derive a general perturbation theory. In this approach, we can include the effects of all known forces on a body during a small time interval. The Special Perturbations approach suffers from the issues of round-off error in the numerical integration and the selection of an appropriate integration method. Additionally, Special Perturbations do not usually provide clear insight into the fundamental behavior of the system due to the perturbations.

J. M. Longuski et al., *Introduction to Orbital Perturbations*, Space Technology Library 40, https://doi.org/10.1007/978-3-030-89758-1_3

3.2   General Perturbations (Theory of Perturbations)

Our development in this book focuses on General Perturbations. The *Theory of Perturbations* allows us to examine the effect of perturbations on the orbital elements. While the numerical techniques of Special Perturbations may yield high accuracy, such approaches do not provide much information about the qualitative behavior of the particle's orbit because the output is usually position and velocity vectors—which change rapidly. If we instead express the orbital behavior in terms of the orbital elements, we benefit from the ability to study the slowly changing effects that the perturbations have on the size, shape, and orientation of the orbit over time. See Vallado (2013).

3.3   Variation of Parameters

We assume that the Cartesian coordinates defining the position of the body of interest (a planet, spacecraft, etc.) are

$$x = x\,(t, c_1, c_2, c_3, c_4, c_5, c_6)$$

$$y = y\,(t, c_1, c_2, c_3, c_4, c_5, c_6)$$

$$z = z\,(t, c_1, c_2, c_3, c_4, c_5, c_6)$$

or, in vector form,

$$\mathbf{r} = \mathbf{r}\,(t, c_1, c_2, c_3, c_4, c_5, c_6)$$

$$= x\mathbf{i} + y\mathbf{j} + z\mathbf{k}, \tag{3.1}$$

where the c_k ($k = 1, 2, \ldots, 6$) are slowly varying functions due to the disturbance, for example, the Keplerian orbital elements: a, e, i, ω, Ω, and T or M_0. At any instant, we can consider the c_k to be fixed, which defines the osculating orbit. At this instant, the body has the same orbital state in both the unperturbed model and the perturbed orbit.

3.4   Example: Two-Body Motion Disturbed by a Third Body

We now consider the motion of a mass, m_2 associated with P_2, around a central mass, m_1 associated with P_1, under the perturbing influence of a third mass, m_3 associated with P_3. From Eqs. 1.77 and 1.78, we have

$$\ddot{\mathbf{r}}^{P_1 P_2} - \nabla \frac{G\left(m_1 + m_2\right)}{\left|\mathbf{r}^{P_1 P_2}\right|} = \nabla \mathcal{R}_3, \tag{3.2}$$

where $\mathcal{R}_3$ is the disturbing function due to the effect of the third body. The second term on the left-hand side is the gradient of the potential function for the two-body problem, whereas the term on the right-hand side is the gradient of the disturbing function associated with the third body. It follows from Eq. 1.66 (with $i = 2$) that

$$\ddot{\mathbf{r}}^{P_1 P_2} + \frac{G\left(m_1 + m_2\right) \mathbf{r}^{P_1 P_2}}{\left|\mathbf{r}^{P_1 P_2}\right|^3} = \nabla \mathcal{R}_3, \tag{3.3}$$

where from Eq. 1.67 (with $i = 2$ and $j = 3$), we have

$$\mathcal{R}_3 = G m_3 \left(\frac{1}{\left|\mathbf{r}^{P_2 P_3}\right|} - \frac{\mathbf{r}^{P_1 P_2} \cdot \mathbf{r}^{P_1 P_3}}{\left|\mathbf{r}^{P_1 P_3}\right|^3} \right). \tag{3.4}$$

Let us simplify the notation as follows:

$$\mathbf{r} = \mathbf{r}^{P_1 P_2} = x\mathbf{i} + y\mathbf{j} + z\mathbf{k}$$
$$\mathbf{r}_3 = \mathbf{r}^{P_1 P_3} = x_3\mathbf{i} + y_3\mathbf{j} + z_3\mathbf{k}, \tag{3.5}$$
$$\boldsymbol{\rho} = \mathbf{r}^{P_2 P_3} = (x_3 - x)\mathbf{i} + (y_3 - y)\mathbf{j} + (z_3 - z)\mathbf{k}$$

and

$$r^3 = |\mathbf{r}|^3 = \left|\mathbf{r}^{P_1 P_2}\right|^3$$
$$r_3^3 = |\mathbf{r}_3|^3 = \left|\mathbf{r}^{P_1 P_3}\right|^3 \tag{3.6}$$
$$\rho^3 = |\boldsymbol{\rho}|^3 = \left|\mathbf{r}^{P_2 P_3}\right|^3,$$

where we have placed the origin of the inertial **ijk**-frame at the center of the central mass, m_1. Making the notational changes from Eqs. 3.5 and 3.6 in Eq. 3.3, we obtain the equation of motion of our body of interest, m_2, with respect to the central body, m_1, including the disturbance (from m_3):

$$\ddot{\mathbf{r}} + \frac{G\left(m_1 + m\right)\mathbf{r}}{r^3} = \nabla \mathcal{R}_3, \tag{3.7}$$

where we have dropped the subscripts denoting $i = 2$. Equation 3.4 can be written as

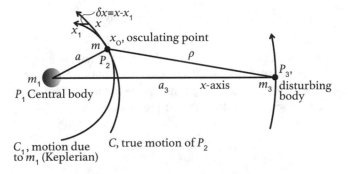

Fig. 3.1 The osculating orbit of m about the central body, with disturbing body m_3. A typical example involves the perturbation of an asteroid's orbit around the Sun by Jupiter. Adapted from McCuskey (1963)

$$R_3 = Gm_3 \left[\frac{1}{\rho} - \frac{(x_3 x + y_3 y + z_3 z)}{r_3^3} \right]. \tag{3.8}$$

3.4.1 Simple Order-of-Magnitude Calculation

Prior to undertaking a general discussion of Eq. 3.7, we consider the case where m and m_3 are in circular orbits with radii a and a_3 ($a_3 > a$) around the central body m_1 (Fig. 3.1).

In this example, we consider only the x-component of the motion of m. For a short time interval we have

$$x = x_0 + \left(\frac{dx}{dt} \right)_0 \tau + \left(\frac{d^2 x}{dt^2} \right)_0 \frac{\tau^2}{2} + \ldots, \quad \text{(true)}$$

$$x_1 = x_0 + \left(\frac{\partial x}{\partial t} \right)_0 \tau + \left(\frac{\partial^2 x}{\partial t^2} \right)_0 \frac{\tau^2}{2} + \ldots, \quad \text{(osculating)}, \tag{3.9}$$

where we use the subscript 0 to denote a term that is evaluated at the osculating point. If the perturbing mass m_3 were not present, the body of interest m would move on the osculating orbit, C_1. The perturbation due to m_3 is

$$x - x_1 \equiv \delta x = \left[\left(\frac{dx}{dt} \right)_0 - \left(\frac{\partial x}{\partial t} \right)_0 \right] \tau + \left[\left(\frac{d^2 x}{dt^2} \right)_0 - \left(\frac{\partial^2 x}{\partial t^2} \right)_0 \right] \frac{\tau^2}{2}, \tag{3.10}$$

where we have dropped terms higher than second order. To distinguish the total derivatives from the partial derivatives, we note that the total derivatives refer to the true orbit and take into account the changing orbital elements, $c_k(t)$:

$$\frac{dx}{dt} = \frac{\partial x}{\partial t} + \sum_{k=1}^{6} \frac{\partial x}{\partial c_k} \dot{c}_k, \tag{3.11}$$

whereas the partial derivatives refer to the osculating orbit (where the c_k are regarded as constants). In other words, the c_k are treated as constants in the partial derivatives for the osculating orbit, but in the true orbit we take them to be live functions of time.

By the definition of the osculating orbit, at the osculating point, x_0, the velocities are the same

$$\left(\frac{dx}{dt}\right)_0 = \left(\frac{\partial x}{\partial t}\right)_0. \tag{3.12}$$

From Eq. 3.7, we have for the x-component as follows true orbit

$$\left(\frac{d^2x}{dt^2}\right)_0 + \frac{\mu x_0}{r_0^3} = \left(\frac{\partial \mathcal{R}}{\partial x}\right)_0, \tag{3.13}$$

and for the osculating orbit

$$\left(\frac{\partial^2 x}{\partial t^2}\right)_0 + \frac{\mu x_0}{r_0^3} = 0 \tag{3.14}$$

(because the osculating orbit is assumed not to be influenced by the disturbing function), where we define

$$\mu = G(m_1 + m). \tag{3.15}$$

Subtracting Eq. 3.14 from Eq. 3.13 yields

$$\left(\frac{d^2x}{dt^2}\right)_0 - \left(\frac{\partial^2 x}{\partial t^2}\right)_0 = \left(\frac{\partial \mathcal{R}}{\partial x}\right)_0. \tag{3.16}$$

We can evaluate the right-hand side of Eq. 3.16 as follows. From Eq. 3.8, we have

$$\frac{\partial \mathcal{R}}{\partial x} = Gm_3 \frac{\partial}{\partial x} \left[\left(\rho^2\right)^{-1/2} - \frac{(x_3 x + y_3 y + z_3 z)}{r_3^3} \right]$$

$$= Gm_3 \left[-\frac{1}{2} \left(\rho^2\right)^{-3/2} \frac{\partial \left(\rho^2\right)}{\partial x} - \frac{x_3}{r_3^3} \right]$$

$$= Gm_3 \left\{ -\frac{1}{2\rho^3} \frac{\partial}{\partial x} \left[(x_3 - x)^2 + (y_3 - y)^2 + (z_3 - z)^2 \right] - \frac{x_3}{r_3^3} \right\}$$

$$= Gm_3 \left[-\frac{1}{2\rho^3} 2 (x_3 - x) (-1) - \frac{x_3}{r_3^3} \right]$$

$$= Gm_3 \left[\frac{(x_3 - x)}{\rho^3} - \frac{x_3}{r_3^3} \right]. \tag{3.17}$$

Equation 3.17 can be verified using the right-hand side of Eq. 1.66. Substituting Eqs. 3.12, 3.16, and 3.17 into the perturbation equation (Eq. 3.10) provides

$$\delta x = \frac{1}{2}\tau^2 \left(\frac{\partial \mathcal{R}}{\partial x} \right)_0 = \frac{1}{2}\tau^2 Gm_3 \left[\frac{(x_3 - x)}{\rho^3} - \frac{x_3}{r_3^3} \right]_0. \tag{3.18}$$

Let us assume that the masses are approximately aligned so that

$$x_3 - x \approx a_3 - a, \tag{3.19}$$

and we can ignore perturbations in the y and z components. By Kepler's third law for the orbital period, P:

$$P^2 = \frac{4\pi^2 a^3}{Gm_1}, \qquad (m_1 \gg m), \tag{3.20}$$

so we can write the universal gravitational constant as

$$G = \frac{4\pi^2 a^3}{P^2 m_1}. \tag{3.21}$$

Substituting Eqs. 3.19 and 3.21 into Eq. 3.18, we have

$$\delta x = 2\pi^2 a^3 \left(\frac{m_3}{m_1} \right) \left(\frac{\tau}{P} \right)^2 \left[\frac{1}{(a_3 - a)^2} - \frac{1}{a_3^2} \right] \tag{3.22}$$

for the perturbation in this simplified approach.

3.4.2   Illustration of Simplified Approach

As a numerical example, we consider the perturbation produced by Jupiter on an asteroid that has a period one-half that of Jupiter's period. We make the following definitions:

$$m_1 = m_{\mathrm{Sun}}$$

$$m = m_{asteroid}$$

$$m_3 = m_{Jupiter} \qquad (3.23)$$

$$P = \frac{1}{2} P_{Jupiter}.$$

From Kepler's Law for orbital periods (Eq. 3.20),

$$\frac{P^2}{P_{Jupiter}^2} = \frac{1}{4} = \frac{a^3}{a_{Jupiter}^3}. \qquad (3.24)$$

From Eq. 3.24, we have

$$a = (1/4)^{1/3} a_{Jupiter} = 0.63 \, (5.2 \, \text{AU}) = 3.3 \, \text{AU} \qquad (3.25)$$

and

$$\frac{m_3}{m_1} = \frac{m_{Jupiter}}{m_{Sun}} = 0.001. \qquad (3.26)$$

For simplicity, we assume that the effective contact for the perturbation lasts for 0.1 period of the asteroid:

$$\tau = 0.1 \, P. \qquad (3.27)$$

Substituting Eqs. 3.25–3.27 into Eq. 3.22 gives

$$\delta x = 2\pi^2 \, (3.3 \, \text{AU})^3 \, (0.001) \, (0.1)^2 \left[\frac{1}{(5.2 \, \text{AU} - 3.3 \, \text{AU})^2} - \frac{1}{(5.2 \, \text{AU})^2} \right]$$

$$\approx 1.7 \times 10^{-3} \, \text{AU}$$

$$\approx 250,000 \, \text{km}, \qquad (3.28)$$

where

$$1 \, \text{AU} = 149.6 \times 10^6 \, \text{km}. \qquad (3.29)$$

This example illustrates that perturbations from a distant planet can give rise to significant deviations in position for an asteroid. Perturbations such as this from Jupiter cause the Kirkwood gaps in the asteroid belt. The gravity of Mimas causes the Cassini division in Saturn's rings in a similar fashion.

3.5   General Discussion of the Perturbation Equation

We now recall our perturbation equation, Eq. 3.7, and we generalize where $\mathbf{F}$ is the force per unit mass:

$$\ddot{\mathbf{r}} + \frac{G\left(m_1 + m\right)\mathbf{r}}{r^3} = \mathbf{F}, \qquad (3.30)$$

and also Eq. 3.1

$$\mathbf{r} = \mathbf{r}\left(t, c_1, c_2, c_3, c_4, c_5, c_6\right),$$

where the c_k are the orbital elements.

In employing the method of variation of parameters and following Fitzpatrick (1970), we assume that the c_k are slowly varying functions of time. From Eq. 3.1, we obtain for the total time derivative:

$$\dot{\mathbf{r}} = \frac{\partial \mathbf{r}}{\partial t} + \sum_{k=1}^{6} \frac{\partial \mathbf{r}}{\partial c_k} \dot{c}_k \qquad \text{(true orbit)}, \qquad (3.31)$$

which describes the behavior of the true orbit at any instant, say t_0. However, at that same instant t_0, there is a particular osculating orbit (with particular values of the c_k) such that

$$\dot{\mathbf{r}}_{t_0} = \frac{\partial \mathbf{r}}{\partial t} \qquad \text{(osculating orbit)}. \qquad (3.32)$$

The velocity in the osculating orbit at the point corresponding to t_0 is the same as the velocity in the true orbit at that same point. For each point on the true orbit, there is an osculating orbit at the same point for which Eq. 3.32 must hold. Since Eqs. 3.31 and 3.32 must hold simultaneously for each instant of time, we must have for all time

$$\sum_{k=1}^{6} \frac{\partial \mathbf{r}}{\partial c_k} \dot{c}_k = \mathbf{0}. \qquad (3.33)$$

Equation 3.33 is a vector constraint on the differential equations for the orbital elements, c_k. If we take the total time derivative of Eq. 3.31, we obtain

$$\ddot{\mathbf{r}} = \frac{\partial^2 \mathbf{r}}{\partial t^2} + \sum_{k=1}^{6} \frac{\partial^2 \mathbf{r}}{\partial c_k \partial t} \dot{c}_k. \qquad (3.34)$$

Substituting $\ddot{\mathbf{r}}$ from Eq. 3.34 into Eq. 3.30 yields the following equation of motion for the true orbit

$$\frac{\partial^2 \mathbf{r}}{\partial t^2} + \frac{\mu \mathbf{r}}{r^3} + \sum_{k=1}^{6} \frac{\partial^2 \mathbf{r}}{\partial c_k \partial t} \dot{c}_k = \mathbf{F}, \tag{3.35}$$

where we define the gravitational parameter as

$$\mu = G \left(m_1 + m \right). \tag{3.36}$$

For the osculating orbit, however, the disturbing function does not exist, so $\mathbf{F} = 0$ and the c_k are constants, and we have

$$\frac{\partial^2 \mathbf{r}}{\partial t^2} + \frac{\mu \mathbf{r}}{r^3} = \mathbf{0}. \tag{3.37}$$

Subtracting Eq. 3.37 from Eq. 3.35 provides the vector equation:

$$\sum_{k=1}^{6} \frac{\partial^2 \mathbf{r}}{\partial c_k \partial t} \dot{c}_k = \mathbf{F}. \tag{3.38}$$

Equations 3.33 and 3.38 can be used to calculate the time rates of change of the orbital elements, $\dot{c}_k$. In the following development, we show how these equations lead directly to Lagrange's planetary equations (after some algebra).

3.6   The Lagrange Brackets

We can simplify Eq. 3.38 by changing the order of the partial derivatives and writing

$$\frac{\partial^2 \mathbf{r}}{\partial t \partial c_k} = \frac{\partial}{\partial c_k} \left(\frac{\partial \mathbf{r}}{\partial t} \right) = \frac{\partial \dot{\mathbf{r}}}{\partial c_k}, \tag{3.39}$$

so that Eq. 3.38 becomes

$$\sum_{k=1}^{6} \frac{\partial \dot{\mathbf{r}}}{\partial c_k} \dot{c}_k = \mathbf{F}. \tag{3.40}$$

We seek a solution of Eqs. 3.33 and 3.40 for the $\dot{c}_k$. That is, we endeavor to isolate the $\dot{c}_k$ to give six first-order differential equations for the orbital elements. This process can be made easier by a rearrangement that introduces Lagrange brackets.

We complete the rearrangement by taking the scalar product of Eq. 3.40 with $\partial \mathbf{r}/\partial c_j$ and subtracting the scalar product of Eq. 3.33 with $\partial \dot{\mathbf{r}}/\partial c_j$ to obtain six scalar equations:

$$\sum_{k=1}^{6} \left[\frac{\partial \mathbf{r}}{\partial c_j} \cdot \frac{\partial \dot{\mathbf{r}}}{\partial c_k} - \frac{\partial \mathbf{r}}{\partial c_k} \cdot \frac{\partial \dot{\mathbf{r}}}{\partial c_j} \right] \dot{c}_k = \mathbf{F} \cdot \frac{\partial \mathbf{r}}{\partial c_j}, \qquad j = 1, 2, \ldots, 6. \qquad (3.41)$$

From Eq. 3.33, we know that the second term in the square brackets in Eq. 3.41 is zero when multiplied by $\dot{c}_k$ and summed. The quantity in square brackets in Eq. 3.41 is known as the Lagrange bracket and denoted by $[c_j, c_k]$.

In terms of its Cartesian components, the Lagrange bracket may be written as

$$[c_j, c_k] = \frac{\partial (x, \dot{x})}{\partial (c_j, c_k)} + \frac{\partial (y, \dot{y})}{\partial (c_j, c_k)} + \frac{\partial (z, \dot{z})}{\partial (c_j, c_k)}, \qquad (3.42)$$

where we use the Jacobian to write the difference in the components of the two scalar products

$$J^x \equiv \frac{\partial (x, \dot{x})}{\partial (c_j, c_k)} \equiv \begin{vmatrix} \frac{\partial x}{\partial c_j} & \frac{\partial x}{\partial c_k} \\ \frac{\partial \dot{x}}{\partial c_j} & \frac{\partial \dot{x}}{\partial c_k} \end{vmatrix}, \qquad (3.43)$$

with analogous terms for y and z. Equation 3.43 is referred to as the "Jacobian of x and $\dot{x}$ with respect to c_j and c_k."

We recall from the definition of the gradient operator in Eq. 1.69 that for Cartesian coordinates

$$\nabla \mathcal{R} = \frac{\partial \mathcal{R}}{\partial x}\mathbf{i} + \frac{\partial \mathcal{R}}{\partial y}\mathbf{j} + \frac{\partial \mathcal{R}}{\partial z}\mathbf{k}, \qquad (3.44)$$

so that

$$\begin{aligned}
\nabla \mathcal{R} \cdot \frac{\partial \mathbf{r}}{\partial c_j} &= \left(\frac{\partial \mathcal{R}}{\partial x}\mathbf{i} + \frac{\partial \mathcal{R}}{\partial y}\mathbf{j} + \frac{\partial \mathcal{R}}{\partial z}\mathbf{k} \right) \cdot \left(\frac{\partial x}{\partial c_j}\mathbf{i} + \frac{\partial y}{\partial c_j}\mathbf{j} + \frac{\partial z}{\partial c_j}\mathbf{k} \right) \\
&= \frac{\partial \mathcal{R}}{\partial x}\frac{\partial x}{\partial c_j} + \frac{\partial \mathcal{R}}{\partial y}\frac{\partial y}{\partial c_j} + \frac{\partial \mathcal{R}}{\partial z}\frac{\partial z}{\partial c_j} \\
&= \frac{\partial \mathcal{R}}{\partial c_j}.
\end{aligned} \qquad (3.45)$$

If the force can be derived from a scalar potential, then $\mathbf{F} = \nabla \mathcal{R}$. Therefore, Eq. 3.41 can be written in terms of Lagrange brackets (using Eqs. 3.45 and 3.42):

$$\sum_{k=1}^{6} \left[c_j, c_k \right] \dot{c}_k = \frac{\partial \mathcal{R}}{\partial c_j}, \qquad j = 1, 2, \ldots, 6. \qquad (3.46)$$

Equation 3.46 provides six equations, which we must solve for the $\dot{c}_k$ to obtain Lagrange's planetary equations.

3.6.1 An Illustrative Example

As an example, we consider a 1-dimensional linear oscillator to demonstrate the use of Lagrange brackets. (For the analysis that follows, see McCuskey (1963) and Szebehely (1989).) The equation of motion for the forced oscillator is

$$\ddot{x} + x = R(t). \qquad (3.47)$$

When the forcing function $R(t) = 0$, the solution is

$$x(t) = c_1 \sin t + c_2 \cos t. \qquad (3.48)$$

Now if we let c_1 and c_2 be functions of time so that the total time derivative is

$$\dot{x} = (c_1 \cos t - c_2 \sin t) - (\dot{c}_1 \sin t + \dot{c}_2 \cos t), \qquad (3.49)$$

we can observe how the same result is obtained using Eq. 3.31:

$$\dot{\mathbf{r}} = \frac{\partial \mathbf{r}}{\partial t} + \sum_{k=1}^{2} \frac{\partial \mathbf{r}}{\partial c_k} \dot{c}_k.$$

The first term on the right-hand side is

$$\frac{\partial r}{\partial t} = \frac{\partial x}{\partial t} = c_1 \cos t - c_2 \sin t, \qquad (3.50)$$

and the second term is

$$\sum_{k=1}^{2} \frac{\partial r}{\partial c_k} \dot{c}_k = \sum_{k=1}^{2} \frac{\partial x}{\partial c_k} \dot{c}_k = \dot{c}_1 \sin t + \dot{c}_2 \cos t, \qquad (3.51)$$

which is identical to the total time derivative in Eq. 3.49. From Eq. 3.33,

$$\sum_{k=1}^{2} \frac{\partial r}{\partial c_k} \dot{c}_k = 0,$$

so Eq. 3.51 is equal to zero (i.e., the second parenthetical term of Eq. 3.49 is zero):

$$\dot{c}_1 \sin t + \dot{c}_2 \cos t = 0. \tag{3.52}$$

Equation 3.52 can be viewed as analogous to the osculating orbit, with no perturbations.

Equation 3.49 reduces to

$$\dot{x} = (c_1 \cos t - c_2 \sin t), \tag{3.53}$$

and differentiating once more directly provides

$$\ddot{x} = (-c_1 \sin t - c_2 \cos t) + (\dot{c}_1 \cos t - \dot{c}_2 \sin t). \tag{3.54}$$

If we now use Eq. 3.34 with $x = r$, we obtain the same result as in Eq. 3.54:

$$\ddot{\mathbf{x}} = \frac{\partial^2 \mathbf{x}}{\partial t^2} + \sum_{k=1}^{6} \frac{\partial^2 \mathbf{x}}{\partial c_k \partial t} \dot{c}_k = (-c_1 \sin t - c_2 \cos t) + (\dot{c}_1 \cos t - \dot{c}_2 \sin t),$$

where

$$\frac{\partial^2 \mathbf{x}}{\partial t^2} = (-c_1 \sin t - c_2 \cos t) \tag{3.55}$$

$$\sum_{k=1}^{6} \frac{\partial^2 \mathbf{x}}{\partial c_k \partial t} \dot{c}_k = (\dot{c}_1 \cos t - \dot{c}_2 \sin t). \tag{3.56}$$

Substituting Eqs. 3.48 and 3.54 into our differential equation in Eq. 3.47 yields

$$(-c_1 \sin t - c_2 \cos t) + (\dot{c}_1 \cos t - \dot{c}_2 \sin t) + (c_1 \sin t + c_2 \cos t) = R(t) \tag{3.57}$$

or

$$\dot{c}_1 \cos t - \dot{c}_2 \sin t = R(t). \tag{3.58}$$

(Equation 3.58 is analogous to the true orbit.) Our problem is to simultaneously solve for $\dot{c}_1$ and $\dot{c}_2$ in Eqs. 3.52 and 3.58:

$$\dot{c}_1 \sin t + \dot{c}_2 \cos t = 0$$

$$\dot{c}_1 \cos t - \dot{c}_2 \sin t = R(t). \tag{3.59}$$

We now rewrite Eqs. 3.59 before using Lagrange brackets to solve them. From Eq. 3.39 and Eq. 3.49 (or Eq. 3.53) we have

$$\frac{\partial^2 x}{\partial t \partial c_1} = \frac{\partial}{\partial c_1}\left(\frac{\partial x}{\partial t}\right) = \frac{\partial \dot{x}}{\partial c_1} = \cos t$$

$$\frac{\partial^2 x}{\partial t \partial c_2} = \frac{\partial}{\partial c_2}\left(\frac{\partial x}{\partial t}\right) = \frac{\partial \dot{x}}{\partial c_2} = -\sin t, \tag{3.60}$$

and we note that from Eq. 3.48 that

$$\frac{\partial x}{\partial c_1} = \sin t$$

$$\frac{\partial x}{\partial c_2} = \cos t. \tag{3.61}$$

For purposes of comparison with Eqs. 3.33 and 3.40, Eqs. 3.59 can be written as follows. From Eq. 3.33,

$$\sum_{k=1}^{2} \frac{\partial x}{\partial c_k}\dot{c}_k = \frac{\partial x}{\partial c_1}\dot{c}_1 + \frac{\partial x}{\partial c_2}\dot{c}_2 = 0 \tag{3.62}$$

and from Eq. 3.40,

$$\sum_{k=1}^{2} \frac{\partial \dot{x}}{\partial c_k}\dot{c}_k = \frac{\partial \dot{x}}{\partial c_1}\dot{c}_1 + \frac{\partial \dot{x}}{\partial c_2}\dot{c}_2 = R(t). \tag{3.63}$$

We can see, by inspection, that substitution of the partial derivatives in Eqs. 3.60 and 3.61 into Eqs. 3.62 and 3.63 recovers Eqs. 3.59.

Multiplying Eq. 3.63 by $\partial x/\partial c_j$, multiplying Eq. 3.62 by $-\partial \dot{x}/\partial c_j$, and adding the results, we obtain

$$\dot{c}_1\left[\frac{\partial x}{\partial c_j}\frac{\partial \dot{x}}{\partial c_1} - \frac{\partial x}{\partial c_1}\frac{\partial \dot{x}}{\partial c_j}\right] + \dot{c}_2\left[\frac{\partial x}{\partial c_j}\frac{\partial \dot{x}}{\partial c_2} - \frac{\partial x}{\partial c_2}\frac{\partial \dot{x}}{\partial c_j}\right] = \frac{\partial x}{\partial c_j}R(t), \qquad j = 1, 2. \tag{3.64}$$

Introducing Lagrange brackets (Eq. 3.42), we may write Eq. 3.64 more compactly as

$$[c_j, c_1]\dot{c}_1 + [c_j, c_2]\dot{c}_2 = R(t)\frac{\partial x}{\partial c_j}, \qquad j = 1, 2 \tag{3.65}$$

or

$$\sum_{k=1}^{2} \left[c_j, c_k \right] \dot{c}_k = R\left(t \right) \frac{\partial x}{\partial c_j}, \qquad j = 1, 2, \tag{3.66}$$

which are analogous to Eq. 3.41.

Now let us demonstrate that the Lagrange brackets admit the following properties for this particular problem:

$$[c_1, c_1] = [c_2, c_2] = 0 \tag{3.67}$$

$$[c_1, c_2] = -1 \tag{3.68}$$

$$[c_2, c_1] = 1. \tag{3.69}$$

To verify Eq. 3.67, let us set $i = j = k$ in Eqs. 3.42 and 3.43:

$$[c_i, c_i] = \frac{\partial x}{\partial c_i} \frac{\partial \dot{x}}{\partial c_i} - \frac{\partial \dot{x}}{\partial c_i} \frac{\partial x}{\partial c_i} = 0. \tag{3.70}$$

The Jacobian terms are all equal to zero for $[c_i, c_i]$, and Eq. 3.70 verifies Eq. 3.67 for $i = 1$ and for $i = 2$. For Eq. 3.68, we have

$$[c_1, c_2] = \frac{\partial x}{\partial c_1} \frac{\dot{x}}{\partial c_2} - \frac{\partial x}{\partial c_2} \frac{\partial \dot{x}}{\partial c_1}. \tag{3.71}$$

Substituting Eqs. 3.60 and 3.61 into Eq. 3.71, we find

$$[c_1, c_2] = \sin t \left(-\sin t \right) - \cos t \left(\cos t \right) = -1 \tag{3.72}$$

and similarly for Eq. 3.69:

$$[c_1, c_2] = \frac{\partial x}{\partial c_2} \frac{\dot{x}}{\partial c_1} - \frac{\partial x}{\partial c_1} \frac{\partial \dot{x}}{\partial c_2} = \cos t \left(\cos t \right) - \sin t \left(-\sin t \right) = 1. \tag{3.73}$$

Equations 3.72 and 3.73 verify Eqs. 3.68 and 3.69, respectively. We also note that

$$[c_1, c_2] = -[c_2, c_1]. \tag{3.74}$$

Thus, having evaluated the four Lagrange brackets in Eqs. 3.67–3.69, we find that Eq. 3.65 for $j = 1, 2$

$$[c_1, c_1] \dot{c}_1 + [c_1, c_2] \dot{c}_2 = \frac{\partial x}{\partial c_1} R\left(t \right) \tag{3.75}$$

$$[c_2, c_1] \dot{c}_1 + [c_2, c_2] \dot{c}_2 = \frac{\partial x}{\partial c_2} R\left(t \right) \tag{3.76}$$

reduces to

$$0 + (-1)\, \dot{c}_2 = (\sin t) R(t)$$
$$(1)\, \dot{c}_1 + 0 = (\cos t) R(t).$$

So the solution is given by McCuskey (1963) and Szebehely (1989)

$$\dot{c}_1 = R(t) \cos t \tag{3.77}$$

$$\dot{c}_2 = R(t) \sin t \tag{3.78}$$

and can provide the exact solution for the functions $c_1(t)$ and $c_2(t)$ directly upon integration when $R(t)$ is known. Later, we make the simplifying assumption that the perturbations are at least an order of magnitude smaller than the gravity of the central body, but in this example we have made no such assumption.

3.6.2   Properties of the Lagrange Brackets

These first two properties hold for any perturbing force. We can show that by their definition the Lagrange brackets have the following properties:

$$\left[c_j, c_j\right] = 0 \tag{3.79}$$

$$\left[c_k, c_j\right] = -\left[c_j, c_k\right], \tag{3.80}$$

where we recall that the c_k are the 6 orbital elements, which are assumed to vary slowly with time. From Eq. 3.42, we have the definition of the Lagrange bracket in terms of the inertial, Cartesian coordinates

$$\left[c_j, c_k\right] \equiv \frac{\partial(x, \dot{x})}{\partial(c_j, c_k)} + \frac{\partial(y, \dot{y})}{\partial(c_j, c_k)} + \frac{\partial(z, \dot{z})}{\partial(c_j, c_k)}. \tag{3.81}$$

To verify Eq. 3.79, we set $k = j$ in Eq. 3.81:

$$\left[c_j, c_j\right] = \frac{\partial(x, \dot{x})}{\partial(c_j, c_j)} + \frac{\partial(y, \dot{y})}{\partial(c_j, c_j)} + \frac{\partial(z, \dot{z})}{\partial(c_j, c_j)}, \tag{3.82}$$

where from Eq. 3.43 we recall the Jacobian of x and $\dot{x}$ with respect to c_j and c_k is

$$\frac{\partial(x, \dot{x})}{\partial(c_j, c_k)} \equiv \begin{vmatrix} \frac{\partial x}{\partial c_j} & \frac{\partial x}{\partial c_k} \\ \frac{\partial \dot{x}}{\partial c_j} & \frac{\partial \dot{x}}{\partial c_k} \end{vmatrix}, \tag{3.83}$$

with similar expressions for y and z. Evaluating the first term on the right-hand side of Eq. 3.82, we obtain

$$\frac{\partial\,(x,\dot{x})}{\partial\,(c_j,c_j)} = \frac{\partial x}{\partial c_j}\frac{\partial \dot{x}}{\partial c_j} - \frac{\partial x}{\partial c_j}\frac{\partial \dot{x}}{\partial c_j} = 0. \tag{3.84}$$

The remaining terms for y and z in Eq. 3.82 are also zero so that

$$[c_j, c_j] = 0,$$

and Eq. 3.79 is verified.

To demonstrate Eq. 3.80, we apply the definition of the Lagrange bracket and note that

$$\frac{\partial\,(x,\dot{x})}{\partial\,(c_k,c_j)} = \frac{\partial x}{\partial c_k}\frac{\partial \dot{x}}{\partial c_j} - \frac{\partial \dot{x}}{\partial c_k}\frac{\partial x}{\partial c_j} \tag{3.85}$$

and that

$$\frac{\partial\,(x,\dot{x})}{\partial\,(c_j,c_k)} = \frac{\partial x}{\partial c_j}\frac{\partial \dot{x}}{\partial c_k} - \frac{\partial \dot{x}}{\partial c_j}\frac{\partial x}{\partial c_k}. \tag{3.86}$$

Comparing Eqs. 3.85 and 3.86, we see that

$$\frac{\partial\,(x,\dot{x})}{\partial\,(c_k,c_j)} = -\frac{\partial\,(x,\dot{x})}{\partial\,(c_j,c_k)}, \tag{3.87}$$

and since there are similar expressions for y and z, we have the property in Eq. 3.80:

$$[c_k, c_j] = -[c_j, c_k],$$

reversing the indices in a Lagrange bracket leads to a sign change.

3.6.3   Time Independence (for Forces Derivable from a Scalar Potential Function)

Another important property of Lagrange brackets applies only to forces that are derivable from a scalar potential function. In such cases, the Lagrange brackets do not depend explicitly on time. We may state this property as

$$\frac{\partial}{\partial t}[c_j, c_k] = 0, \tag{3.88}$$

which implies that the $[c_j, c_k]$ are independent of time explicitly. Equation 3.88 indicates that the Lagrange brackets may be computed at any convenient epoch (for example, at perihelion) and they remain invariant with respect to time thereafter.

To prove Eq. 3.88, let us denote any one bracket involving two general orbital elements by $[p, q]$ and one of the Jacobians (from Eq. 3.83) by $J^x [(x, \dot{x}) / (p, q)]$. Then we have

$$\frac{\partial J^x}{\partial t} = \frac{\partial}{\partial t} \left(\frac{\partial x}{\partial p} \frac{\partial \dot{x}}{\partial q} - \frac{\partial x}{\partial q} \frac{\partial \dot{x}}{\partial p} \right) = \frac{\partial^2 x}{\partial t \partial p} \frac{\partial \dot{x}}{\partial q} - \frac{\partial^2 x}{\partial t \partial q} \frac{\partial \dot{x}}{\partial p} + \frac{\partial^2 \dot{x}}{\partial t \partial q} \frac{\partial x}{\partial p} - \frac{\partial^2 \dot{x}}{\partial t \partial p} \frac{\partial x}{\partial q}.$$

$$(3.89)$$

By the definition of the osculating orbit, we have $\partial x / \partial t = \dot{x}$, so the first two terms on the right-hand side of Eq. 3.89 sum to zero. The surviving terms on the right-hand side of Eq. 3.89 can be written as

$$\frac{\partial}{\partial q} \left(\frac{\partial \dot{x}}{\partial t} \right) \frac{\partial x}{\partial p} - \frac{\partial}{\partial p} \left(\frac{\partial \dot{x}}{\partial t} \right) \frac{\partial x}{\partial q}.$$

$$(3.90)$$

Now, from the equation of motion for the osculating orbit (see Eq. 3.14), we have

$$\frac{\partial \dot{x}}{\partial t} \equiv \frac{\partial^2 x}{\partial t^2} = -\frac{\mu x}{r^3} = \frac{\partial U}{\partial x},$$

$$(3.91)$$

where the potential function is

$$U = \frac{\mu}{r}.$$

$$(3.92)$$

Thus we can use Eqs. 3.90–3.92 to write Eq. 3.89 as

$$\frac{\partial J^x}{\partial t} = \frac{\partial}{\partial q} \left(\frac{\partial U}{\partial x} \right) \frac{\partial x}{\partial p} - \frac{\partial}{\partial p} \left(\frac{\partial U}{\partial x} \right) \frac{\partial x}{\partial q}.$$

$$(3.93)$$

Similar expressions can be written for y and z. The potential function $U(x, y, z)$ is continuous and has continuous derivatives at all points except the origin. Adding the terms for y and z (by analogy with Eq. 3.93), we have the derivative of the entire Lagrange bracket:

$$\frac{\partial J}{\partial t} = \frac{\partial J^x}{\partial t} + \frac{\partial J^y}{\partial t} + \frac{\partial J^z}{\partial t}$$

$$= \frac{\partial}{\partial t} [p, q]$$

$$= \left(\frac{\partial U_q}{\partial x} \frac{\partial x}{\partial p} + \frac{\partial U_q}{\partial y} \frac{\partial y}{\partial p} + \frac{\partial U_q}{\partial z} \frac{\partial z}{\partial p} \right) - \left(\frac{\partial U_p}{\partial x} \frac{\partial x}{\partial q} + \frac{\partial U_p}{\partial y} \frac{\partial y}{\partial q} + \frac{\partial U_p}{\partial z} \frac{\partial z}{\partial q} \right),$$

$$(3.94)$$

where

$$U_p \equiv \frac{\partial U}{\partial p}$$

$$U_q \equiv \frac{\partial U}{\partial q}.$$ (3.95)

From Eq. 3.94, it is clear that

$$\frac{\partial J}{\partial t} = \frac{\partial U_q}{\partial p} - \frac{\partial U_p}{\partial q} = \frac{\partial [p,q]}{\partial t} = 0$$ (3.96)

because the order of the partial derivatives does not matter. Equation 3.96 confirms the Lagrange bracket property stated in Eq. 3.88: The Lagrange bracket is independent of time for forces derivable from a scalar potential function.

References

Fitzpatrick, P. M. (1970). *Principles of celestial mechanics*. New York: Academic Press.
McCuskey, S. W. (1963). *Introduction to celestial mechanics*. Reading: Addison-Wesley Publishing Company.
Szebehely, V. G. (1989). *Adventures in celestial mechanics, a first course in the theory of orbits*. Austin: University of Texas Press.
Vallado, D. A. (2013). *Fundamentals of astrodynamics and applications* (4th ed.). El Segundo: Microcosm Press.

Chapter 4
Evaluation of the Lagrange Brackets

The Lagrange brackets, $[c_j, c_k]$, must be evaluated in terms of the orbital elements and a particular coordinate frame so that

$$\sum_{k=1}^{6} [c_j, c_k] \dot{c}_k = \mathbf{F} \cdot \frac{\partial \mathbf{r}}{\partial c_j}, \qquad j = 1, 2, \ldots, 6 \qquad (4.1)$$

may be solved for the $\dot{c}_k$ (i.e., to find the differential equations). (See Eq. 3.41.)

Other authors develop formulas for the Lagrange brackets by assuming the brackets do not depend on time explicitly. Such time independence allows evaluation at any convenient point such as periapsis, which introduces considerable simplification and easy derivation of the Lagrange brackets.

It can be proven that the brackets do not contain time explicitly if the perturbing force is derivable from a scalar potential. However, brackets derived under such a restrictive assumption cannot validly be used to develop perturbation equations for a non-conservative force. Nevertheless, McCuskey (1963, p. 145) applies his potential-based Lagrange brackets to general forces without any justification for doing so. Additionally, Battin (1999, p. 484) states, "Although Lagrange's variational equations were derived for the special case in which the disturbing acceleration was represented as the gradient of the disturbing function, this restriction is wholly unnecessary." Unfortunately, he does not cite any reference or give any justification for this claim.

We do not know of any published result that justifies such an assumption by previous authors. Therefore, we carefully and fully develop a general derivation of the Lagrange brackets demonstrating that they are free of any explicit time dependence.

© The Author(s), under exclusive license to Springer Nature Switzerland AG 2022
J. M. Longuski et al., *Introduction to Orbital Perturbations*, Space Technology
Library 40, https://doi.org/10.1007/978-3-030-89758-1_4

4.1   Perifocal Coordinate System, PQW

One of the most convenient coordinate frames for describing the motion of a satellite is the perifocal coordinate system, denoted by PQW, and introduced in Chap. 2. From Eqs. 2.50 and 2.51, $\mathbf{u}_P$, $\mathbf{u}_Q$, $\mathbf{u}_W$ depend only upon the orbital elements Ω, ω, and i, which orient the orbit in inertial space and not upon a (or n), e, and M_0, which define the orbit's *shape* and the *position* of the satellite in the orbital plane.

The orbital elements, c_k, exist in two groups:

$$\alpha_1 = n$$
$$\alpha_2 = e \qquad\qquad\qquad (4.2)$$
$$\alpha_3 = M_0,$$

which give the shape of the orbit and the position of the satellite in the orbit, and

$$\beta_1 = \Omega$$
$$\beta_2 = \omega \qquad\qquad\qquad (4.3)$$
$$\beta_3 = i,$$

which give the orientation of the orbit in inertial space. The Lagrange brackets, then, can be grouped into three categories: $[\alpha_r, \alpha_s]$, $[\alpha_r, \beta_s]$, and $[\beta_r, \beta_s]$, where $r, s = 1, 2, 3$ and for the first and third categories, $r \neq s$. These categories yield a total of 15 unique equations that need to be examined.

4.1.1   Dependent and Independent Variables

One of the most important concepts that one must understand is that of dependent and independent variables. We have shown that the motion of a satellite can be described by six independent variables, e.g., the rectangular components of position and velocity or some set of orbital elements.

If we select the set of orbital elements just identified, then we must be intensely aware that only those six elements are independent of one another. As we use this set in our development, we may need to introduce auxiliary variables such as the semimajor axis and the angular anomalies: mean, eccentric, and true.

In dealing with the appearance of these variables, we must always remember they are dependent on one or more of our independent set of variables. Before proceeding with any further development steps, one must clearly determine the dependence of any auxiliary variables likely to be used.

For the set of independent variables we are using, we consider Kepler's equation
(Eq. 2.65)

$$M = M_0 + nt = E - e \sin E.$$

We can see that any of the auxiliary angular variables have a dependence on both
the mean motion and the eccentricity. We identify the following dependencies of
our auxiliary variables:

$$a = f_1(n)$$

$$E = f_2(n, e, M_0)$$

$$M = f_3(n, e, M_0)$$

$$f = f_4(n, e, M_0),$$

where the f_i functions are determined by the definitions given in Eqs. 2.61, 2.65,
and 2.74.

There is a reason why we selected the mean motion rather than the semimajor
axis as one of six independent variables. The natural choice would seem to be the
semimajor axis. It has a clear physical meaning and is one of a pair of parameters
describing the size and shape of the ellipse. But we should always be keeping in
mind the big picture of why we are developing these perturbation equations. We
want to apply them to model a real-world perturbation, and we are hopeful that we
can develop an analytical solution for the set of six first-order differential equations.

Suppose that we use the semimajor axis for our independent variable. Further
suppose that we are able to develop an explicit expression for the behavior of the
semimajor axis as a function of time. Let us imagine that expression is a polynomial
in time with perhaps some trigonometric terms in the true anomaly of the satellite.

Having a solution for the semimajor axis, we now turn our attention to finding a
solution for the mean anomaly. The very first term in the mean anomaly rate is the
mean motion. So we must integrate the mean motion with respect to time. Oops!
When we substitute the closed-form equation for semimajor axis into the mean
motion, we have a time polynomial and trigonometric terms all raised to the $-3/2$
power. Good luck with finding that in the integral tables.

In our development, we use the mean motion rather than the semimajor axis as
our independent variable, so we would have an easier task integrating it with respect
to time. A simple choice for what independent variable to use can have significant
payoff down the road. One should always be thinking ahead when embarking on a
perturbation problem. For this reason, we diverge from other authors and choose the
mean motion as our independent variable in the Lagrange planetary equations.

4.2   Evaluation of the $[\alpha_r, \alpha_s]$ Lagrange Bracket

We now evaluate the first category of Lagrange bracket, $[\alpha_r, \alpha_s]$, involving the three orbital elements that provide the shape of the orbit and position of the body in the orbit. The radius vector can be written as

$$\mathbf{r} = x\mathbf{i} + y\mathbf{j} + z\mathbf{k} = \xi\mathbf{u}_P + \eta\mathbf{u}_Q, \tag{4.4}$$

where ξ and η are the perifocal Cartesian coordinates in the orbital plane. The inertial velocity is

$$\dot{\mathbf{r}} = \dot{x}\mathbf{i} + \dot{y}\mathbf{j} + \dot{z}\mathbf{k} = \dot{\xi}\mathbf{u}_P + \dot{\eta}\mathbf{u}_Q + \xi\dot{\mathbf{u}}_P + \eta\dot{\mathbf{u}}_Q. \tag{4.5}$$

But, by definition of the osculating orbit:

$$\dot{\mathbf{r}} = \frac{\partial \mathbf{r}}{\partial t}$$

so that

$$\xi\dot{\mathbf{u}}_P + \eta\dot{\mathbf{u}}_Q = 0, \tag{4.6}$$

and Eq. 4.5 reduces to

$$\dot{\mathbf{r}} = \dot{\xi}\mathbf{u}_P + \dot{\eta}\mathbf{u}_Q. \tag{4.7}$$

By another line of reasoning, we can observe that $\dot{\mathbf{u}}_P = \dot{\mathbf{u}}_Q = 0$ because the P_i and Q_i are constant functions of the orientation angles Ω, ω, and i, all of which are constants on the osculating orbit.

From Eq. 3.42, the Lagrange bracket, $[\alpha_r, \alpha_s]$, is

$$
\begin{aligned}
[\alpha_r, \alpha_s] &= \frac{\partial\,(x, \dot{x})}{\partial\,(\alpha_r, \alpha_s)} + \frac{\partial\,(y, \dot{y})}{\partial\,(\alpha_r, \alpha_s)} + \frac{\partial\,(z, \dot{z})}{\partial\,(\alpha_r, \alpha_s)} \\[2mm]
&= \begin{vmatrix} \frac{\partial x}{\partial \alpha_r} & \frac{\partial x}{\partial \alpha_s} \\ \frac{\partial \dot{x}}{\partial \alpha_r} & \frac{\partial \dot{x}}{\partial \alpha_s} \end{vmatrix} + \begin{vmatrix} \frac{\partial y}{\partial \alpha_r} & \frac{\partial y}{\partial \alpha_s} \\ \frac{\partial \dot{y}}{\partial \alpha_r} & \frac{\partial \dot{y}}{\partial \alpha_s} \end{vmatrix} + \begin{vmatrix} \frac{\partial z}{\partial \alpha_r} & \frac{\partial z}{\partial \alpha_s} \\ \frac{\partial \dot{z}}{\partial \alpha_r} & \frac{\partial \dot{z}}{\partial \alpha_s} \end{vmatrix}.
\end{aligned} \tag{4.8}
$$

But in place of $(\partial x/\partial \alpha_r)$, $(\partial \dot{x}/\partial \alpha_r)$, etc., using Eqs. 4.4 and 4.7 and the chain rule, we can write

$$\frac{\partial x}{\partial \alpha_r} = \frac{\partial x}{\partial \xi}\frac{\partial \xi}{\partial \alpha_r} + \frac{\partial x}{\partial \eta}\frac{\partial \eta}{\partial \alpha_r},$$

$$\frac{\partial \dot{x}}{\partial \alpha_r} = \frac{\partial \dot{x}}{\partial \dot{\xi}}\frac{\partial \dot{\xi}}{\partial \alpha_r} + \frac{\partial \dot{x}}{\partial \dot{\eta}}\frac{\partial \dot{\eta}}{\partial \alpha_r}, \tag{4.9}$$

along with corresponding terms in y and z (and for the partial derivatives with respect to α_s). Furthermore, from Eqs. 4.4 and 2.50

$$\mathbf{r} = \xi \, (P_1\mathbf{i} + P_2\mathbf{j} + P_3\mathbf{k}) + \eta \, (Q_1\mathbf{i} + Q_2\mathbf{j} + Q_3\mathbf{k}), \tag{4.10}$$

so we have

$$\frac{\partial x}{\partial \xi} = P_1$$

$$\frac{\partial x}{\partial \eta} = Q_1, \tag{4.11}$$

and from Eqs. 4.7 and 2.50

$$\dot{\mathbf{r}} = \dot{\xi} \, (P_1\mathbf{i} + P_2\mathbf{j} + P_3\mathbf{k}) + \dot{\eta} \, (Q_1\mathbf{i} + Q_2\mathbf{j} + Q_3\mathbf{k}), \tag{4.12}$$

we have

$$\frac{\partial \dot{x}}{\partial \dot{\xi}} = P_1$$

$$\frac{\partial \dot{x}}{\partial \dot{\eta}} = Q_1. \tag{4.13}$$

We can write similar terms for y and z corresponding to Eqs. 4.11 and 4.13. Thus, we obtain

$$\frac{\partial x}{\partial \alpha_r} = P_1 \frac{\partial \xi}{\partial \alpha_r} + Q_1 \frac{\partial \eta}{\partial \alpha_r} \tag{4.14}$$

$$\frac{\partial y}{\partial \alpha_r} = P_2 \frac{\partial \xi}{\partial \alpha_r} + Q_2 \frac{\partial \eta}{\partial \alpha_r} \tag{4.15}$$

$$\frac{\partial z}{\partial \alpha_r} = P_3 \frac{\partial \xi}{\partial \alpha_r} + Q_3 \frac{\partial \eta}{\partial \alpha_r} \tag{4.16}$$

$$\frac{\partial \dot{x}}{\partial \alpha_r} = P_1 \frac{\partial \dot{\xi}}{\partial \alpha_r} + Q_1 \frac{\partial \dot{\eta}}{\partial \alpha_r} \tag{4.17}$$

$$\frac{\partial \dot{y}}{\partial \alpha_r} = P_2 \frac{\partial \dot{\xi}}{\partial \alpha_r} + Q_2 \frac{\partial \dot{\eta}}{\partial \alpha_r} \tag{4.18}$$

$$\frac{\partial \dot{z}}{\partial \alpha_r} = P_3 \frac{\partial \dot{\xi}}{\partial \alpha_r} + Q_3 \frac{\partial \dot{\eta}}{\partial \alpha_r}. \tag{4.19}$$

For the first determinant of Eq. 4.8, we have

$$\begin{vmatrix} \dfrac{\partial x}{\partial \alpha_r} & \dfrac{\partial x}{\partial \alpha_s} \\[2mm] \dfrac{\partial \dot{x}}{\partial \alpha_r} & \dfrac{\partial \dot{x}}{\partial \alpha_s} \end{vmatrix} = \begin{vmatrix} \left(P_1 \dfrac{\partial \xi}{\partial \alpha_r} + Q_1 \dfrac{\partial \eta}{\partial \alpha_r} \right) & \left(P_1 \dfrac{\partial \xi}{\partial \alpha_s} + Q_1 \dfrac{\partial \eta}{\partial \alpha_s} \right) \\[3mm] \left(P_1 \dfrac{\partial \dot{\xi}}{\partial \alpha_r} + Q_1 \dfrac{\partial \dot{\eta}}{\partial \alpha_r} \right) & \left(P_1 \dfrac{\partial \dot{\xi}}{\partial \alpha_s} + Q_1 \dfrac{\partial \dot{\eta}}{\partial \alpha_s} \right) \end{vmatrix}. \tag{4.20}$$

To obtain the analogous determinants for y and z, we simply change the subscript 1 to 2 and 3, respectively. Multiplying out the determinant in Eq. 4.20, we get

$$P_1^2 \frac{\partial \xi}{\partial \alpha_r} \frac{\partial \dot{\xi}}{\partial \alpha_s} + Q_1^2 \frac{\partial \eta}{\partial \alpha_r} \frac{\partial \dot{\eta}}{\partial \alpha_s} + P_1 Q_1 \left(\frac{\partial \xi}{\partial \alpha_r} \frac{\partial \dot{\eta}}{\partial \alpha_s} + \frac{\partial \eta}{\partial \alpha_r} \frac{\partial \dot{\xi}}{\partial \alpha_s} \right)$$

$$- P_1^2 \frac{\partial \dot{\xi}}{\partial \alpha_r} \frac{\partial \xi}{\partial \alpha_s} - Q_1^2 \frac{\partial \dot{\eta}}{\partial \alpha_r} \frac{\partial \eta}{\partial \alpha_s} - P_1 Q_1 \left(\frac{\partial \dot{\xi}}{\partial \alpha_r} \frac{\partial \eta}{\partial \alpha_s} + \frac{\partial \dot{\eta}}{\partial \alpha_r} \frac{\partial \xi}{\partial \alpha_s} \right). \tag{4.21}$$

We get similar terms for the determinants involving y and z except the subscript 1 is replaced by 2 and 3, respectively.

We note that the elements of the direction cosine matrix have the following properties, which allow us to simplify the sum of the three determinants:

$$P_1^2 + P_2^2 + P_3^2 = 1 \tag{4.22}$$

$$Q_1^2 + Q_2^2 + Q_3^2 = 1 \tag{4.23}$$

$$P_1 Q_1 + P_2 Q_2 + P_3 Q_3 = 0. \tag{4.24}$$

Thus, we can add the three determinants corresponding to x, y, and z in Eq. 4.8 to obtain

$$\begin{aligned} [\alpha_r, \alpha_s] &= \frac{\partial \xi}{\partial \alpha_r} \frac{\partial \dot{\xi}}{\partial \alpha_s} - \frac{\partial \dot{\xi}}{\partial \alpha_r} \frac{\partial \xi}{\partial \alpha_s} + \frac{\partial \eta}{\partial \alpha_r} \frac{\partial \dot{\eta}}{\partial \alpha_s} - \frac{\partial \dot{\eta}}{\partial \alpha_r} \frac{\partial \eta}{\partial \alpha_s} \\[3mm] &= \begin{vmatrix} \dfrac{\partial \xi}{\partial \alpha_r} & \dfrac{\partial \xi}{\partial \alpha_s} \\[2mm] \dfrac{\partial \dot{\xi}}{\partial \alpha_r} & \dfrac{\partial \dot{\xi}}{\partial \alpha_s} \end{vmatrix} + \begin{vmatrix} \dfrac{\partial \eta}{\partial \alpha_r} & \dfrac{\partial \eta}{\partial \alpha_s} \\[2mm] \dfrac{\partial \dot{\eta}}{\partial \alpha_r} & \dfrac{\partial \dot{\eta}}{\partial \alpha_s} \end{vmatrix} \\[3mm] &= \frac{\partial(\xi, \dot{\xi})}{\partial(\alpha_r, \alpha_s)} + \frac{\partial(\eta, \dot{\eta})}{\partial(\alpha_r, \alpha_s)}, \end{aligned} \tag{4.25}$$

where all the PQ terms vanish in the summation due to Eq. 4.24.

To evaluate the brackets in Eq. 4.25, we need several partial derivatives. Because of the dependencies just mentioned, we begin by developing chain rule relationships for those dependencies. The relationship between the mean motion and semimajor axis is given by

$$n^2 a^3 = \mu. \tag{4.26}$$

Then

$$2na^3 \frac{\partial n}{\partial n} + 3n^2 a^2 \frac{\partial a}{\partial n} = 0$$

so that

$$\frac{\partial a}{\partial n} = -\frac{2a}{3n}. \tag{4.27}$$

We introduce an auxiliary variable

$$\beta^2 = 1 - e^2. \tag{4.28}$$

Then

$$\frac{\partial}{\partial e}\beta^2 = -2e \tag{4.29}$$

and

$$\frac{\partial \beta}{\partial e} = -\frac{e}{\beta}. \tag{4.30}$$

To determine the dependent relationships, we begin with Kepler's equation

$$M = M_0 + nt = E - e \sin E. \tag{4.31}$$

Then

$$\frac{\partial M}{\partial n}\frac{\partial n}{\partial} + \frac{\partial M}{\partial M_0}\frac{\partial M_0}{\partial} = \frac{\partial E}{\partial} - \frac{\partial e}{\partial}\sin E - e \cos E \frac{\partial E}{\partial},$$

which simplifies to

$$t\frac{\partial n}{\partial} + \frac{\partial M_0}{\partial} = \frac{\partial E}{\partial} - \frac{\partial e}{\partial}\sin E - e \cos E \frac{\partial E}{\partial}. \tag{4.32}$$

Equation 4.32 provides a general formula for partial derivatives with respect to any of our chosen set of independent variables. Careful development and application of this equation is the key to properly handling dependent and independent variables.

We apply Eq. 4.32 to partial derivatives with respect to e. Keeping in mind our independent variables, we find

$$0 = \frac{\partial E}{\partial e} - \sin E - e \cos E \frac{\partial E}{\partial e}.$$

Then we solve for $\partial E / \partial e$ to obtain

$$\frac{\partial E}{\partial e} = \frac{\sin E}{(1 - e \cos E)}.$$

(4.33)

We apply Eq. 4.32 to partial derivatives with respect to n to find

$$t = \frac{\partial E}{\partial n} - e \cos E \frac{\partial E}{\partial n}.$$

Then we solve for $\partial E / \partial n$ to obtain

$$\frac{\partial E}{\partial n} = t \frac{1}{(1 - e \cos E)}.$$

(4.34)

We apply Eq. 4.32 to partial derivatives with respect to M_0 to find

$$1 = \frac{\partial E}{\partial M_0} - e \cos E \frac{\partial E}{\partial M_0}.$$

Then we solve for $\partial E / \partial M_0$ to obtain

$$\frac{\partial E}{\partial M_0} = \frac{1}{(1 - e \cos E)}.$$

(4.35)

Equation 4.35 completes the primary derivation of partial derivatives based on the assumed set of six independent variables. If one wishes to use a different set of six variables, then the equivalent derivation must be accomplished for the new choice of the independent variables.

The coordinates in the orbital plane are given by Eqs. 2.52 and 2.56

$$\xi = a (\cos E - e)$$

(4.36)

$$\eta = a\beta \sin E = a \sin E \sqrt{1 - e^2}.$$

(4.37)

Equations 4.33, 4.34, and 4.35 enable the development of the position and velocity partial derivatives.

The partial of ξ with respect to n is

$$\frac{\partial \xi}{\partial n} = -\frac{2a}{3n} (\cos E - e) - a \sin E \frac{\partial E}{\partial n},$$

which reduces to

$$\frac{\partial \xi}{\partial n} = -\frac{2a}{3n} (\cos E - e) - at \frac{\sin E}{(1 - e \cos E)}$$

$$= -\frac{2r}{3n} \cos f - \frac{a}{\beta} t \sin f.$$

(4.38)

The partial of ξ with respect to e is

$$\frac{\partial \xi}{\partial e} = -a \sin E \frac{\partial E}{\partial e} - a,$$

which reduces to

$$\frac{\partial \xi}{\partial e} = -a \left[1 + \left(\frac{\sin^2 E}{1 - e \cos E} \right) \right]$$

$$= -a - \frac{r}{\beta^2} \sin^2 f. \tag{4.39}$$

The partial of ξ with respect to M_0 is

$$\frac{\partial \xi}{\partial M_0} = -a \sin E \frac{\partial E}{\partial M_0},$$

which reduces to

$$\frac{\partial \xi}{\partial M_0} = -\frac{a \sin E}{(1 - e \cos E)}$$

$$= -\frac{a}{\beta} \sin f. \tag{4.40}$$

The partial of η with respect to n is

$$\frac{\partial \eta}{\partial n} = -\frac{2a\beta}{3n} \sin E + a\beta \cos E \frac{\partial E}{\partial n},$$

which reduces to

$$\frac{\partial \eta}{\partial n} = -\frac{2a\beta}{3n} \sin E + (a\beta) t \frac{\cos E}{(1 - e \cos E)}$$

$$= -\frac{2r}{3n} \sin f + \frac{a}{\beta} t (\cos f + e). \tag{4.41}$$

The partial of η with respect to e is

$$\frac{\partial \eta}{\partial e} = -\frac{ae}{\beta} \sin E + a\beta \cos E \frac{\partial E}{\partial e},$$

which reduces to

$$\frac{\partial \eta}{\partial e} = \frac{a}{\beta} \frac{\sin E}{(1 - e \cos E)} (\cos E - e)$$

$$= \frac{r}{\beta^2} \sin f \cos f. \tag{4.42}$$

The partial of η with respect to M_0 is

$$\frac{\partial \eta}{\partial M_0} = a\beta \cos E \frac{\partial E}{\partial M_0},$$

which reduces to

$$\frac{\partial \eta}{\partial M_0} = a\beta \frac{\cos E}{(1 - e \cos E)}$$

$$= \frac{a}{\beta} (\cos f + e). \tag{4.43}$$

The velocity components in the orbital plane are obtained by differentiating Eqs. 4.36 and 4.37 to get

$$\dot{\xi} = -a \sin E \left(\dot{E} \right) \tag{4.44}$$

$$\dot{\eta} = a\beta \cos E \left(\dot{E} \right). \tag{4.45}$$

We can determine the eccentric anomaly rate by differentiating

$$\sin E = \frac{\beta \sin f}{1 + e \cos f}$$

and then

$$\cos E \left(\dot{E} \right) = \beta \frac{\cos E}{(1 + e \cos f)} \left(\dot{f} \right). \tag{4.46}$$

Using Eqs. 2.17 and 2.28, and substituting expressions for $\sqrt{\mu a}$ and $\sqrt{(1 - e^2)}$ obtained from Eqs. 2.61 and 4.28, respectively, we can write

$$\dot{f} = \frac{na^2 \beta}{r^2}. \tag{4.47}$$

Substituting into Eq. 4.46 and solving for $\dot{E}$ give

$$\dot{E} = \frac{na}{r}. \tag{4.48}$$

Substituting Eq. 4.48 into Eqs. 4.44 and 4.45, and using Eqs. 2.57, 2.76, and 2.77, gives

$$\dot{\xi} = -na\frac{\sin E}{(1 - e\cos E)} = -\frac{na}{\beta}\sin f \tag{4.49}$$

$$\dot{\eta} = na\beta\frac{\cos E}{(1 - e\cos E)} = \frac{na}{\beta}(\cos f + e). \tag{4.50}$$

The partial of $\dot{\xi}$ with respect to n is

$$\frac{\partial \dot{\xi}}{\partial n} = \left(-\frac{a}{3}\right)\frac{\sin E}{1 - e\cos E} - na\frac{(1 - e\cos E)\cos E - e\sin^2 E}{(1 - e\cos E)^2}\frac{\partial E}{\partial n},$$

which reduces to

$$\frac{\partial \dot{\xi}}{\partial n} = -\frac{a}{3\beta}\sin f - \frac{na^3}{r^2}t\cos f, \tag{4.51}$$

and we note the *explicit appearance of the time*, t. In the following pages, we prove the claim that the time terms all vanish and the Lagrange brackets are, in fact, time independent.

The partial of $\dot{\xi}$ with respect to e is

$$\frac{\partial \dot{\xi}}{\partial e} = -na\frac{(1 - e\cos E)\cos E\dfrac{\partial E}{\partial e} - \sin E\left(e\sin E\dfrac{\partial E}{\partial e} - \cos E\right)}{(1 - e\cos E)^2},$$

which reduces to

$$\frac{\partial \dot{\xi}}{\partial e} = -\frac{na}{\beta^3}\sin f\left(e + 2\cos f + e\cos^2 f\right). \tag{4.52}$$

The partial of $\dot{\xi}$ with respect to M_0 is

$$\frac{\partial \dot{\xi}}{\partial M_0} = -na\frac{(1 - e\cos E)\cos E - e\sin^2 E}{(1 - e\cos E)^2}\frac{\partial E}{\partial M_0},$$

which reduces to

$$\frac{\partial \dot{\xi}}{\partial M_0} = -\frac{na^3}{r^2}\cos f. \tag{4.53}$$

The partial of $\dot{\eta}$ with respect to n is

$$\frac{\partial \dot{\eta}}{\partial n} = \frac{a}{3\beta}\frac{\cos E}{(1 - e\cos E)} + na\beta\frac{(-\sin E + e\cos E\sin E - e\cos E\sin E)}{(1 - e\cos E)^2}\frac{\partial E}{\partial n},$$

which reduces to

$$\frac{\partial \dot{\eta}}{\partial n} = \frac{a}{3\beta}(e + \cos f) - \frac{na^3}{r^2}t \sin f. \tag{4.54}$$

The partial of $\dot{\eta}$ with respect to e is

$$\frac{\partial \dot{\eta}}{\partial e} = -na\frac{e}{\beta}\frac{\cos E}{(1 - e\cos E)} + na\beta\frac{-(1 - e\cos E)\sin E\frac{\partial E}{\partial e} - \cos E\left(-\cos E + e\sin E\frac{\partial E}{\partial e}\right)}{(1 - e\cos E)^2},$$

which reduces to

$$\frac{\partial \dot{\eta}}{\partial e} = \frac{na}{\beta^3}\left(1 - 2\sin^2 f + e\cos^3 f\right). \tag{4.55}$$

The partial of $\dot{\eta}$ with respect to M_0 is

$$\frac{\partial \dot{\eta}}{\partial M_0} = na\beta\frac{-(1 - e\cos E)\sin E - e\cos E\sin E}{(1 - e\cos E)^2}\frac{\partial E}{\partial M_0},$$

which reduces to

$$\frac{\partial \dot{\eta}}{\partial M_0} = -\frac{na^3}{r^2}\sin f. \tag{4.56}$$

This completes the derivation of the partial derivatives of the ellipse components of position and velocity with respect to the independent variables. They are valid at any point in the orbit.

4.2.1   Evaluation of $[n, e]$

Let us evaluate the Lagrange bracket for the mean motion and the eccentricity:

$$[\alpha_r, \alpha_s] = [n, e]. \tag{4.57}$$

We can now evaluate the Jacobian

$$[n, e] = \frac{\partial \xi}{\partial n}\frac{\partial \dot{\xi}}{\partial e} - \frac{\partial \xi}{\partial e}\frac{\partial \dot{\xi}}{\partial n} + \frac{\partial \eta}{\partial n}\frac{\partial \dot{\eta}}{\partial e} - \frac{\partial \eta}{\partial e}\frac{\partial \dot{\eta}}{\partial n}$$

using Eqs. 4.38, 4.39, 4.41, 4.42, 4.51, 4.52, 4.54, and 4.55 to obtain

$$[n, e] = \left[-\frac{2r}{3n} \cos f - \frac{a}{\beta} t \sin f \right] \left[-\frac{na}{\beta^3} \sin f \left(e + 2 \cos f + e\cos^2 f \right) \right]$$

$$- \left[-a - \frac{r}{\beta^2} \sin^2 f \right] \left[-\frac{a}{3\beta} \sin f - \frac{na^3}{r^2} t \cos f \right]$$

$$+ \left[-\frac{2r}{3n} \sin f + \frac{a}{\beta} t (\cos f + e) \right] \left[\frac{na}{\beta^3} \left(1 - 2\sin^2 f + e\cos^3 f \right) \right]$$

$$- \left[\frac{r}{\beta^2} \sin f \cos f \right] \left[\frac{a}{3\beta} (e + \cos f) - \frac{na^3}{r^2} t \sin f \right]. \tag{4.58}$$

The reader is encouraged to verify that the secular terms vanish, proving that the Lagrange bracket, $[n, e]$, is constant. After a considerable amount of algebra, we find

$$[n, e] = 0. \tag{4.59}$$

4.2.2 Evaluation of $[n, M_0]$

Let us evaluate the Lagrange bracket for the mean motion and the mean anomaly:

$$[\alpha_r, \alpha_s] = [n, M_0]. \tag{4.60}$$

We can now evaluate the Jacobian

$$[n, M_0] = \frac{\partial \xi}{\partial n} \frac{\partial \dot{\xi}}{\partial M_0} - \frac{\partial \xi}{\partial M_0} \frac{\partial \dot{\xi}}{\partial n} + \frac{\partial \eta}{\partial n} \frac{\partial \dot{\eta}}{\partial M_0} - \frac{\partial \eta}{\partial M_0} \frac{\partial \dot{\eta}}{\partial n}$$

using Eqs. 4.38, 4.40, 4.41, 4.43, 4.51, 4.53, 4.54, and 4.56 to obtain

$$[n, M_0] = \left[\frac{2r}{3n} \cos f + \frac{a}{\beta} t \sin f \right] \left[\frac{na^3}{r^2} \cos f \right]$$

$$- \left[\frac{a}{\beta} \sin f \right] \left[\frac{a}{3\beta} \sin f + \frac{na^3}{r^2} t \cos f \right]$$

$$+ \left[-\frac{2r}{3n} \sin f + \frac{a}{\beta} t (\cos f + e) \right] \left[-\frac{na^3}{r^2} \sin f \right]$$

$$- \left[\frac{a}{\beta} (\cos f + e) \right] \left[\frac{a}{3\beta} (e + \cos f) - \frac{na^3}{r^2} t \sin f \right]. \tag{4.61}$$

After a considerable amount of algebra, we find

$$[n, M_0] = \frac{a^2}{3}. \tag{4.62}$$

4.2.3   Evaluation of $[e, M_0]$

Let us evaluate the Lagrange bracket for the eccentricity and the mean anomaly:

$$[\alpha_r, \alpha_s] = [e, M_0]. \tag{4.63}$$

We can now evaluate the Jacobian

$$[e, M_0] = \frac{\partial \xi}{\partial e} \frac{\partial \dot{\xi}}{\partial M_0} - \frac{\partial \xi}{\partial M_0} \frac{\partial \dot{\xi}}{\partial e} + \frac{\partial \eta}{\partial e} \frac{\partial \dot{\eta}}{\partial M_0} - \frac{\partial \eta}{\partial M_0} \frac{\partial \dot{\eta}}{\partial e}$$

using Eqs. 4.39, 4.40, 4.42, 4.43, 4.52, 4.53, 4.55, and 4.56 to obtain

$$[e, M_0] = \left[-a - \frac{r}{\beta^2} \sin^2 f \right] \left[-\frac{na^3}{r^2} \cos f \right]$$

$$- \left[-\frac{a}{\beta} \sin f \right] \left[-\frac{na}{\beta^3} \sin f \left(e + 2\cos f + e\cos^2 f \right) \right]$$

$$- \left[\frac{r}{\beta^2} \sin f \cos f \right] \left[-\frac{na^3}{r^2} \sin f \right]$$

$$- \left[\frac{a}{\beta} (\cos f + e) \right] \left[\frac{na}{\beta^3} \left(1 - 2\sin^2 f + e\cos^3 f \right) \right]. \tag{4.64}$$

After a considerable amount of algebra, we find

$$[e, M_0] = 0. \tag{4.65}$$

4.3   Evaluation of the $[\beta_r, \beta_s]$ Lagrange Brackets

Given:

$$x = \xi P_1 + \eta Q_1$$
$$y = \xi P_2 + \eta Q_2 \tag{4.66}$$
$$z = \xi P_3 + \eta Q_3$$

and the corresponding expressions for $\dot{x}$, $\dot{y}$, and $\dot{z}$, we will show that

$$[\beta_r, \beta_s] = (\xi\dot{\eta} - \eta\dot{\xi})\left[\frac{\partial\,(P_1, Q_1)}{\partial\,(\beta_r, \beta_s)} + \frac{\partial\,(P_2, Q_2)}{\partial\,(\beta_r, \beta_s)} + \frac{\partial\,(P_3, Q_3)}{\partial\,(\beta_r, \beta_s)}\right], \qquad (4.67)$$

where we note that these Jacobians (in contrast to the earlier ones for the $[\alpha_r, \alpha_s]$ Lagrange brackets) involve partial derivatives with respect to the orbital elements Ω, ω, and i. These three elements provide the orientation of the orbit in inertial space.

By the definition of the Lagrange bracket:

$$[\beta_r, \beta_s] = \frac{\partial\,(x, \dot{x})}{\partial\,(\beta_r, \beta_s)} + \frac{\partial\,(y, \dot{y})}{\partial\,(\beta_r, \beta_s)} + \frac{\partial\,(z, \dot{z})}{\partial\,(\beta_r, \beta_s)}, \qquad (4.68)$$

where

$$\frac{\partial\,(x, \dot{x})}{\partial\,(\beta_r, \beta_s)} = \begin{vmatrix} \frac{\partial x}{\partial \beta_r} & \frac{\partial x}{\partial \beta_s} \\ \frac{\partial \dot{x}}{\partial \beta_r} & \frac{\partial \dot{x}}{\partial \beta_s} \end{vmatrix}, \qquad (4.69)$$

and similar expressions follow for y and z. We assume that the direction cosines are functions of the orientation, $P(\Omega, \omega, i)$, $Q(\Omega, \omega, i)$ and that $\beta_r = \Omega, \omega$, or i.

Now we can write, via the chain rule and using Eq. 4.4:

$$\frac{\partial x}{\partial \beta_r} = \frac{\partial x}{\partial P_1}\frac{\partial P_1}{\partial \beta_r} + \frac{\partial x}{\partial Q_1}\frac{\partial Q_1}{\partial \beta_r} = \xi\frac{\partial P_1}{\partial \beta_r} + \eta\frac{\partial Q_1}{\partial \beta_r}, \qquad (4.70)$$

and using Eq. 4.7,

$$\frac{\partial \dot{x}}{\partial \beta_r} = \frac{\partial \dot{x}}{\partial P_1}\frac{\partial P_1}{\partial \beta_r} + \frac{\partial \dot{x}}{\partial Q_1}\frac{\partial Q_1}{\partial \beta_r} = \dot{\xi}\frac{\partial P_1}{\partial \beta_r} + \dot{\eta}\frac{\partial Q_1}{\partial \beta_r}, \qquad (4.71)$$

with analogous terms for the partial derivatives with respect to β_s. Thus, from Eqs. 4.69–4.71, we have

$$\frac{\partial\,(x, \dot{x})}{\partial\,(\beta_r, \beta_s)} = \begin{vmatrix} \left(\xi\frac{\partial P_1}{\partial \beta_r} + \eta\frac{\partial Q_1}{\partial \beta_r}\right) & \left(\xi\frac{\partial P_1}{\partial \beta_s} + \eta\frac{\partial Q_1}{\partial \beta_s}\right) \\ \left(\dot{\xi}\frac{\partial P_1}{\partial \beta_r} + \dot{\eta}\frac{\partial Q_1}{\partial \beta_r}\right) & \left(\dot{\xi}\frac{\partial P_1}{\partial \beta_s} + \dot{\eta}\frac{\partial Q_1}{\partial \beta_s}\right) \end{vmatrix}$$

$$= \xi\dot{\xi}\frac{\partial P_1}{\partial \beta_r}\frac{\partial P_1}{\partial \beta_s} + \eta\dot{\eta}\frac{\partial Q_1}{\partial \beta_r}\frac{\partial Q_1}{\partial \beta_s} + \xi\dot{\eta}\frac{\partial P_1}{\partial \beta_r}\frac{\partial Q_1}{\partial \beta_s} + \eta\dot{\xi}\frac{\partial Q_1}{\partial \beta_r}\frac{\partial P_1}{\partial \beta_s}$$

$$- \xi\dot{\xi}\frac{\partial P_1}{\partial \beta_s}\frac{\partial P_1}{\partial \beta_r} - \eta\dot{\eta}\frac{\partial Q_1}{\partial \beta_s}\frac{\partial Q_1}{\partial \beta_r} - \xi\dot{\eta}\frac{\partial P_1}{\partial \beta_s}\frac{\partial Q_1}{\partial \beta_r} - \eta\dot{\xi}\frac{\partial Q_1}{\partial \beta_s}\frac{\partial P_1}{\partial \beta_r}. \qquad (4.72)$$

We note that the $\xi\dot{\xi}$ and $\eta\dot{\eta}$ terms cancel, but the mixed terms do not cancel because, for example $\partial P_1/\partial \beta_r \neq \partial P_1/\partial \beta_s$ as $r \neq s$.

The right-hand side of Eq. 4.72 becomes

$$(\xi\dot{\eta} - \dot{\xi}\eta)\frac{\partial P_1}{\partial \beta_r}\frac{\partial Q_1}{\partial \beta_s} + (\eta\dot{\xi} - \xi\dot{\eta})\frac{\partial P_1}{\partial \beta_s}\frac{\partial Q_1}{\partial \beta_r} = (\xi\dot{\eta} - \eta\dot{\xi})\left[\frac{\partial P_1}{\partial \beta_r}\frac{\partial Q_1}{\partial \beta_s} - \frac{\partial P_1}{\partial \beta_s}\frac{\partial Q_1}{\partial \beta_r}\right]$$

$$= (\xi\dot{\eta} - \eta\dot{\xi})\begin{vmatrix} \frac{\partial P_1}{\partial \beta_r} & \frac{\partial P_1}{\partial \beta_s} \\ \frac{\partial Q_1}{\partial \beta_r} & \frac{\partial Q_1}{\partial \beta_s} \end{vmatrix}. \tag{4.73}$$

From Eq. 4.73, we can write

$$\frac{\partial(x,\dot{x})}{\partial(\beta_r,\beta_s)} = (\xi\dot{\eta} - \eta\dot{\xi})\frac{\partial(P_1,Q_1)}{\partial(\beta_r,\beta_s)}. \tag{4.74}$$

Since the expressions for y and z are found by changing the subscript from 1 to 2 and from 1 to 3, respectively, after adding the three expressions, we obtain

$$[\beta_r, \beta_s] = (\xi\dot{\eta} - \eta\dot{\xi})\left[\frac{\partial(P_1,Q_1)}{\partial(\beta_r,\beta_s)} + \frac{\partial(P_2,Q_2)}{\partial(\beta_r,\beta_s)} + \frac{\partial(P_3,Q_3)}{\partial(\beta_r,\beta_s)}\right], \tag{4.75}$$

which verifies Eq. 4.67.

Next, we evaluate the leading term, $(\xi\dot{\eta} - \eta\dot{\xi})$. From Eqs. 4.36, 4.37, 4.49, and 4.50, we have

$$\xi\dot{\eta} - \eta\dot{\xi} = a(\cos E - e)\frac{na^2\beta}{r}\cos E + a\beta\sin E\frac{na^2}{r}\sin E$$

$$= \frac{na^3\beta}{r}(\cos E - e)\cos E + \frac{na^3\beta}{r}\sin E\sin E$$

$$= \frac{na^3\beta}{r}\left(\cos^2 E - e\cos E + \sin^2 E\right)$$

$$= \frac{na^3\beta}{a(1 - e\cos E)}(1 - e\cos E)$$

$$= na^2\beta = na^2\sqrt{1 - e^2}. \tag{4.76}$$

The remaining calculations for the $[\beta_r, \beta_s]$ brackets require partial derivatives of direction cosines with respect to Ω, ω, i. Once we compute those derivatives, the brackets can be computed with simple substitution and simplification.

From Eq. 2.51

$$P_1 = \cos\omega\cos\Omega - \cos i\sin\omega\sin\Omega, \tag{4.77}$$

and we have

$$\frac{\partial P_1}{\partial \Omega} = -\cos\omega\sin\Omega - \cos i\sin\omega\cos\Omega \tag{4.78}$$

$$\frac{\partial P_1}{\partial \omega} = -\sin \omega \cos \Omega - \cos i \cos \omega \sin \Omega \tag{4.79}$$

$$\frac{\partial P_1}{\partial i} = \sin i \sin \omega \sin \Omega. \tag{4.80}$$

From Eq. 2.51

$$P_2 = \cos \omega \sin \Omega + \cos i \sin \omega \cos \Omega, \tag{4.81}$$

and we obtain

$$\frac{\partial P_2}{\partial \Omega} = \cos \omega \cos \Omega - \cos i \sin \omega \sin \Omega \tag{4.82}$$

$$\frac{\partial P_2}{\partial \omega} = -\sin \omega \sin \Omega + \cos i \cos \omega \cos \Omega \tag{4.83}$$

$$\frac{\partial P_2}{\partial i} = -\sin i \sin \omega \cos \Omega. \tag{4.84}$$

From Eq. 2.51

$$P_3 = \sin i \sin \omega, \tag{4.85}$$

and we have

$$\frac{\partial P_3}{\partial \Omega} = 0 \tag{4.86}$$

$$\frac{\partial P_3}{\partial \omega} = \sin i \cos \omega \tag{4.87}$$

$$\frac{\partial P_3}{\partial i} = \cos i \sin \omega. \tag{4.88}$$

From Eq. 2.51

$$Q_1 = -\sin \omega \cos \Omega - \cos i \cos \omega \sin \Omega, \tag{4.89}$$

and we have

$$\frac{\partial Q_1}{\partial \Omega} = \sin \omega \sin \Omega - \cos i \cos \omega \cos \Omega \tag{4.90}$$

$$\frac{\partial Q_1}{\partial \omega} = -\cos \omega \cos \Omega + \cos i \sin \omega \sin \Omega \qquad (4.91)$$

$$\frac{\partial Q_1}{\partial i} = \sin i \cos \omega \sin \Omega. \qquad (4.92)$$

From Eq. 2.51

$$Q_2 = -\sin \omega \sin \Omega + \cos i \cos \omega \cos \Omega, \qquad (4.93)$$

and we obtain

$$\frac{\partial Q_2}{\partial \Omega} = -\sin \omega \cos \Omega - \cos i \cos \omega \sin \Omega \qquad (4.94)$$

$$\frac{\partial Q_2}{\partial \omega} = -\cos \omega \sin \Omega - \cos i \sin \omega \cos \Omega \qquad (4.95)$$

$$\frac{\partial Q_2}{\partial i} = -\sin i \cos \omega \cos \Omega. \qquad (4.96)$$

From Eq. 2.51

$$Q_3 = \sin i \cos \omega, \qquad (4.97)$$

and we have

$$\frac{\partial Q_3}{\partial \Omega} = 0 \qquad (4.98)$$

$$\frac{\partial Q_3}{\partial \omega} = -\sin i \sin \omega \qquad (4.99)$$

$$\frac{\partial Q_3}{\partial i} = \cos i \cos \omega. \qquad (4.100)$$

4.3.1   Evaluation of $[\Omega, i]$

Next, we evaluate the Lagrange bracket $[\Omega, i]$. We see from Eq. 4.75 that we need to evaluate three Jacobians, the first of which is

$$\frac{\partial (P_1, Q_1)}{\partial (\Omega, i)} = \begin{vmatrix} \frac{\partial P_1}{\partial \Omega} & \frac{\partial P_1}{\partial i} \\ \frac{\partial Q_1}{\partial \Omega} & \frac{\partial Q_1}{\partial i} \end{vmatrix}. \qquad (4.101)$$

Using Eqs. 4.78, 4.92, 4.80, and 4.90, we find

$$\frac{\partial (P_1, Q_1)}{\partial (\Omega, i)} = -\cos^2 \omega \sin^2 \Omega \sin i - \sin \omega \cos \omega \sin \Omega \cos \Omega \sin i \cos i$$

$$-\sin^2 \omega \sin^2 \Omega \sin i + \sin \omega \cos \omega \sin \Omega \cos \Omega \sin i \cos i, \qquad (4.102)$$

which reduces to

$$\frac{\partial (P_1, Q_1)}{\partial (\Omega, i)} = -\sin^2 \Omega \sin i. \qquad (4.103)$$

Next we must evaluate the second Jacobian:

$$\frac{\partial (P_2, Q_2)}{\partial (\Omega, i)} = \begin{vmatrix} \frac{\partial P_2}{\partial \Omega} & \frac{\partial P_2}{\partial i} \\ \frac{\partial Q_2}{\partial \Omega} & \frac{\partial Q_2}{\partial i} \end{vmatrix}. \qquad (4.104)$$

Using Eqs. 4.82, 4.96, 4.84, and 4.94, we find

$$\frac{\partial (P_2, Q_2)}{\partial (\Omega, i)} = -\cos^2 \omega \cos^2 \Omega \sin i + \sin \omega \cos \omega \sin \Omega \cos \Omega \sin i \cos i$$

$$-\sin^2 \omega \cos^2 \Omega \sin i - \sin \omega \cos \omega \sin \Omega \cos \Omega \sin i \cos i, \qquad (4.105)$$

which reduces to

$$\frac{\partial (P_2, Q_2)}{\partial (\Omega, i)} = -\cos^2 \Omega \sin i. \qquad (4.106)$$

In a similar manner, the third Jacobian is

$$\frac{\partial (P_3, Q_3)}{\partial (\Omega, i)} = \begin{vmatrix} \frac{\partial P_3}{\partial \Omega} & \frac{\partial P_3}{\partial i} \\ \frac{\partial Q_3}{\partial \Omega} & \frac{\partial Q_3}{\partial i} \end{vmatrix}. \qquad (4.107)$$

Using Eqs. 4.86, 4.100, 4.88, and 4.98, we find

$$\frac{\partial (P_3, Q_3)}{\partial (\Omega, i)} = 0. \qquad (4.108)$$

From Eqs. 4.76, 4.103, 4.106, and 4.108, we can finally write the Lagrange bracket of Ω with i:

$$[\Omega, i] = a^2 n \sqrt{1 - e^2} \left[-\sin^2 \Omega \sin i - \cos^2 \Omega \sin i + 0 \right]. \qquad (4.109)$$

After simplification, we obtain

$$[\Omega, i] = -na^2\sqrt{1 - e^2} \sin i = -na^2\beta \sin i, \qquad (4.110)$$

our second Lagrange bracket.

4.3.2   Evaluation of $[\Omega, \omega]$

The Lagrange bracket $[\Omega, \omega]$ (the bracket of the ascending node angle with the argument of periapsis) is of the type $[\beta_r, \beta_s]$ involving partial derivatives with respect to the orbital elements that provide the orientation of the orbit in space (i.e., Ω, ω, and i). As we see from Eq. 4.75, we need to evaluate three Jacobians, the first of which is

$$\frac{\partial (P_1, Q_1)}{\partial (\Omega, \omega)} = \begin{vmatrix} \frac{\partial P_1}{\partial \Omega} & \frac{\partial P_1}{\partial \omega} \\ \frac{\partial Q_1}{\partial \Omega} & \frac{\partial Q_1}{\partial \omega} \end{vmatrix}. \qquad (4.111)$$

Using Eqs. 4.78, 4.91, 4.79, and 4.90, we find

$$\begin{aligned}
\frac{\partial (P_1, Q_1)}{\partial (\Omega, \omega)} &= \cos^2 \omega \sin \Omega \cos \Omega - \cos \omega \sin \omega \sin^2 \Omega \cos i \\
&\quad + \sin \omega \cos \omega \cos^2 \Omega \cos i - \sin^2 \omega \cos \Omega \sin \Omega \cos^2 i \\
&\quad + \sin^2 \omega \sin \Omega \cos \Omega + \sin \omega \cos \omega \sin^2 \Omega \cos i \\
&\quad - \cos \omega \sin \omega \cos^2 \Omega \cos i - \cos^2 \omega \cos \Omega \sin \Omega \cos^2 i \\
&= \sin \Omega \cos \Omega - \cos \omega \sin \omega \cos i + \cos \omega \sin \omega \cos i - \cos \Omega \sin \Omega \cos^2 i \\
&= \sin \Omega \cos \Omega \left(1 - \cos^2 i\right), \qquad (4.112)
\end{aligned}$$

so that the first Jacobian is

$$\frac{\partial (P_1, Q_1)}{\partial (\Omega, \omega)} = \sin \Omega \cos \Omega \sin^2 i. \qquad (4.113)$$

In a similar manner, we evaluate the second Jacobian

$$\frac{\partial (P_2, Q_2)}{\partial (\Omega, \omega)} = \begin{vmatrix} \frac{\partial P_2}{\partial \Omega} & \frac{\partial P_2}{\partial \omega} \\ \frac{\partial Q_2}{\partial \Omega} & \frac{\partial Q_2}{\partial \omega} \end{vmatrix}. \qquad (4.114)$$

Using Eqs. 4.82, 4.95, 4.83, and 4.94, we find

$$\frac{\partial (P_2, Q_2)}{\partial (\Omega, \omega)} = -\cos^2 \omega \cos \Omega \sin \Omega - \cos \omega \sin \omega \cos^2 \Omega \cos i$$

$$+ \sin \omega \cos \omega \sin^2 \Omega \cos i + \sin^2 \omega \sin \Omega \cos \Omega \cos^2 i$$

$$- \sin^2 \omega \cos \Omega \sin \Omega + \sin \omega \cos \omega \cos^2 \Omega \cos i$$

$$- \sin \omega \cos \omega \sin^2 \Omega \cos i + \cos^2 \omega \sin \Omega \cos \Omega \cos^2 i$$

$$= -\cos \Omega \sin \Omega + \cos \Omega \sin \Omega \cos^2 i$$

$$= -\cos \Omega \sin \Omega \left(1 - \cos^2 i \right), \tag{4.115}$$

so the second Jacobian is

$$\frac{\partial (P_2, Q_2)}{\partial (\Omega, \omega)} = -\sin \Omega \cos \Omega \sin^2 i. \tag{4.116}$$

The third Jacobian is

$$\frac{\partial (P_3, Q_3)}{\partial (\Omega, \omega)} = \begin{vmatrix} \frac{\partial P_3}{\partial \Omega} & \frac{\partial P_3}{\partial \omega} \\ \frac{\partial Q_3}{\partial \Omega} & \frac{\partial Q_3}{\partial \omega} \end{vmatrix}. \tag{4.117}$$

Using Eqs. 4.86, 4.99, 4.87, and 4.98, we find

$$\frac{\partial (P_3, Q_3)}{\partial (\Omega, \omega)} = 0. \tag{4.118}$$

We can now add the three Jacobians in Eqs. 4.113, 4.116, and 4.118

$$\left[\frac{\partial (P_1, Q_1)}{\partial (\Omega, \omega)} + \frac{\partial (P_2, Q_2)}{\partial (\Omega, \omega)} + \frac{\partial (P_3, Q_3)}{\partial (\Omega, \omega)} \right] = \sin \Omega \cos \Omega \sin^2 i - \sin \Omega \cos \Omega \sin^2 i + 0$$

$$= 0 \tag{4.119}$$

and substitute this result into Eq. 4.75 to calculate $[\Omega, \omega]$:

$$[\Omega, \omega] = 0. \tag{4.120}$$

4.3.3   Evaluation of $[i, \omega]$

As before, we must evaluate the three Jacobians of Eq. 4.75, the first of which is

$$\frac{\partial (P_1, Q_1)}{\partial (i, \omega)} = \begin{vmatrix} \frac{\partial P_1}{\partial i} & \frac{\partial P_1}{\partial \omega} \\ \frac{\partial Q_1}{\partial i} & \frac{\partial Q_1}{\partial \omega} \end{vmatrix}. \tag{4.121}$$

Using Eqs. 4.80, 4.91, 4.79, and 4.92, we find

$$\frac{\partial (P_1, Q_1)}{\partial (i, \omega)} = - \sin \omega \cos \omega \sin \Omega \cos \Omega \sin i + \sin^2 \omega \sin^2 \Omega \sin i \cos i$$

$$+ \sin \omega \cos \omega \sin \Omega \cos \Omega \sin i + \cos^2 \omega \sin^2 \Omega \sin i \cos i, \tag{4.122}$$

which simplifies to

$$\frac{\partial (P_1, Q_1)}{\partial (i, \omega)} = \sin^2 \Omega \sin i \cos i. \tag{4.123}$$

To compute the second Jacobian, we use Eqs. 4.84, 4.95, 4.83, and 4.96 to find

$$\begin{vmatrix} \frac{\partial P_2}{\partial i} & \frac{\partial P_2}{\partial \omega} \\ \frac{\partial Q_2}{\partial i} & \frac{\partial Q_2}{\partial \omega} \end{vmatrix} = \sin \omega \cos \omega \cos \Omega \sin \Omega \sin i + \sin^2 \omega \cos^2 \Omega \sin i \cos i$$

$$- \sin \omega \cos \omega \cos \Omega \sin \Omega \sin i + \cos^2 \omega \cos^2 \Omega \sin i \cos i, \tag{4.124}$$

so we have

$$\frac{\partial (P_2, Q_2)}{\partial (i, \omega)} = \cos^2 \Omega \sin i \cos i. \tag{4.125}$$

To evaluate the third Jacobian, we use Eqs. 4.88, 4.99, 4.87, and 4.100 to find

$$\begin{vmatrix} \frac{\partial P_3}{\partial i} & \frac{\partial P_3}{\partial \omega} \\ \frac{\partial Q_3}{\partial i} & \frac{\partial Q_3}{\partial \omega} \end{vmatrix} = - \sin^2 \omega \cos i \sin i - \cos^2 \omega \cos i \sin i, \tag{4.126}$$

which simplifies to

$$\frac{\partial (P_3, Q_3)}{\partial (i, \omega)} = - \cos i \sin i. \tag{4.127}$$

Adding Eqs. 4.123, 4.125, and 4.127, we obtain

$$\frac{\partial (P_1, Q_1)}{\partial (i, \omega)} + \frac{\partial (P_2, Q_2)}{\partial (i, \omega)} + \frac{\partial (P_3, Q_3)}{\partial (i, \omega)} = \sin^2 \Omega \sin i \cos i + \cos^2 \Omega \sin i \cos i$$

$$- \sin i \cos i$$

$$= \sin i \cos i - \sin i \cos i = 0, \quad (4.128)$$

so we have

$$[i, \omega] = 0. \quad (4.129)$$

4.4 Evaluation of Lagrange Brackets: $[\alpha_r, \beta_s]$

The $[\alpha_r, \beta_s]$ Lagrange brackets include an orbital element from the α_r group (n, e, M_0) and an orbital element from the β_s group (Ω, ω, i). From the definition of the Lagrange bracket (Eq. 3.42),

$$[\alpha_r, \beta_s] = \frac{\partial(x, \dot{x})}{\partial(\alpha_r, \beta_s)} + \frac{\partial(y, \dot{y})}{\partial(\alpha_r, \beta_s)} + \frac{\partial(z, \dot{z})}{\partial(\alpha_r, \beta_s)}, \quad (4.130)$$

where

$$\frac{\partial(x, \dot{x})}{\partial(\alpha_r, \beta_s)} = \begin{vmatrix} \frac{\partial x}{\partial \alpha_r} & \frac{\partial x}{\partial \beta_s} \\ \frac{\partial \dot{x}}{\partial \alpha_r} & \frac{\partial \dot{x}}{\partial \beta_s} \end{vmatrix}, \quad (4.131)$$

and where there are similar terms for y and z. We recall from Eqs. 4.14, 4.70, 4.17, and 4.71:

$$\begin{aligned}
\frac{\partial x}{\partial \alpha_r} &= P_1 \frac{\partial \xi}{\partial \alpha_r} + Q_1 \frac{\partial \eta}{\partial \alpha_r} \\
\frac{\partial x}{\partial \beta_s} &= \xi \frac{\partial P_1}{\partial \beta_s} + \eta \frac{\partial Q_1}{\partial \beta_s} \\
\frac{\partial \dot{x}}{\partial \alpha_r} &= P_1 \frac{\partial \dot{\xi}}{\partial \alpha_r} + Q_1 \frac{\partial \dot{\eta}}{\partial \alpha_r} \\
\frac{\partial \dot{x}}{\partial \beta_s} &= \dot{\xi} \frac{\partial P_1}{\partial \beta_s} + \dot{\eta} \frac{\partial Q_1}{\partial \beta_s}.
\end{aligned} \quad (4.132)$$

Substituting Eqs. 4.132 into Eq. 4.131, we have

$$\frac{\partial(x, \dot{x})}{\partial(\alpha_r, \beta_s)} = \begin{vmatrix} \left(P_1 \frac{\partial \xi}{\partial \alpha_r} + Q_1 \frac{\partial \eta}{\partial \alpha_r} \right) & \left(\xi \frac{\partial P_1}{\partial \beta_s} + \eta \frac{\partial Q_1}{\partial \beta_s} \right) \\ \left(P_1 \frac{\partial \dot{\xi}}{\partial \alpha_r} + Q_1 \frac{\partial \dot{\eta}}{\partial \alpha_r} \right) & \left(\dot{\xi} \frac{\partial P_1}{\partial \beta_s} + \dot{\eta} \frac{\partial Q_1}{\partial \beta_s} \right) \end{vmatrix}$$

$$= P_1 \dot{\xi} \frac{\partial \xi}{\partial \alpha_r} \frac{\partial P_1}{\partial \beta_s} + P_1 \dot{\eta} \frac{\partial \xi}{\partial \alpha_r} \frac{\partial Q_1}{\partial \beta_s} + Q_1 \dot{\xi} \frac{\partial \eta}{\partial \alpha_r} \frac{\partial P_1}{\partial \beta_s} + Q_1 \dot{\eta} \frac{\partial \eta}{\partial \alpha_r} \frac{\partial Q_1}{\partial \beta_s}$$

$$- P_1 \xi \frac{\partial \dot{\xi}}{\partial \alpha_r} \frac{\partial P_1}{\partial \beta_s} - P_1 \eta \frac{\partial \dot{\xi}}{\partial \alpha_r} \frac{\partial Q_1}{\partial \beta_s} - Q_1 \xi \frac{\partial \dot{\eta}}{\partial \alpha_r} \frac{\partial P_1}{\partial \beta_s} - Q_1 \eta \frac{\partial \dot{\eta}}{\partial \alpha_r} \frac{\partial Q_1}{\partial \beta_s}.$$

(4.133)

We have similar terms for y and z that are given by changing the subscripts to 2 and 3, respectively. By summing the terms corresponding to the 1, 2, 3 subscripts in Eq. 4.133, we obtain the expression for $[\alpha_r, \beta_s]$ in Eq. 4.130:

$$[\alpha_r, \beta_s] = \frac{\partial (x, \dot{x})}{\partial (\alpha_r, \beta_s)} + \frac{\partial (y, \dot{y})}{\partial (\alpha_r, \beta_s)} + \frac{\partial (z, \dot{z})}{\partial (\alpha_r, \beta_s)}$$

$$= \left(\sum_{i=1}^{3} P_i \frac{\partial P_i}{\partial \beta_s} \right) \left(\dot{\xi} \frac{\partial \xi}{\partial \alpha_r} - \xi \frac{\partial \dot{\xi}}{\partial \alpha_r} \right)$$

$$+ \left(\sum_{i=1}^{3} P_i \frac{\partial Q_i}{\partial \beta_s} \right) \left(\dot{\eta} \frac{\partial \xi}{\partial \alpha_r} - \eta \frac{\partial \dot{\xi}}{\partial \alpha_r} \right)$$

$$+ \left(\sum_{i=1}^{3} Q_i \frac{\partial P_i}{\partial \beta_s} \right) \left(\dot{\xi} \frac{\partial \eta}{\partial \alpha_r} - \xi \frac{\partial \dot{\eta}}{\partial \alpha_r} \right)$$

$$+ \left(\sum_{i=1}^{3} Q_i \frac{\partial Q_i}{\partial \beta_s} \right) \left(\dot{\eta} \frac{\partial \eta}{\partial \alpha_r} - \eta \frac{\partial \dot{\eta}}{\partial \alpha_r} \right).$$

(4.134)

For convenience in evaluating these Lagrange brackets, let us label the terms in Eq. 4.134 as follows:

$$[\alpha_r, \beta_s] = \left(\sum_{i=1}^{3} P_i \frac{\partial P_i}{\partial \beta_s} \right) \left(\dot{\xi} \frac{\partial \xi}{\partial \alpha_r} - \xi \frac{\partial \dot{\xi}}{\partial \alpha_r} \right) \qquad \text{Term 1}$$

$$+ \left(\sum_{i=1}^{3} P_i \frac{\partial Q_i}{\partial \beta_s} \right) \left(\dot{\eta} \frac{\partial \xi}{\partial \alpha_r} - \eta \frac{\partial \dot{\xi}}{\partial \alpha_r} \right) \qquad \text{Term 2}$$

$$+ \left(\sum_{i=1}^{3} Q_i \frac{\partial P_i}{\partial \beta_s} \right) \left(\dot{\xi} \frac{\partial \eta}{\partial \alpha_r} - \xi \frac{\partial \dot{\eta}}{\partial \alpha_r} \right) \qquad \text{Term 3}$$

$$+ \left(\sum_{i=1}^{3} Q_i \frac{\partial Q_i}{\partial \beta_s} \right) \left(\dot{\eta} \frac{\partial \eta}{\partial \alpha_r} - \eta \frac{\partial \dot{\eta}}{\partial \alpha_r} \right) \qquad \text{Term 4.}$$

(4.135)

Each of the terms is the product of a β element and an α element, and we refer to the elements using this terminology.

To compute the β elements for partial derivatives with respect to Ω of Term 1, we use Eqs. 4.77, 4.78, 4.81, 4.82, 4.85, and 4.86 to find

$$\sum_{i=1}^{3} P_i \frac{\partial P_i}{\partial \Omega} = (\cos \omega \cos \Omega - \cos i \sin \omega \sin \Omega)(-\cos \omega \sin \Omega - \cos i \sin \omega \cos \Omega)$$

$$+ (\cos \omega \sin \Omega + \cos i \sin \omega \cos \Omega)(\cos \omega \cos \Omega - \cos i \sin \omega \sin \Omega)$$

$$= -\cos^2 \omega \sin \Omega \cos \Omega + \cos i \sin \omega \cos \omega \sin^2 \Omega - \cos i \sin \omega \cos \omega \cos^2 \Omega$$

$$+ \cos^2 i \sin^2 \omega \sin \Omega \cos \Omega + \cos^2 \omega \sin \Omega \cos \Omega - \cos i \sin \omega \cos \omega \sin^2 \Omega$$

$$+ \cos i \sin \omega \cos \omega \cos^2 \Omega - \cos^2 i \sin^2 \omega \sin \Omega \cos \Omega$$

$$= 0. \tag{4.136}$$

To compute the β elements for partial derivatives with respect to Ω of Term 2, we use Eqs. 4.77, 4.81, 4.85, 4.90, 4.94, and 4.98 to find

$$\sum_{i=1}^{3} P_i \frac{\partial Q_i}{\partial \Omega} = (\cos \omega \cos \Omega - \cos i \sin \omega \sin \Omega)(\sin \omega \sin \Omega - \cos i \cos \omega \cos \Omega)$$

$$+ (\cos \omega \sin \Omega + \cos i \sin \omega \cos \Omega)(-\sin \omega \cos \Omega - \cos i \cos \omega \sin \Omega)$$

$$+ (\sin i \sin \omega)(0)$$

$$= \sin \omega \cos \omega \sin \Omega \cos \Omega - \cos i \sin^2 \omega \sin^2 \Omega$$

$$- \cos i \cos^2 \omega \cos^2 \Omega + \cos^2 i \sin \omega \cos \omega \sin \Omega \cos \Omega$$

$$- \sin \omega \cos \omega \sin \Omega \cos \Omega - \cos i \sin^2 \omega \cos^2 \Omega$$

$$- \cos i \cos^2 \omega \sin^2 \Omega - \cos^2 i \sin \omega \cos \omega \sin \Omega \cos \Omega$$

$$= -\cos i. \tag{4.137}$$

To compute the β elements for partial derivatives with respect to Ω of Term 3, we use Eqs. 4.89, 4.93, 4.97, 4.78, 4.82, and 4.86 to find

$$\sum_{i=1}^{3} Q_i \frac{\partial P_i}{\partial \Omega} = (-\sin \omega \cos \Omega - \cos i \cos \omega \sin \Omega)(-\cos \omega \sin \Omega - \cos i \sin \omega \cos \Omega)$$

$$+ (-\sin \omega \sin \Omega + \cos i \cos \omega \cos \Omega)(\cos \omega \cos \Omega - \cos i \sin \omega \sin \Omega)$$

$$= \sin \omega \cos \omega \sin \Omega \cos \Omega + \cos i \sin^2 \omega \cos^2 \Omega + \cos i \cos^2 \omega \sin^2 \Omega$$

$$+ \cos^2 i \sin \omega \cos \omega \sin \Omega \cos \Omega - \sin \omega \cos \omega \sin \Omega \cos \Omega$$

$$+ \cos i \sin^2 \omega \sin^2 \Omega + \cos i \cos^2 \omega \cos^2 \Omega - \cos^2 i \sin \omega \cos \omega \sin \Omega \cos \Omega$$

$$= \cos i. \tag{4.138}$$

To compute the β elements for partial derivatives with respect to Ω of Term 4, we use Eqs. 4.89, 4.90, 4.93, 4.94, 4.97, and 4.98 to find

$$\sum_{i=1}^{3} Q_i \frac{\partial Q_i}{\partial \Omega} = (-\sin \omega \cos \Omega - \cos i \cos \omega \sin \Omega)(\sin \omega \sin \Omega - \cos i \cos \omega \cos \Omega)$$

$$+ (-\sin \omega \sin \Omega + \cos i \cos \omega \cos \Omega)(-\sin \omega \cos \Omega - \cos i \cos \omega \sin \Omega)$$

$$= -\sin^2\omega \sin \Omega \cos \Omega + \cos i \sin \omega \cos \omega \cos^2\Omega - \cos i \sin \omega \cos \omega \sin^2\Omega$$

$$+ \cos^2 i \cos^2\omega \sin \Omega \cos \Omega + \sin^2\omega \sin \Omega \cos \Omega + \cos i \sin \omega \cos \omega \sin^2\Omega$$

$$- \cos i \sin \omega \cos \omega \cos^2\Omega - \cos^2 i \cos^2\omega \sin \Omega \cos \Omega$$

$$= 0. \tag{4.139}$$

To compute the β elements for partial derivatives with respect to ω of Term 1, we use Eqs. 4.77, 4.79, 4.81, 4.83, 4.85, and 4.87 to find

$$\sum_{i=1}^{3} P_i \frac{\partial P_i}{\partial \omega} = (\cos \omega \cos \Omega - \cos i \sin \omega \sin \Omega)(-\sin \omega \cos \Omega - \cos i \cos \omega \sin \Omega)$$

$$+ (\cos \omega \sin \Omega + \cos i \sin \omega \cos \Omega)(-\sin \omega \sin \Omega + \cos i \cos \omega \cos \Omega)$$

$$+ \sin \omega \sin i (\cos \omega \sin i)$$

$$= -\sin \omega \cos \omega \cos^2\Omega - \cos i \cos^2\omega \sin \Omega \cos \Omega + \cos i \sin^2\omega \sin \Omega \cos \Omega$$

$$+ \cos^2 i \sin \omega \cos \omega \sin^2\Omega - \sin \omega \cos \omega \sin^2\Omega + \cos i \cos^2\omega \sin \Omega \cos \Omega$$

$$- \cos i \sin^2\omega \sin \Omega \cos \Omega + \cos^2 i \sin \omega \cos \omega \cos^2\Omega + \sin \omega \cos \omega \sin^2 i$$

$$= 0. \tag{4.140}$$

To compute the β elements for partial derivatives with respect to ω of Term 2, we use Eqs. 4.77, 4.81, 4.85, 4.91, 4.95, and 4.99 to find

$$\sum_{i=1}^{3} P_i \frac{\partial Q_i}{\partial \omega} = (\cos \omega \cos \Omega - \cos i \sin \omega \sin \Omega)(-\cos \omega \cos \Omega + \cos i \sin \omega \sin \Omega)$$

$$+ (\cos \omega \sin \Omega + \cos i \sin \omega \cos \Omega)(-\cos \omega \sin \Omega - \cos i \sin \omega \cos \Omega)$$

$$+ \sin i \sin \omega (-\sin \omega \sin i)$$

$$= -\cos^2\omega \cos^2\Omega + \cos i \sin \omega \cos \omega \sin \Omega \cos \Omega$$

$$+ \cos i \sin \omega \cos \omega \sin \Omega \cos \Omega - \cos^2 i \sin^2\omega \sin^2\Omega$$

$$- \cos^2\omega \sin^2\Omega - \cos i \sin \omega \cos \omega \sin \Omega \cos \Omega$$

$$- \cos i \sin \omega \cos \omega \sin \Omega \cos \Omega - \cos^2 i \sin^2\omega \cos^2\Omega - \sin^2 i \sin^2\omega$$

$$= -1. \tag{4.141}$$

To compute the β elements for partial derivatives with respect to ω of Term 3, we use Eqs. 4.89, 4.93, 4.97, 4.79, 4.83, and 4.87 to find

$$\sum_{i=1}^{3} Q_i \frac{\partial P_i}{\partial \omega} = (-\sin \omega \cos \Omega - \cos i \cos \omega \sin \Omega)(-\sin \omega \cos \Omega - \cos i \cos \omega \sin \Omega)$$

$$+ (-\sin \omega \sin \Omega + \cos i \cos \omega \cos \Omega)(-\sin \omega \sin \Omega + \cos i \cos \omega \cos \Omega)$$

$$+ \sin i \cos \omega (\cos \omega \sin i)$$

$$= \sin^2\omega \cos^2\Omega$$

$$+ \cos i \sin \omega \cos \omega \sin \Omega \cos \Omega + \cos i \sin \omega \cos \omega \sin \Omega \cos \Omega$$

$$+ \cos^2 i \cos^2\omega \sin^2\Omega + \sin^2\omega \sin^2\Omega - \cos i \sin \omega \cos \omega \sin \Omega \cos \Omega$$

$$- \cos i \sin \omega \cos \omega \sin \Omega \cos \Omega + \cos^2 i \cos^2\omega \cos^2\Omega + \sin^2 i \cos^2\omega$$

$$= 1. \tag{4.142}$$

To compute the β elements for partial derivatives with respect to ω of Term 4, we use Eqs. 4.89, 4.91, 4.93, 4.95, 4.97, and 4.99 to find

$$\sum_{i=1}^{3} Q_i \frac{\partial Q_i}{\partial \omega} = (-\sin \omega \cos \Omega - \cos i \cos \omega \sin \Omega)(-\cos \omega \cos \Omega + \cos i \sin \omega \sin \Omega)$$

$$+ (-\sin \omega \sin \Omega + \cos i \cos \omega \cos \Omega)(-\cos \omega \sin \Omega - \cos i \sin \omega \cos \Omega)$$

$$+ \cos \omega \sin i (-\sin \omega \sin i)$$

$$= \sin \omega \cos \omega \cos^2\Omega - \cos i \sin^2\omega \sin \Omega \cos \Omega + \cos i \cos^2\omega \sin \Omega \cos \Omega$$

$$- \cos^2 i \sin \omega \cos \omega \sin^2\Omega + \sin \omega \cos \omega \sin^2\Omega + \cos i \sin^2\omega \sin \Omega \cos \Omega$$

$$- \cos i \cos^2\omega \sin \Omega \cos \Omega - \cos^2 i \sin \omega \cos \omega \cos^2\Omega - \sin \omega \cos \omega \sin^2 i$$

$$= 0. \tag{4.143}$$

To compute the β elements for partial derivatives with respect to i of Term 1, we use Eqs. 4.77, 4.80, 4.81, 4.84, 4.85, and 4.88 to find

$$\sum_{i=1}^{3} P_i \frac{\partial P_i}{\partial i} = (\cos \omega \cos \Omega - \cos i \sin \omega \sin \Omega)(\sin i \sin \omega \sin \Omega)$$

$$+ (\cos \omega \sin \Omega + \cos i \sin \omega \cos \Omega)(-\sin i \sin \omega \cos \Omega)$$

$$+ (\sin i \sin \omega)(\cos i \sin \omega)$$

$$= \sin i \sin \omega \cos \omega \sin \Omega \cos \Omega - \sin i \cos i \sin^2 \omega \sin^2 \Omega$$

$$- \sin i \sin \omega \cos \omega \sin \Omega \cos \Omega - \sin i \cos i \sin^2 \omega \cos^2 \Omega + \sin i \cos i \sin^2 \omega$$

$$= 0. \tag{4.144}$$

To compute the β elements for partial derivatives with respect to i of Term 2, we use Eqs. 4.77, 4.81, 4.85, 4.92, 4.96, and 4.100 to find

$$\sum_{i=1}^{3} P_i \frac{\partial Q_i}{\partial i} = (\cos \omega \cos \Omega - \cos i \sin \omega \sin \Omega)(\sin i \cos \omega \sin \Omega)$$

$$+ (\cos \omega \sin \Omega + \cos i \sin \omega \cos \Omega)(-\sin i \cos \omega \cos \Omega)$$

$$+ (\sin i \sin \omega)(\cos i \cos \omega)$$

$$= \sin i \cos^2 \omega \sin \Omega \cos \Omega - \sin i \cos i \sin \omega \cos \omega \sin^2 \Omega$$

$$- \sin i \cos^2 \omega \sin \Omega \cos \Omega - \sin i \cos i \sin \omega \cos \omega \cos^2 \Omega$$

$$+ \sin i \cos i \sin \omega \cos \omega$$

$$= 0. \tag{4.145}$$

To compute the β elements for partial derivatives with respect to i of Term 3, we use Eqs. 4.89, 4.93, 4.97, 4.80, 4.84, and 4.88 to find

$$\sum_{i=1}^{3} Q_i \frac{\partial P_i}{\partial i} = (-\sin \omega \cos \Omega - \cos i \cos \omega \sin \Omega)(\sin i \sin \omega \sin \Omega)$$

$$+ (-\sin \omega \sin \Omega + \cos i \cos \omega \cos \Omega)(-\sin i \sin \omega \cos \Omega)$$

$$+ (\sin i \cos \omega)(\cos i \sin \omega)$$

$$= -\sin i \sin^2 \omega \sin \Omega \cos \Omega - \sin i \cos i \sin \omega \cos \omega \sin^2 \Omega$$

$$+ \sin i \sin^2 \omega \sin \Omega \cos \Omega - \sin i \cos i \sin \omega \cos \omega \cos^2 \Omega$$

$$+ \sin i \cos i \sin \omega \cos \omega$$

$$= 0. \tag{4.146}$$

To compute the β elements for partial derivatives with respect to i of Term 4, we use Eqs. 4.89, 4.92, 4.93, 4.96, 4.97, and 4.100 to find

$$\sum_{i=1}^{3} Q_i \frac{\partial Q_i}{\partial i} = (-\sin\omega\cos\Omega - \cos i \cos\omega \sin\Omega)(\sin i \cos\omega \sin\Omega)$$

$$+ (-\sin\omega\sin\Omega + \cos i \cos\omega\cos\Omega)(-\sin i \cos\omega\cos\Omega)$$

$$+ (\sin i \cos\omega)(\cos i \cos\omega)$$

$$= -\sin i \sin\omega\cos\omega\sin\Omega\cos\Omega - \sin i \cos i \cos^2\omega \sin^2\Omega$$

$$+ \sin i \sin\omega\cos\omega\sin\Omega\cos\Omega - \sin i \cos i \cos^2\omega\cos^2\Omega$$

$$+ \sin i \cos i \cos^2\omega$$

$$= 0. \tag{4.147}$$

4.4.1   Evaluation of $[n, \Omega]$

Using Eq. 4.135, let us evaluate the bracket

$$[\alpha_r, \beta_s] = [n, \Omega]. \tag{4.148}$$

From Eqs. 4.136, 4.137, 4.138, and 4.139, we have the β element of each of the terms:

Term 1Ω $\displaystyle\sum_{i=1}^{3} P_i \frac{\partial P_i}{\partial \Omega} = 0$

Term 2Ω $\displaystyle\sum_{i=1}^{3} P_i \frac{\partial Q_i}{\partial \Omega} = -\cos i$

Term 3Ω $\displaystyle\sum_{i=1}^{3} Q_i \frac{\partial P_i}{\partial \Omega} = \cos i$

Term 4Ω $\displaystyle\sum_{i=1}^{3} Q_i \frac{\partial Q_i}{\partial \Omega} = 0$

so that

$$[n, \Omega] = (\text{Term } 2\Omega)(\text{Term } 2n) + (\text{Term } 3\Omega)(\text{Term } 3n), \tag{4.149}$$

where

$$\text{Term } 2n = \left(\dot{\eta} \frac{\partial \xi}{\partial n} - \eta \frac{\partial \dot{\xi}}{\partial n} \right)$$

$$\text{Term } 3n = \left(\dot{\xi} \frac{\partial \eta}{\partial n} - \xi \frac{\partial \dot{\eta}}{\partial n} \right)$$

and

$$[n, \Omega] = -\cos i \,(\text{Term } 2n) + \cos i \,(\text{Term } 3n)$$

$$[n, \Omega] = -\cos i \,[(\text{Term } 2n) - (\text{Term } 3n)]. \tag{4.150}$$

Next we have

$$\text{Term } 2n = \left(\dot{\eta} \frac{\partial \xi}{\partial n} - \eta \frac{\partial \dot{\xi}}{\partial n} \right).$$

Using Eqs. 2.56, 4.38, 4.50, and 4.51, we can write

$$\left(\dot{\eta} \frac{\partial \xi}{\partial n} - \eta \frac{\partial \dot{\xi}}{\partial n} \right) = \frac{na}{\beta} (\cos f + e) \left[-\frac{2r}{3n} \cos f - \frac{a}{\beta} t \sin f \right]$$

$$- r \sin f \left[-\frac{a}{3\beta} \sin f - t \frac{na^3}{r^2} \cos f \right],$$

which simplifies to

$$\left(\dot{\eta} \frac{\partial \xi}{\partial n} - \eta \frac{\partial \dot{\xi}}{\partial n} \right) = -\frac{2ar}{3\beta} (\cos f + e) \cos f - \frac{na^2}{\beta^2} t \sin f (\cos f + e) + \frac{ar}{3\beta} \sin^2 f$$

$$+ \frac{na^3}{r} t \sin f \cos f. \tag{4.151}$$

And

$$\text{Term } 3n = \left(\dot{\xi} \frac{\partial \eta}{\partial n} - \xi \frac{\partial \dot{\eta}}{\partial n} \right),$$

so that using Eqs. 2.52, 4.41, 4.49, and 4.54, we can write

$$\left(\dot{\xi} \frac{\partial \eta}{\partial n} - \xi \frac{\partial \dot{\eta}}{\partial n} \right) = -\left(\frac{na}{\beta} \sin f \right) \left(-\frac{2r}{3n} \sin f + \frac{a}{\beta} t (\cos f + e) \right)$$

$$- r \cos f \left(\frac{a}{3\beta} (e + \cos f) - \frac{na^3}{r^2} t \sin f \right),$$

which simplifies to

$$\left(\dot{\xi}\frac{\partial \eta}{\partial n} - \xi\frac{\partial \dot{\eta}}{\partial n}\right) = \frac{2ar}{3\beta}\sin^2 f - \frac{na^2}{\beta^2}t\sin f\,(\cos f + e) - \frac{ar}{3\beta}(e + \cos f)\cos f$$

$$+ \frac{na^3}{r}t\sin f\cos f. \tag{4.152}$$

Using Eqs. 4.151 and 4.152, we find

$$(\text{Term }2n - \text{Term }3n) = -\frac{2ar}{3\beta}(\cos f + e)\cos f - \frac{na^2}{\beta^2}t\sin f\,(\cos f + e)$$

$$+ \frac{ar}{3\beta}\sin^2 f + \frac{na^3}{r}t\sin f\cos f - \frac{2ar}{3\beta}\sin^2 f$$

$$+ \frac{na^2}{\beta^2}t\sin f\,(\cos f + e) + \frac{ar}{3\beta}(e + \cos f)\cos f$$

$$- \frac{na^3}{r}t\sin f\cos f,$$

which simplifies to

$$(\text{Term }2n - \text{Term }3n) = -\frac{a^2\beta}{3}. \tag{4.153}$$

Using Eq. 4.153 in Eq. 4.150, we find

$$[n, \Omega] = \frac{a^2\beta}{3}\cos i. \tag{4.154}$$

4.4.2   Evaluation of $[e, \Omega]$

Using Eq. 4.135, let us evaluate the bracket

$$[\alpha_r, \beta_s] = [e, \Omega]. \tag{4.155}$$

From Eq. 4.149, we see that the only difference between $[e, \Omega]$ and $[n, \Omega]$ is the α element, so we can immediately write

$$[e, \Omega] = (\text{Term }2\Omega)\,(\text{Term }2e) + (\text{Term }3\Omega)\,(\text{Term }3e),$$

where

$$\text{Term } 2e = \left(\dot{\eta} \frac{\partial \xi}{\partial e} - \eta \frac{\partial \dot{\xi}}{\partial e} \right)$$

$$\text{Term } 3e = \left(\dot{\xi} \frac{\partial \eta}{\partial e} - \xi \frac{\partial \dot{\eta}}{\partial e} \right)$$

and

$$[e, \Omega] = - \cos i \, (\text{Term } 2e) + \cos i \, (\text{Term } 3e)$$
$$= - \cos i \, [(\text{Term } 2e) - (\text{Term } 3e)] \, . \tag{4.156}$$

We first examine

$$\text{Term } 2e = \left(\dot{\eta} \frac{\partial \xi}{\partial e} - \eta \frac{\partial \dot{\xi}}{\partial e} \right) \, .$$

Using Eqs. 2.56, 4.39, 4.50, and 4.52, we can write

$$\left(\dot{\eta} \frac{\partial \xi}{\partial e} - \eta \frac{\partial \dot{\xi}}{\partial e} \right) = \frac{na}{\beta} \, (\cos f + e) \left(-a - \frac{r}{\beta^2} \sin^2 f \right)$$
$$+ r \sin f \frac{na}{\beta^3} \sin f \left(e + 2 \cos f + e\cos^2 f \right) ,$$

which simplifies to

$$\text{Term } 2e = - \frac{na^2}{\beta} \left(e + \cos^3 f \right) . \tag{4.157}$$

And

$$\text{Term } 3e = \left(\dot{\xi} \frac{\partial \eta}{\partial e} - \xi \frac{\partial \dot{\eta}}{\partial e} \right) ,$$

so that using Eqs. 2.52, 4.42, 4.49, and 4.55, we can write

$$\left(\dot{\xi} \frac{\partial \eta}{\partial e} - \xi \frac{\partial \dot{\eta}}{\partial e} \right) = \left[- \frac{na}{\beta} \sin f \right] \left[\frac{r}{\beta^2} \sin f \cos f \right]$$
$$- [r \cos f] \left[\frac{na}{\beta^3} \left(1 - 2\sin^2 f + e\cos^3 f \right) \right] ,$$

which simplifies to

$$\text{Term } 3e = - \frac{na^2}{\beta} \cos^3 f . \tag{4.158}$$

Using Eqs. 4.157 and 4.158, we find

$$\text{(Term } 2e - \text{Term } 3e) = -\frac{na^2}{\beta}\left(e + \cos^3 f\right) + \frac{na^2}{\beta}\cos^3 f,$$

which simplifies to

$$\text{(Term } 2e - \text{Term } 3e) = -\frac{na^2 e}{\beta}. \tag{4.159}$$

Using Eq. 4.159 in Eq. 4.156, we find

$$[e, \Omega] = \frac{na^2 e}{\beta}\cos i. \tag{4.160}$$

4.4.3 Evaluation of $[M_0, \Omega]$

Using Eq. 4.135, let us evaluate the bracket

$$[\alpha_r, \beta_s] = [M_0, \Omega]. \tag{4.161}$$

From Eq. 4.149, we see that the only difference between $[M_0, \Omega]$ and $[n, \Omega]$ is the α element. We can immediately write

$$[M_0, \Omega] = \text{(Term } 2\Omega)\,\text{(Term } 2M_0) + \text{(Term } 3\Omega)\,\text{(Term } 3M_0),$$

where

$$\text{Term } 2M_0 = \dot{\eta}\frac{\partial \xi}{\partial M_0} - \eta\frac{\partial \dot{\xi}}{\partial M_0}$$

$$\text{Term } 3M_0 = \dot{\xi}\frac{\partial \eta}{\partial M_0} - \xi\frac{\partial \dot{\eta}}{\partial M_0},$$

then

$$[M_0, \Omega] = -\cos i\,\text{(Term } 2M_0) + \cos i\,\text{(Term } 3M_0)$$
$$= -\cos i\,[\text{(Term } 2M_0) - \text{(Term } 3M_0)], \tag{4.162}$$

and

$$\text{Term } 2M_0 = \dot{\eta}\frac{\partial \xi}{\partial M_0} - \eta\frac{\partial \dot{\xi}}{\partial M_0}.$$

Using Eqs. 2.56, 4.40, 4.50, and 4.53, we can write

$$\dot{\eta}\frac{\partial \xi}{\partial M_0} - \eta\frac{\partial \dot{\xi}}{\partial M_0} = -\frac{na^2}{\beta^2}(\cos f + e)\sin f + \frac{na^3}{r}\cos f \sin f,$$

which simplifies to

$$\text{Term } 2M_0 = -\frac{na^2 e}{\beta^2}\sin^3 f. \tag{4.163}$$

And

$$\text{Term } 3M_0 = \dot{\xi}\frac{\partial \eta}{\partial M_0} - \xi\frac{\partial \dot{\eta}}{\partial M_0},$$

so that using Eqs. 2.52, 4.43, 4.49, and 4.56, we can write

$$\dot{\xi}\frac{\partial \eta}{\partial M_0} - \xi\frac{\partial \dot{\eta}}{\partial M_0} = -\frac{na^2}{\beta^2}\sin f(\cos f + e) + \frac{na^3}{r}\sin f \cos f,$$

which simplifies to

$$\text{Term } 3M_0 = -\frac{na^2 e}{\beta^2}\sin^3 f. \tag{4.164}$$

Using Eqs. 4.163 and 4.164, we find

$$(\text{Term } 2M_0 - \text{Term } 3M_0) = -\cos i\left[-\frac{na^2 e}{\beta^2}\sin^3 f + \frac{na^2 e}{\beta^2}\sin^3 f\right],$$

which simplifies to

$$(\text{Term } 2M_0 - \text{Term } 3M_0) = 0. \tag{4.165}$$

Using Eq. 4.165 in Eq. 4.162, we find

$$[M_0, \Omega] = 0. \tag{4.166}$$

4.4.4   Evaluation of $[n, \omega]$

Using Eq. 4.135, let us evaluate the bracket

$$[\alpha_r, \beta_s] = [n, \omega]. \qquad (4.167)$$

From Eqs. 4.140, 4.141, 4.142, and 4.143, we have the β element of each of the terms:

Term 1ω
$$\sum_{i=1}^{3} P_i \frac{\partial P_i}{\partial \omega} = 0$$

Term 2ω
$$\sum_{i=1}^{3} P_i \frac{\partial Q_i}{\partial \omega} = -1$$

Term 3ω
$$\sum_{1}^{3} Q_i \frac{\partial P_i}{\partial \omega} = 1$$

Term 4ω
$$\sum_{1}^{3} Q_i \frac{\partial Q_i}{\partial \omega} = 0$$

and

$$[n, \omega] = (\text{Term 2}\omega)\,(\text{Term 2}n) + (\text{Term 3}\omega)\,(\text{Term 3}n)\,, \qquad (4.168)$$

where

$$\text{Term 2}n = \left(\dot{\eta}\frac{\partial \xi}{\partial n} - \eta\frac{\partial \dot{\xi}}{\partial n} \right)$$

$$\text{Term 3}n = \left(\dot{\xi}\frac{\partial \eta}{\partial n} - \xi\frac{\partial \dot{\eta}}{\partial n} \right)$$

and

$$[n, \omega] = -\,(\text{Term 2}n) + (\text{Term 3}n)$$
$$= -\,[(\text{Term 2}n) - (\text{Term 3}n)]\,. \qquad (4.169)$$

Using Eq. 4.153 in Eq. 4.169, we find

$$[n, \omega] = \frac{a^2 \beta}{3}. \qquad (4.170)$$

4.4.5 Evaluation of [e, ω]

Using Eq. 4.135, let us evaluate the bracket

$$[\alpha_r, \beta_s] = [e, \omega].$$ (4.171)

From Eq. 4.168, we see that the only difference between $[e, \omega]$ and $[n, \omega]$ is the α element, so we can immediately write

$$[e, \omega] = (\text{Term } 2\omega)\,(\text{Term } 2e) + (\text{Term } 3\omega)\,(\text{Term } 3e)\,,$$

where

$$\text{Term } 2e = \left(\dot{\eta}\frac{\partial \xi}{\partial e} - \eta\frac{\partial \dot{\xi}}{\partial e} \right)$$

$$\text{Term } 3e = \left(\dot{\xi}\frac{\partial \eta}{\partial e} - \xi\frac{\partial \dot{\eta}}{\partial e} \right)$$

and

$$[e, \omega] = -\,(\text{Term } 2e) + (\text{Term } 3e)$$
$$= -\,[(\text{Term } 2e) - (\text{Term } 3e)]\,.$$ (4.172)

Using Eq. 4.159 in Eq. 4.172, we find

$$[e, \omega] = \frac{na^2 e}{\beta}.$$ (4.173)

4.4.6 Evaluation of [M₀, ω]

Using Eq. 4.135, let us evaluate the bracket

$$[\alpha_r, \beta_s] = [M_0, \omega].$$ (4.174)

From Eq. 4.168, we see that the only difference between $[M_0, \omega]$ and $[n, \omega]$ is the α element, so we can immediately write

$$[M_0, \omega] = (\text{Term } 2\omega)\,(\text{Term } 2M_0) + (\text{Term } 3\omega)\,(\text{Term } 3M_0)\,,$$

where

$$\text{Term } 2M_0 = \dot{\eta}\frac{\partial \xi}{\partial M_0} - \eta\frac{\partial \dot{\xi}}{\partial M_0}$$

$$\text{Term } 3M_0 = \dot{\eta}\frac{\partial \xi}{\partial M_0} - \eta\frac{\partial \dot{\xi}}{\partial M_0}$$

and

$$[M_0, \omega] = -(\text{Term } 2M_0) + (\text{Term } 3M_0)$$
$$= -[(\text{Term } 2M_0) - (\text{Term } 3M_0)]. \tag{4.175}$$

Using Eq. 4.165 in Eq. 4.175, we find

$$[M_0, \omega] = 0. \tag{4.176}$$

4.4.7   Evaluation of $[n, i]$, $[e, i]$, $[M_0, i]$

Using Eq. 4.135, let us evaluate the bracket

$$[\alpha_r, \dot{\beta}_s] = [n, i]. \tag{4.177}$$

From Eqs. 4.144, 4.145, 4.146, and 4.147, we have the β element of each of the terms:

$$\text{Term } 1i \qquad \sum_{i=1}^{3} P_i\frac{\partial P_i}{\partial i} = 0$$

$$\text{Term } 2i \qquad \sum_{i=1}^{3} P_i\frac{\partial Q_i}{\partial i} = 0$$

$$\text{Term } 3i \qquad \sum_{i=1}^{3} Q_i\frac{\partial P_i}{\partial i} = 0$$

$$\text{Term } 4i \qquad \sum_{i=1}^{3} Q_i\frac{\partial Q_i}{\partial i} = 0,$$

so that

$$[n, i] = 0. \tag{4.178}$$

Similarly,

$$[e, i] = 0 \tag{4.179}$$

and

$$[M_0, i] = 0. \tag{4.180}$$

4.5   The Nonvanishing Lagrange Brackets

Collecting the nonzero brackets from Eqs. 4.154, 4.160, 4.170, 4.173, 4.62, and 4.110, we have six nonzero Lagrange brackets:

$$[n, \Omega] = \frac{a^2 \beta}{3} \cos i$$

$$[e, \Omega] = \frac{na^2 e}{\beta} \cos i$$

$$[n, \omega] = \frac{a^2 \beta}{3}$$

$$[e, \omega] = \frac{na^2 e}{\beta} \tag{4.181}$$

$$[n, M_0] = \frac{a^2}{3}$$

$$[\Omega, i] = -na^2 \beta \sin i.$$

For our choice of perifocal coordinates and the classical orbital elements, these six are the only nonvanishing brackets. These brackets are also given (but not always derived) by Battin (1999), Fitzpatrick (1970), Kaula (1966), Moulton (1914), Taff (1985), and Vallado (2013). These brackets differ slightly from the cited references due to our choice of mean motion rather than semimajor axis as the independent variable.

4.6   The Vanishing Lagrange Brackets

Nine of the 15 unique Lagrange brackets vanish. These vanishing brackets are listed by type below. The vanishing brackets of the type $[\alpha_r, \alpha_s]$ are:

$$[n, e] = 0$$

$$[e, M_0] = 0, \tag{4.182}$$

which were proven in Eqs. 4.59 and 4.65. The type $[\beta_r, \beta_s]$ admits the following vanishing brackets:

$$[\omega, \Omega] = 0$$

$$[i, \omega] = 0, \tag{4.183}$$

which were proven in Eqs. 4.120 and 4.129. The type $[\alpha_r, \beta_s]$ admits the following vanishing brackets:

$$[M_0, \Omega] = 0$$

$$[M_0, \omega] = 0$$

$$[n, i] = 0 \tag{4.184}$$

$$[e, i] = 0$$

$$[M_0, i] = 0,$$

which we have shown in Eqs. 4.166, 4.176, 4.178, 4.179, and 4.180.

We have evaluated all 15 unique Lagrange brackets. *We have not assumed at any step that the brackets are free of time explicitly. But we have, in fact, proven that they are free of time explicitly.* Furthermore, we have not placed any restriction on the nature of the perturbing force. Thus, this set of Lagrange brackets can be used to develop the Lagrange planetary equations for *both conservative and non-conservative forces.*

References

Battin, R. H. (1999). *An introduction to the mathematics and methods of astrodynamics* (Revised ed.). Reston: American Institute of Aeronautics and Astronautics, Inc.

Fitzpatrick, P. M. (1970). *Principles of celestial mechanics.* New York: Academic Press.

Kaula, W. M. (1966). *Theory of satellite geodesy.* Waltham: Blaisdell Publishing Company.

McCuskey, S. W. (1963). *Introduction to celestial mechanics.* Reading: Addison-Wesley Publishing Company.

Moulton, F. R. (1914). *An introduction to celestial mechanics* (2nd ed.). New York: The Macmillan Company.

Taff, L. G. (1985). *Celestial mechanics, a computational guide for the practitioner.* New York: John Wiley & Sons.

Vallado, D. A. (2013). *Fundamentals of astrodynamics and applications* (4th ed.). El Segundo: Microcosm Press.

Chapter 5
The Lagrange Planetary Equations for a General Perturbing Force

5.1 Derivation of Lagrange's Planetary Equations

Now that we have evaluated all of the Lagrange brackets (Eqs. 4.181–4.184), we can turn our attention to formulating the Lagrange planetary equations. The equations take on a different form depending on the nature of the perturbing force. If the perturbing force is due to a gravitational potential function $\mathcal{R}$, then from Eq. 3.7, $\mathbf{F} = \nabla \mathcal{R}$, and from Eq. 3.46

$$\sum_{k=1}^{6} \left[c_j, c_k \right] \dot{c}_k = \frac{\partial \mathcal{R}}{\partial c_j}, \qquad j = 1, 2, \ldots, 6. \tag{5.1}$$

For the purpose of expanding Eq. 5.1, let

$$\mathbf{c}^T = n, e, i, \Omega, \omega, M_0, \tag{5.2}$$

where the superscript T denotes the transpose operation.

Using the Lagrange bracket for the first element of $\mathbf{c}$, we find

$$\sum_{k=1}^{6} [n, c_k] \dot{c}_k = \frac{\partial \mathcal{R}}{\partial n}. \tag{5.3}$$

From Eq. 4.181, we have

$$\sum_{k=1}^{6} [n, c_k] \dot{c}_k = [n, n] \dot{n} + [n, e] \dot{e} + [n, i] \frac{di}{dt} + [n, \Omega] \dot{\Omega} + [n, \omega] \dot{\omega} + [n, M_0] \dot{M}_0$$

$$= [n, \Omega] \dot{\Omega} + [n, \omega] \dot{\omega} + [n, M_0] \dot{M}_0, \tag{5.4}$$

J. M. Longuski et al., *Introduction to Orbital Perturbations*, Space Technology Library 40, https://doi.org/10.1007/978-3-030-89758-1_5

99

so that

$$\left(\frac{a^2\beta}{3}\cos i\right)\dot{\Omega} + \left(\frac{a^2\beta}{3}\right)\dot{\omega} + \left(\frac{a^2}{3}\right)\dot{M}_0 = \frac{\partial\mathcal{R}}{\partial n}. \tag{5.5}$$

Using the Lagrange bracket for the second element of **c**, we find

$$\sum_{k=1}^{6}[e, c_k]\dot{c}_k = \frac{\partial\mathcal{R}}{\partial e}. \tag{5.6}$$

From Eq. 4.181, we have

$$\sum_{k=1}^{6}[e, c_k]\dot{c}_k = [e, n]\dot{n} + [e, e]\dot{e} + [e, i]\frac{di}{dt} + [e, \Omega]\dot{\Omega} + [e, \omega]\dot{\omega} + [e, M_0]\dot{M}_0$$

$$= [e, \Omega]\dot{\Omega} + [e, \omega]\dot{\omega}, \tag{5.7}$$

so that

$$\left(\frac{na^2 e}{\beta}\cos i\right)\dot{\Omega} + \left(\frac{na^2 e}{\beta}\right)\dot{\omega} = \frac{\partial\mathcal{R}}{\partial e}. \tag{5.8}$$

Using the Lagrange bracket for the third element of **c**, we find

$$\sum_{k=1}^{6}[i, c_k]\dot{c}_k = \frac{\partial\mathcal{R}}{\partial i}, \tag{5.9}$$

and from Eq. 4.181

$$\sum_{k=1}^{6}[i, c_k]\dot{c}_k = [i, n]\dot{n} + [i, e]\dot{e} + [i, i]\frac{di}{dt} + [i, \Omega]\dot{\Omega} + [i, \omega]\dot{\omega} + [i, M_0]\dot{M}_0$$

$$= [\Omega, i]\dot{\Omega}, \tag{5.10}$$

so we have

$$\left(na^2\beta\sin i\right)\dot{\Omega} = \frac{\partial\mathcal{R}}{\partial i}. \tag{5.11}$$

Using the Lagrange bracket for the fourth element of **c**, we find

$$\sum_{k=1}^{6}[\Omega, c_k]\dot{c}_k = \frac{\partial\mathcal{R}}{\partial\Omega}. \tag{5.12}$$

Using Eq. 4.181, we obtain

$$\sum_{k=1}^{6} [\Omega, c_k] \dot{c}_k = [\Omega, n] \dot{n} + [\Omega, e] \dot{e} + [\Omega, i] \frac{di}{dt} + [\Omega, \Omega] \dot{\Omega} + [\Omega, \omega] \dot{\omega} + [\Omega, M_0] \dot{M}_0$$

$$= [\Omega, n] \dot{n} + [\Omega, e] \dot{e} + [\Omega, i] \frac{di}{dt}, \tag{5.13}$$

so that

$$-\left(\frac{a^2 \beta}{3} \cos i\right) \dot{n} - \left(\frac{na^2 e}{\beta} \cos i\right) \dot{e} - \left(na^2 \beta \sin i\right) \frac{di}{dt} = \frac{\partial \mathcal{R}}{\partial \Omega}. \tag{5.14}$$

Using the Lagrange bracket for the fifth element of **c**, we find

$$\sum_{k=1}^{6} [\omega, c_k] \dot{c}_k = \frac{\partial \mathcal{R}}{\partial \omega}, \tag{5.15}$$

and from Eq. 4.181

$$\sum_{k=1}^{6} [\omega, c_k] \dot{c}_k = [\omega, n] \dot{n} + [\omega, e] \dot{e} + [\omega, i] \frac{di}{dt} + [\omega, \Omega] \dot{\Omega} + [\omega, \omega] \dot{\omega} + [\omega, M_0] \dot{M}_0$$

$$= [\omega, n] \dot{n} + [\omega, e] \dot{e}, \tag{5.16}$$

so we have

$$-\left(\frac{a^2 \beta}{3}\right) \dot{n} - \left(\frac{na^2 e}{\beta}\right) \dot{e} = \frac{\partial \mathcal{R}}{\partial \omega}. \tag{5.17}$$

Using the Lagrange bracket for the sixth element of **c**, we find

$$\sum_{k=1}^{6} [M_0, c_k] \dot{c}_k = \frac{\partial \mathcal{R}}{\partial M_0}, \tag{5.18}$$

and using Eq. 4.181 gives

$$\sum_{k=1}^{6} [M_0, c_k] \dot{c}_k = [M_0, n] \dot{n} + [M_0, e] \dot{e} + [M_0, i] \frac{di}{dt} + [M_0, \Omega] \dot{\Omega}$$

$$+ [M_0, \omega] \dot{\omega} + [M_0, M_0] \dot{M}_0$$

$$= [M_0, n] \dot{n} \tag{5.19}$$

and

$$-\frac{a^2}{3}\dot{n} = \frac{\partial \mathcal{R}}{\partial M_0}. \tag{5.20}$$

Next, we find the simultaneous solution of Eqs. 5.5, 5.8, 5.11, 5.14, 5.17, and 5.20 for the time derivatives of the orbital elements. In other words, we need to find separate differential equations for $\dot{n}$, $\dot{e}$, di/dt, $\dot{\Omega}$, $\dot{\omega}$, $\dot{M}_0$.

5.1.1   Solution for the Time Derivatives of the Orbital Elements

From Eq. 5.20, we have $\dot{n}$ directly:

$$\dot{n} = -\frac{3}{a^2} \frac{\partial \mathcal{R}}{\partial M_0}. \tag{5.21}$$

The $\dot{e}$ equation can be found by substituting Eq. 5.21 into Eq. 5.17 for $\dot{n}$, giving

$$-\left(\frac{a^2 \beta}{3}\right)\left(-\frac{3}{a^2}\frac{\partial \mathcal{R}}{\partial M_0}\right) - \left(\frac{na^2 e}{\beta}\right)\dot{e} = \frac{\partial \mathcal{R}}{\partial \omega}$$

$$\left(\frac{na^2 e}{\beta}\right)\dot{e} = \beta \frac{\partial \mathcal{R}}{\partial M_0} - \frac{\partial \mathcal{R}}{\partial \omega},$$

which can be rearranged to provide

$$\dot{e} = \left(\frac{\beta^2}{na^2 e}\right)\frac{\partial \mathcal{R}}{\partial M_0} - \left(\frac{\beta}{na^2 e}\right)\frac{\partial \mathcal{R}}{\partial \omega}. \tag{5.22}$$

We have $\dot{\Omega}$ directly from Eq. 5.11:

$$\dot{\Omega} = \left(\frac{1}{na^2 \beta \sin i}\right)\frac{\partial \mathcal{R}}{\partial i}. \tag{5.23}$$

We obtain the $\dot{\omega}$ equation by substituting Eq. 5.23 into Eq. 5.8 to obtain

$$\left(\frac{na^2 e}{\beta}\right)\dot{\omega} = \frac{\partial \mathcal{R}}{\partial e} - \left(\frac{na^2 e}{\beta}\cos i\right)\left(\frac{1}{na^2 \beta \sin i}\right)\frac{\partial \mathcal{R}}{\partial i}$$

$$\left(\frac{na^2 e}{\beta}\right)\dot{\omega} = \frac{\partial \mathcal{R}}{\partial e} - \left(\frac{e \cos i}{\beta^2 \sin i}\right)\frac{\partial \mathcal{R}}{\partial i},$$

which can be rearranged to give

$$\dot{\omega} = \frac{\beta}{na^2e} \frac{\partial R}{\partial e} - \left(\frac{\cos i}{na^2\beta \sin i} \right) \frac{\partial R}{\partial i}. \tag{5.24}$$

To find $\dot{M}_0$, we substitute Eqs. 5.23 and 5.24 into Eq. 5.5:

$$\frac{\partial R}{\partial n} = \left(\frac{a^2\beta}{3} \cos i \right) \left(\frac{1}{na^2\beta \sin i} \right) \frac{\partial R}{\partial i} + \left(\frac{a^2\beta}{3} \right) \left[\frac{\beta}{na^2e} \frac{\partial R}{\partial e} - \left(\frac{\cos i}{na^2\beta \sin i} \right) \frac{\partial R}{\partial i} \right]$$

$$+ \left(\frac{a^2}{3} \right) \dot{M}_0$$

so that

$$\left(\frac{a^2}{3} \right) \dot{M}_0 = \frac{\partial R}{\partial n} - \frac{1}{3} \left(\frac{\cos i}{n \sin i} \right) \frac{\partial R}{\partial i} - \frac{1}{3} \frac{\beta^2}{ne} \frac{\partial R}{\partial e} + \frac{1}{3} \left(\frac{\cos i}{n \sin i} \right) \frac{\partial R}{\partial i}$$

or, after solving for $\dot{M}_0$:

$$\dot{M}_0 = \frac{3}{a^2} \frac{\partial R}{\partial n} - \frac{\beta^2}{na^2e} \frac{\partial R}{\partial e}. \tag{5.25}$$

Finally, we find di/dt by substituting Eqs. 5.21 and 5.22 into Eq. 5.14:

$$\frac{\partial R}{\partial \Omega} = - \left(\frac{a^2\beta}{3} \cos i \right) \left(-\frac{3}{a^2} \right) \frac{\partial R}{\partial M_0} - \left(\frac{na^2e}{\beta} \cos i \right) \left[\left(\frac{\beta^2}{na^2e} \right) \frac{\partial R}{\partial M_0} - \left(\frac{\beta}{na^2e} \right) \frac{\partial R}{\partial \omega} \right]$$

$$- \left(na^2\beta \sin i \right) \frac{di}{dt}$$

$$\frac{\partial R}{\partial \Omega} = (\beta \cos i) \frac{\partial R}{\partial M_0} - (\beta \cos i) \frac{\partial R}{\partial M_0} + (\cos i) \frac{\partial R}{\partial \omega} - \left(na^2\beta \sin i \right) \frac{di}{dt}$$

yielding

$$\left(na^2\beta \sin i \right) \frac{di}{dt} = (\cos i) \frac{\partial R}{\partial \omega} - \frac{\partial R}{\partial \Omega},$$

which simplifies to

$$\frac{di}{dt} = \left(\frac{\cos i}{na^2\beta \sin i} \right) \frac{\partial R}{\partial \omega} - \left(\frac{1}{na^2\beta \sin i} \right) \frac{\partial R}{\partial \Omega}. \tag{5.26}$$

Equations 5.21, 5.22, 5.23, 5.24, 5.25, and 5.26 are known as Lagrange's planetary equations, which we summarize in their classical form as follows:

$$\dot{n} = -\frac{3}{a^2} \frac{\partial R}{\partial M_0}$$

$$\dot{e} = \left(\frac{\beta^2}{na^2e}\right)\frac{\partial\mathcal{R}}{\partial M_0} - \left(\frac{\beta}{na^2e}\right)\frac{\partial\mathcal{R}}{\partial\omega}$$

$$\frac{di}{dt} = \left(\frac{\cos i}{na^2\beta\sin i}\right)\frac{\partial\mathcal{R}}{\partial\omega} - \left(\frac{1}{na^2\beta\sin i}\right)\frac{\partial\mathcal{R}}{\partial\Omega}$$

$$\dot{\Omega} = \left(\frac{1}{na^2\beta\sin i}\right)\frac{\partial\mathcal{R}}{\partial i} \tag{5.27}$$

$$\dot{\omega} = \frac{\beta}{na^2e}\frac{\partial\mathcal{R}}{\partial e} - \left(\frac{\cos i}{na^2\beta\sin i}\right)\frac{\partial\mathcal{R}}{\partial i}$$

$$\dot{M}_0 = \frac{3}{a^2}\frac{\partial\mathcal{R}}{\partial n} - \frac{\beta^2}{na^2e}\frac{\partial\mathcal{R}}{\partial e}.$$

Lagrange's planetary equations provide the first-order rates of change of the orbital elements due to a conservative disturbing function $\mathcal{R}$. Nearly identical (equivalent) forms appear in Battin (1999); Moulton (1914); Roy (2005); Vallado (2013). These equations differ slightly from the cited references due to our choice of mean motion rather than semimajor axis as the independent variable.

5.1.2   Lagrange's Planetary Equations in Terms of Mean Anomaly

In some situations, it may be more convenient to use the mean anomaly

$$M = M_0 + nt \tag{5.28}$$

instead of M_0 in Lagrange's planetary equations, in which case the perturbation can be written as $\mathcal{R}(n, e, i, \Omega, \omega, M)$. We note that the time derivative of the mean anomaly is

$$\dot{M} = \dot{M}_0 + n + \dot{n}t. \tag{5.29}$$

From Eq. 5.25, we see that we need an expression for $\partial\mathcal{R}/\partial n$. We can write

$$\frac{\partial\mathcal{R}}{\partial n} = \left(\frac{\partial\mathcal{R}}{\partial n}\right)_M + \frac{\partial R}{\partial M}\frac{\partial M}{\partial n}, \tag{5.30}$$

where $(\partial\mathcal{R}/\partial n)_M$ denotes that the partial derivative of the disturbing function is to be taken at constant M. From Eq. 5.28, we see that

$$\frac{\partial M}{\partial n} = t. \tag{5.31}$$

Using Eq. 5.31 in Eq. 5.30 gives

$$\frac{\partial \mathcal{R}}{\partial n} = \left(\frac{\partial \mathcal{R}}{\partial n}\right)_M + t\frac{\partial R}{\partial M},$$ (5.32)

so that Eq. 5.25 becomes

$$\dot{M}_0 = \left(\frac{3}{a^2}\right)\left(\frac{\partial \mathcal{R}}{\partial n}\right)_M + \left(\frac{3}{a^2}\right)t\frac{\partial R}{\partial M} - \left(\frac{\beta^2}{na^2e}\right)\frac{\partial \mathcal{R}}{\partial e}.$$ (5.33)

Next using Eqs. 5.21 and 5.33 in Eq. 5.29, we get

$$\dot{M} = \left(\frac{3}{a^2}\right)\left(\frac{\partial \mathcal{R}}{\partial n}\right)_M + \left(\frac{3}{a^2}\right)t\frac{\partial R}{\partial M} - \left(\frac{\beta^2}{na^2e}\right)\frac{\partial \mathcal{R}}{\partial e} + n - \frac{3}{a^2}\frac{\partial \mathcal{R}}{\partial M_0}t.$$ (5.34)

From Eq. 5.28, we can see that

$$\frac{\partial \mathcal{R}}{\partial M} = \frac{\partial \mathcal{R}}{\partial M_0}.$$ (5.35)

Simplifying Eq. 5.34 gives

$$\dot{M} = n + \left(\frac{3}{a^2}\right)\left(\frac{\partial \mathcal{R}}{\partial n}\right)_M - \left(\frac{\beta^2}{na^2e}\right)\frac{\partial \mathcal{R}}{\partial e}.$$ (5.36)

When the mean anomaly is used in place of M_0 as the sixth orbital element, the new set of Lagrange's planetary equations is

$$\dot{n} = -\frac{3}{a^2}\frac{\partial \mathcal{R}}{\partial M}$$

$$\dot{e} = \left(\frac{\beta^2}{na^2e}\right)\frac{\partial \mathcal{R}}{\partial M} - \left(\frac{\beta}{na^2e}\right)\frac{\partial \mathcal{R}}{\partial \omega}$$

$$\frac{di}{dt} = \left(\frac{\cos i}{na^2\beta \sin i}\right)\frac{\partial \mathcal{R}}{\partial \omega} - \left(\frac{1}{na^2\beta \sin i}\right)\frac{\partial \mathcal{R}}{\partial \Omega}$$

$$\dot{\Omega} = \left(\frac{1}{na^2\beta \sin i}\right)\frac{\partial \mathcal{R}}{\partial i}$$ (5.37)

$$\dot{\omega} = \frac{\beta}{na^2e}\frac{\partial \mathcal{R}}{\partial e} - \left(\frac{\cos i}{na^2\beta \sin i}\right)\frac{\partial \mathcal{R}}{\partial i}$$

$$\dot{M} = n - \left(\frac{\beta^2}{na^2e}\right)\frac{\partial \mathcal{R}}{\partial e} + \left(\frac{3}{a^2}\right)\left(\frac{\partial \mathcal{R}}{\partial n}\right)_M.$$

5.2   The Disturbing Function

In astrodynamics, there are many disturbing functions such as atmospheric drag and solar radiation pressure that are not conservative. Thus, we seek a more general development of the Lagrange planetary equations that are valid for any disturbing force. An important component of the development is the use of the Lagrange brackets. *Since we did not make any assumption about the nature of the disturbing function in Chap. 4, we were able to derive a general formulation of the Lagrange brackets that is valid for any disturbing force.*

5.2.1   The Force System (Leading to Gauss's Form)

From Eq. 3.41, we found

$$\sum_{k=1}^{6} [c_j, c_k]\dot{c}_k = \mathbf{F} \cdot \frac{\partial \mathbf{r}}{\partial c_j} \qquad j = 1, 2, \dots, 6. \qquad (5.38)$$

Since we have already obtained the Lagrange brackets in Chap, 4, we simply focus on computing the right-hand side of Eq. 5.38 for each of the six orbital elements. Once that is accomplished, we just need to substitute the Lagrange brackets and solve a set of six linear equations for the time rates of change of the orbital elements.

There are many applications in which the perturbing forces are aligned along a particular direction and may be non-conservative. We consider two cases for such applications of Lagrange's planetary equations.

Case I: Forces are directed along the radius vector, transverse to the radius vector, and perpendicular to the orbital plane.

Case II: Forces are tangent to the orbit, normal to the orbit (in the plane), and perpendicular to the orbital plane. (For example, atmospheric drag acts tangent to the orbit in the direction opposite to the instantaneous velocity vector. Also, low thrust is frequently applied along the velocity vector.)

We note here that in the development of the Force System, we frequently refer to force per unit mass, i.e., the perturbing acceleration. The resulting planetary equations are sometimes referred to as Gauss's form of the planetary equations.

Fig. 5.1 The coordinate
frame for Case I:
radial–transverse–orthogonal
system where $\mathbf{u}_W$ is
orthogonal to the orbital
plane

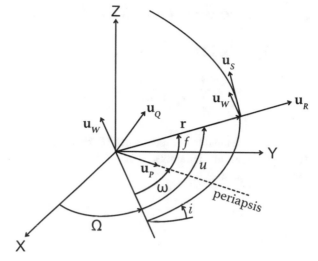

5.2.2   Force System Case I: Radial-Transverse-Orthogonal (RSW)

In the first case, we use the radial–transverse–orthogonal coordinate system, a
satellite-centered coordinate system, RSW. This coordinate system is also called the
Gaussian coordinate system or the local-vertical, local-horizontal (LVLH) frame in
some references. A depiction of the reference frame is given in Fig. 5.1.

Let $\mathbf{u}_R, \mathbf{u}_S, \mathbf{u}_W$ be a right-handed orthogonal set of unit vectors where $\mathbf{u}_R$ is
directed along $\mathbf{r}$ (the radius vector), $\mathbf{u}_S$ is perpendicular to $\mathbf{r}$ in the direction of
motion in the orbital plane, and $\mathbf{u}_W = \mathbf{u}_R \times \mathbf{u}_S$ is perpendicular to the orbital plane.

We define the true argument of latitude as the angle from the line of nodes to $\mathbf{r}$:

$$u = \omega + f \tag{5.39}$$

given by the sum of the argument of periapsis and the true anomaly. We write the
perturbing acceleration (or force per unit mass) as

$$\mathbf{F} = R\mathbf{u}_R + S\mathbf{u}_S + W\mathbf{u}_W, \tag{5.40}$$

where R, S, and W denote the scalar components of the perturbing acceleration in
each of the three principal directions of the RSW system.

5.2.3 Perturbation Formulation for α_j Elements

Returning to Eq. 5.38

$$\sum_{k=1}^{6} [c_j, c_k]\dot{c}_k = \mathbf{F} \cdot \frac{\partial \mathbf{r}}{\partial c_j} \qquad j = 1, 2, \ldots, 6,$$

we see that we need to compute partial derivatives of the position vector with respect to the plane orbital elements n, e, M_0. From the definition of the perifocal coordinate system, we have

$$\mathbf{r} = \xi \mathbf{u}_P + \eta \mathbf{u}_Q,$$

so that

$$\frac{\partial \mathbf{r}}{\partial \alpha_j} = \frac{\partial \xi}{\partial \alpha_j}\mathbf{u}_P + \frac{\partial \eta}{\partial \alpha_j}\mathbf{u}_Q \tag{5.41}$$

and

$$\mathbf{F} \cdot \frac{\partial \mathbf{r}}{\partial c_j} = (R\mathbf{u}_R + S\mathbf{u}_S + W\mathbf{u}_W) \cdot \left(\frac{\partial \xi}{\partial \alpha_j}\mathbf{u}_P + \frac{\partial \eta}{\partial \alpha_j}\mathbf{u}_Q \right). \tag{5.42}$$

By examining Fig. 2.5 in Chap. 2, it is clear that

$$\mathbf{u}_R \cdot \mathbf{u}_P = \cos f$$
$$\mathbf{u}_R \cdot \mathbf{u}_Q = \sin f$$
$$\mathbf{u}_S \cdot \mathbf{u}_P = -\sin f$$
$$\mathbf{u}_S \cdot \mathbf{u}_Q = \cos f \tag{5.43}$$
$$\mathbf{u}_W \cdot \mathbf{u}_P = 0$$
$$\mathbf{u}_W \cdot \mathbf{u}_Q = 0.$$

Using Eq. 5.43 in Eq. 5.42, we obtain

$$\mathbf{F} \cdot \frac{\partial \mathbf{r}}{\partial c_j} = R \left(\frac{\partial \xi}{\partial \alpha_j} \cos f + \frac{\partial \eta}{\partial \alpha_j} \sin f \right) + S \left(-\frac{\partial \xi}{\partial \alpha_j} \sin f + \frac{\partial \eta}{\partial \alpha_j} \cos f \right). \tag{5.44}$$

5.2.4   Perturbation Formulation for Mean Motion

Applying Eq. 5.44 to the mean motion gives

$$\mathbf{F} \cdot \frac{\partial \mathbf{r}}{\partial n} = R \left(\cos f \frac{\partial \xi}{\partial n} + \sin f \frac{\partial \eta}{\partial n} \right) + S \left(-\sin f \frac{\partial \xi}{\partial n} n + \cos f \frac{\partial \eta}{\partial n} \right). \qquad (5.45)$$

Using Eqs. 4.38 and 4.41 in the first term of Eq. 5.45, we obtain

$$\left(\cos f \frac{\partial \xi}{\partial n} + \sin f \frac{\partial \eta}{\partial n} \right) = \cos f \left[\frac{r}{a} \cos f + \frac{3n}{2\beta} t \sin f \right]$$

$$+ \sin f \left[\frac{r}{a} \sin f - \frac{3n}{2\beta} t (\cos f + e) \right]$$

$$= \frac{r}{a} \cos^2 f + \frac{3n}{2\beta} t \sin f \cos f$$

$$+ \frac{r}{a} \sin^2 f - \frac{3n}{2\beta} t \sin f \cos f - \frac{3n}{2\beta} t (e \sin f)$$

$$= \frac{r}{a} - \frac{3n}{2\beta} t (e \sin f). \qquad (5.46)$$

Using Eqs. 4.38 and 4.41 in the second term of Eq. 5.45, we obtain

$$\left(-\sin f \frac{\partial \xi}{\partial n} + \cos f \frac{\partial \eta}{\partial n} \right) = -\sin f \left[\frac{r}{a} \cos f + \frac{3n}{2\beta} t \sin f \right]$$

$$+ \cos f \left[\frac{r}{a} \sin f - \frac{3n}{2\beta} t (\cos f + e) \right]$$

$$= -\frac{r}{a} \sin f \cos f - \frac{3n}{2\beta} t \sin^2 f$$

$$+ \frac{r}{a} \sin f \cos f - \frac{3n}{2\beta} t \cos^2 f - \frac{3n}{2\beta} t (e \cos f)$$

$$= -\frac{3n}{2\beta} t (1 + e \cos f). \qquad (5.47)$$

Using Eqs. 5.46 and 5.47 in Eq. 5.45 yields

$$\mathbf{F} \cdot \frac{\partial \mathbf{r}}{\partial n} = R \left[\frac{r}{a} - \frac{3n}{2\beta} t (e \sin f) \right] + S \left[-\frac{3n}{2\beta} t (1 + e \cos f) \right]. \qquad (5.48)$$

5.2.5 Perturbation Formulation for Eccentricity

Applying Eq. 5.44 to the eccentricity gives

$$\mathbf{F} \cdot \frac{\partial \mathbf{r}}{\partial e} = R \left(\cos f \frac{\partial \xi}{\partial e} + \sin f \frac{\partial \eta}{\partial e} \right) + S \left(-\sin f \frac{\partial \xi}{\partial e} + \cos f \frac{\partial \eta}{\partial e} \right). \tag{5.49}$$

Using Eqs. 4.39 and 4.42 in the first term of Eq. 5.49, we obtain

$$\left(\cos f \frac{\partial \xi}{\partial e} + \sin f \frac{\partial \eta}{\partial e} \right) = \cos f \left(-a - \frac{r}{\beta^2} \sin^2 f \right) + \sin f \left(\frac{r}{\beta^2} \sin f \cos f \right)$$

$$= -a \cos f. \tag{5.50}$$

Using Eqs. 4.39 and 4.42 in the second term of Eq. 5.49, we obtain

$$\left(-\sin f \frac{\partial \xi}{\partial e} + \cos f \frac{\partial \eta}{\partial e} \right) = -\sin f \left(-a - \frac{r}{\beta^2} \sin^2 f \right) + \cos f \left(\frac{r}{\beta^2} \sin f \cos f \right)$$

$$= a \sin f + \frac{r}{\beta^2} \sin^3 f + \frac{r}{\beta^2} \sin f \cos^2 f$$

$$= a \sin f + \frac{r}{\beta^2} \sin f \left(1 - \cos^2 f + \cos^2 f \right)$$

$$= a \sin f \left(1 + \frac{r}{a\beta^2} \right). \tag{5.51}$$

Using Eqs. 5.50 and 5.51 in Eq. 5.49 gives

$$\mathbf{F} \cdot \frac{\partial \mathbf{r}}{\partial e} = -R \left(a \cos f \right) + S \left(a \sin f \right) \left(1 + \frac{r}{a\beta^2} \right). \tag{5.52}$$

5.2.6 Perturbation Formulation for Mean Anomaly

Applying Eq. 5.44 to the mean anomaly gives

$$\mathbf{F} \cdot \frac{\partial \mathbf{r}}{\partial M_0} = R \left(\cos f \frac{\partial \xi}{\partial M_0} + \sin f \frac{\partial \eta}{\partial M_0} \right) + S \left(-\sin f \frac{\partial \xi}{\partial M_0} + \cos f \frac{\partial \eta}{\partial M_0} \right). \tag{5.53}$$

Using Eqs. 4.40 and 4.43 in the first term of Eq. 5.53, we obtain

$$\left(\cos f \frac{\partial \xi}{\partial M_0} + \sin f \frac{\partial \eta}{\partial M_0} \right) = \cos f \left(-\frac{a}{\beta} \sin f \right) + \sin f \left[\frac{a}{\beta} (\cos f + e) \right]$$

$$= \frac{a}{\beta} \left(e \sin f \right) . \tag{5.54}$$

Using Eqs. 4.40 and 4.43 in the second term of Eq. 5.53, we obtain

$$\left(-\sin f \frac{\partial \xi}{\partial M_0} + \cos f \frac{\partial \eta}{\partial M_0} \right) = -\sin f \left(-\frac{a}{\beta} \sin f \right) + \cos f \left[\frac{a}{\beta} \left(\cos f + e \right) \right]$$

$$= \frac{a}{\beta} \left(1 + e \cos f \right)$$

$$= \frac{a^2 \beta}{r} . \tag{5.55}$$

Using Eqs. 5.54 and 5.55 in Eq. 5.53 gives

$$\mathbf{F} \cdot \frac{\partial \mathbf{r}}{\partial M_0} = R \left(\frac{a}{\beta} e \sin f \right) + S \left(\frac{a^2 \beta}{r} \right) . \tag{5.56}$$

5.2.7 Perturbation Formulation for β_j Elements

Returning to Eq. 5.38, we now need to compute the quantities

$$\mathbf{F} \cdot \frac{\partial \mathbf{r}}{\partial \beta_j} . \tag{5.57}$$

We need to compute partial derivatives of the position vector with respect to the orbital orientation elements Ω, ω, i. We note that

$$\mathbf{r} = r \mathbf{u}_r \tag{5.58}$$

and

$$\mathbf{r} = r \left(D_{11} \mathbf{i} + D_{12} \mathbf{j} + D_{13} \mathbf{k} \right) , \tag{5.59}$$

where we make use of a coordinate system that is aligned with the radius direction. The RSW system can be obtained from the perifocal PQW system by simply replacing ω with u, the true argument of latitude, as follows:

$$\begin{pmatrix} \mathbf{u}_R \\ \mathbf{u}_S \\ \mathbf{u}_W \end{pmatrix} = \begin{pmatrix} D_{11} & D_{12} & D_{13} \\ D_{21} & D_{22} & D_{23} \\ D_{31} & D_{32} & D_{33} \end{pmatrix} \begin{pmatrix} \mathbf{i} \\ \mathbf{j} \\ \mathbf{k} \end{pmatrix} \tag{5.60}$$

and

$$
\begin{pmatrix} \mathbf{i} \\ \mathbf{j} \\ \mathbf{k} \end{pmatrix} = \begin{pmatrix} D_{11} & D_{21} & D_{31} \\ D_{12} & D_{22} & D_{32} \\ D_{13} & D_{23} & D_{33} \end{pmatrix} \begin{pmatrix} \mathbf{u}_R \\ \mathbf{u}_S \\ \mathbf{u}_W \end{pmatrix}, \tag{5.61}
$$

where

$$
\begin{aligned}
D_{11} &= \cos u \cos \Omega - \sin u \sin \Omega \cos i \\
D_{12} &= \cos u \sin \Omega + \sin u \cos \Omega \cos i \\
D_{13} &= \sin u \sin i \\
D_{21} &= -\sin u \cos \Omega - \cos u \sin \Omega \cos i \\
D_{22} &= -\sin u \sin \Omega + \cos u \cos \Omega \cos i \\
D_{23} &= \cos u \sin i \\
D_{31} &= \sin \Omega \sin i \\
D_{32} &= -\cos \Omega \sin i \\
D_{33} &= \cos i.
\end{aligned} \tag{5.62}
$$

From Eq. 5.59, it follows that

$$
\frac{\partial \mathbf{r}}{\partial \beta_j} = r \frac{\partial D_{11}}{\partial \beta_j} \mathbf{i} + r \frac{\partial D_{12}}{\partial \beta_j} \mathbf{j} + r \frac{\partial D_{13}}{\partial \beta_j} \mathbf{k}.
$$

Using Eq. 5.60 to transform from **ijk** to the RSW system, we find

$$
\begin{aligned}
\frac{\partial \mathbf{r}}{\partial \beta_j} &= r \frac{\partial D_{11}}{\partial \beta_j} D_{11} \mathbf{u}_R + r \frac{\partial D_{11}}{\partial \beta_j} D_{21} \mathbf{u}_S + r \frac{\partial D_{11}}{\partial \beta_j} D_{31} \mathbf{u}_W \\
&\quad + r \frac{\partial D_{12}}{\partial \beta_j} D_{12} \mathbf{u}_R + r \frac{\partial D_{12}}{\partial \beta_j} D_{22} \mathbf{u}_S + r \frac{\partial D_{12}}{\partial \beta_j} D_{32} \mathbf{u}_W \\
&\quad + r \frac{\partial D_{13}}{\partial \beta_j} D_{13} \mathbf{u}_R + r \frac{\partial D_{13}}{\partial \beta_j} D_{23} \mathbf{u}_S + r \frac{\partial D_{13}}{\partial \beta_j} D_{33} \mathbf{u}_W \\
&= r \left(\frac{\partial D_{11}}{\partial \beta_j} D_{11} + \frac{\partial D_{12}}{\partial \beta_j} D_{12} + \frac{\partial D_{13}}{\partial \beta_j} D_{13} \right) \mathbf{u}_R \\
&\quad + r \left(\frac{\partial D_{11}}{\partial \beta_j} D_{21} + \frac{\partial D_{12}}{\partial \beta_j} D_{22} + \frac{\partial D_{13}}{\partial \beta_j} D_{23} \right) \mathbf{u}_S \\
&\quad + r \left(\frac{\partial D_{11}}{\partial \beta_j} D_{31} + \frac{\partial D_{12}}{\partial \beta_j} D_{32} + \frac{\partial D_{13}}{\partial \beta_j} D_{33} \right) \mathbf{u}_W. \tag{5.63}
\end{aligned}
$$

Using Eqs. 5.40 and 5.63 in Eq. 5.57 gives

$$\mathbf{F} \cdot \frac{\partial \mathbf{r}}{\partial \beta_j} = r R \left(\frac{\partial D_{11}}{\partial \beta_j} D_{11} + \frac{\partial D_{12}}{\partial \beta_j} D_{12} + \frac{\partial D_{13}}{\partial \beta_j} D_{13} \right)$$

$$+ r S \left(\frac{\partial D_{11}}{\partial \beta_j} D_{21} + \frac{\partial D_{12}}{\partial \beta_j} D_{22} + \frac{\partial D_{13}}{\partial \beta_j} D_{23} \right)$$

$$+ r W \left(\frac{\partial D_{11}}{\partial \beta_j} D_{31} + \frac{\partial D_{12}}{\partial \beta_j} D_{32} + \frac{\partial D_{13}}{\partial \beta_j} D_{33} \right). \qquad (5.64)$$

5.2.8   Perturbation Formulation for Ω

Applying Eq. 5.64 to the ascending node angle gives

$$\mathbf{F} \cdot \frac{\partial \mathbf{r}}{\partial \beta_j} = r R \left(\frac{\partial D_{11}}{\partial \Omega} D_{11} + \frac{\partial D_{12}}{\partial \Omega} D_{12} + \frac{\partial D_{13}}{\partial \Omega} D_{13} \right)$$

$$+ r S \left(\frac{\partial D_{11}}{\partial \Omega} D_{21} + \frac{\partial D_{12}}{\partial \Omega} D_{22} + \frac{\partial D_{13}}{\partial \Omega} D_{23} \right)$$

$$+ r W \left(\frac{\partial D_{11}}{\partial \Omega} D_{31} + \frac{\partial D_{12}}{\partial \Omega} D_{32} + \frac{\partial D_{13}}{\partial \Omega} D_{33} \right), \qquad (5.65)$$

where we note that

$$\frac{\partial D_{11}}{\partial \Omega} = -\cos u \sin \Omega - \sin u \cos \Omega \cos i = -D_{12}$$

$$\frac{\partial D_{12}}{\partial \Omega} = \cos u \cos \Omega - \sin u \sin \Omega \cos i = D_{11}$$

$$\frac{\partial D_{21}}{\partial \Omega} = \sin u \sin \Omega - \cos u \cos \Omega \cos i = -D_{22}$$

$$\frac{\partial D_{22}}{\partial \Omega} = -\sin u \cos \Omega - \cos u \sin \Omega \cos i = -D_{21} \qquad (5.66)$$

$$\frac{\partial D_{31}}{\partial \Omega} = \cos \Omega \sin i = -D_{32}$$

$$\frac{\partial D_{32}}{\partial \Omega} = \sin \Omega \sin i = D_{31}.$$

Using Eq. 5.66 in the three elements of Eq. 5.65, we find

$$\left(\frac{\partial D_{11}}{\partial \Omega} D_{11} + \frac{\partial D_{12}}{\partial \Omega} D_{12} + \frac{\partial D_{13}}{\partial \Omega} D_{13} \right) = -D_{12} D_{11} + D_{11} D_{12} + (0) D_{13} = 0$$

$$\left(\frac{\partial D_{11}}{\partial \Omega} D_{21} + \frac{\partial D_{12}}{\partial \Omega} D_{22} + \frac{\partial D_{13}}{\partial \Omega} D_{23}\right) = \cos i \tag{5.67}$$

$$\left(\frac{\partial D_{11}}{\partial \Omega} D_{31} + \frac{\partial D_{12}}{\partial \Omega} D_{32} + \frac{\partial D_{13}}{\partial \Omega} D_{33}\right) = -\cos u \sin i.$$

Substituting these results into Eq. 5.65, we get

$$\mathbf{F} \cdot \frac{\partial \mathbf{r}}{\partial \Omega} = (r \cos i) \, S - (r \cos u \sin i) \, W. \tag{5.68}$$

5.2.9   Perturbation Formulation for ω

Applying Eq. 5.64 to the argument of periapsis gives

$$\mathbf{F} \cdot \frac{\partial \mathbf{r}}{\partial \omega} = rR\left(\frac{\partial D_{11}}{\partial \omega} D_{11} + \frac{\partial D_{12}}{\partial \omega} D_{12} + \frac{\partial D_{13}}{\partial \omega} D_{13}\right)$$

$$+ rS\left(\frac{\partial D_{11}}{\partial \omega} D_{21} + \frac{\partial D_{12}}{\partial \omega} D_{22} + \frac{\partial D_{13}}{\partial \omega} D_{23}\right)$$

$$+ rW\left(\frac{\partial D_{11}}{\partial \omega} D_{31} + \frac{\partial D_{12}}{\partial \omega} D_{32} + \frac{\partial D_{13}}{\partial \omega} D_{33}\right), \tag{5.69}$$

where we note that

$$\frac{\partial D_{11}}{\partial \omega} = -\sin u \cos \Omega - \cos u \sin \Omega \cos i = D_{21}$$

$$\frac{\partial D_{12}}{\partial \omega} = -\sin u \sin \Omega + \cos u \cos \Omega \cos i = D_{22}$$

$$\frac{\partial D_{13}}{\partial \omega} = \cos u \sin i = D_{23}$$

$$\frac{\partial D_{21}}{\partial \omega} = -\cos u \cos \Omega + \sin u \sin \Omega \cos i = -D_{11} \tag{5.70}$$

$$\frac{\partial D_{22}}{\partial \omega} = -\cos u \sin \Omega - \sin u \cos \Omega \cos i = -D_{12}$$

$$\frac{\partial D_{23}}{\partial \omega} = -\sin u \sin i = -D_{13}.$$

Using Eq. 5.70 in the three elements of Eq. 5.69, we find

$$\left(\frac{\partial D_{11}}{\partial \omega} D_{11} + \frac{\partial D_{12}}{\partial \omega} D_{12} + \frac{\partial D_{13}}{\partial \omega} D_{13}\right) = D_{21} D_{11} + D_{22} D_{12} + D_{23} D_{13} = 0$$

$$\left(\frac{\partial D_{11}}{\partial \omega} D_{21} + \frac{\partial D_{12}}{\partial \omega} D_{22} + \frac{\partial D_{13}}{\partial \omega} D_{23} \right) = D_{21} D_{21} + D_{22} D_{22} + D_{23} D_{23} = 1$$

$$(5.71)$$

$$\left(\frac{\partial D_{11}}{\partial \omega} D_{31} + \frac{\partial D_{12}}{\partial \omega} D_{32} + \frac{\partial D_{13}}{\partial \omega} D_{33} \right) = D_{21} D_{31} + D_{22} D_{32} + D_{23} D_{33} = 0.$$

We substitute these results into Eq. 5.69 to get

$$\mathbf{F} \cdot \frac{\partial \mathbf{r}}{\partial \omega} = (r)\, S. \tag{5.72}$$

5.2.10   Perturbation Formulation for i

Applying Eq. 5.64 to the inclination gives

$$\mathbf{F} \cdot \frac{\partial \mathbf{r}}{\partial i} = r R \left(\frac{\partial D_{11}}{\partial i} D_{11} + \frac{\partial D_{12}}{\partial i} D_{12} + \frac{\partial D_{13}}{\partial i} D_{13} \right)$$
$$+ r S \left(\frac{\partial D_{11}}{\partial i} D_{21} + \frac{\partial D_{12}}{\partial i} D_{22} + \frac{\partial D_{13}}{\partial i} D_{23} \right)$$
$$+ r W \left(\frac{\partial D_{11}}{\partial i} D_{31} + \frac{\partial D_{12}}{\partial i} D_{32} + \frac{\partial D_{13}}{\partial i} D_{33} \right), \tag{5.73}$$

where we note that

$$\frac{\partial D_{11}}{\partial i} = \sin u \sin \Omega \sin i$$

$$\frac{\partial D_{12}}{\partial i} = -\sin u \cos \Omega \sin i$$

$$\frac{\partial D_{13}}{\partial i} = \sin u \cos i$$

$$\frac{\partial D_{21}}{\partial i} = \cos u \sin \Omega \sin i$$

$$\frac{\partial D_{22}}{\partial i} = -\cos u \cos \Omega \sin i \tag{5.74}$$

$$\frac{\partial D_{23}}{\partial i} = \cos u \cos i$$

$$\frac{\partial D_{31}}{\partial i} = \sin \Omega \cos i$$

$$\frac{\partial D_{32}}{\partial i} = -\cos \Omega \cos i$$

$$\frac{\partial D_{33}}{\partial i} = -\sin i.$$

Using Eq. 5.74 in the three elements of Eq. 5.73, we find

$$\left(\frac{\partial D_{11}}{\partial i} D_{11} + \frac{\partial D_{12}}{\partial i} D_{12} + \frac{\partial D_{13}}{\partial i} D_{13}\right) = \sin u \sin i \left(-\sin u \cos i\right)$$

$$+ (\sin u \cos i) \sin u \sin i$$

$$= 0$$

$$\left(\frac{\partial D_{11}}{\partial i} D_{21} + \frac{\partial D_{12}}{\partial i} D_{22} + \frac{\partial D_{13}}{\partial i} D_{23}\right) = \sin u \sin i \left(-\cos u \cos i\right)$$

$$+ \sin u \cos i \left(\cos u \sin i\right)$$

$$= 0$$

$$\left(\frac{\partial D_{11}}{\partial i} D_{31} + \frac{\partial D_{12}}{\partial i} D_{32} + \frac{\partial D_{13}}{\partial i} D_{33}\right) = (D_{31} \sin u) D_{31} + (D_{32} \sin u) D_{32}$$

$$+ D_{33} D_{33} \sin u$$

$$= \sin u.$$

Substituting these results into Eq. 5.73, we get

$$\mathbf{F} \cdot \frac{\partial \mathbf{r}}{\partial i} = (r \sin u) \, W. \tag{5.75}$$

5.2.11   Lagrange Planetary Equations for Radial–Transverse–Orthogonal Force

We now have everything needed to evaluate

$$\sum_{k=1}^{6} [c_j, c_k] \dot{c}_k = \mathbf{F} \cdot \frac{\partial \mathbf{r}}{\partial c_j}. \tag{5.76}$$

Using the Lagrange brackets for the first element of $\mathbf{c}$, we find

$$\sum_{k=1}^{6} [n, c_k] \dot{c}_k = \mathbf{F} \cdot \frac{\partial \mathbf{r}}{\partial n}. \tag{5.77}$$

From Eqs. 5.4 and 5.5, we obtain

$$\left(\frac{a^2\beta}{3}\cos i\right)\dot{\Omega} + \left(\frac{a^2\beta}{3}\right)\dot{\omega} + \left(\frac{a^2}{3}\right)\dot{M}_0 = \mathbf{F}\cdot\frac{\partial\mathbf{r}}{\partial n}. \tag{5.78}$$

Substituting Eq. 5.48 into Eq. 5.78 yields

$$\left(\frac{a^2\beta}{3}\cos i\right)\dot{\Omega} + \left(\frac{a^2\beta}{3}\right)\dot{\omega} + \left(\frac{a^2}{3}\right)\dot{M}_0 = R\left[\frac{r}{a} - \frac{3n}{2\beta}t\,(e\sin f)\right]$$
$$+ S\left[-\frac{3n}{2\beta}t\,(1 + e\cos f)\right]. \tag{5.79}$$

Using the Lagrange brackets for the second element of $\mathbf{c}$, we find

$$\sum_{k=1}^{6}[e, c_k]\dot{c}_k = \mathbf{F}\cdot\frac{\partial\mathbf{r}}{\partial e}. \tag{5.80}$$

From Eqs. 5.7 and 5.8, we obtain

$$\left(\frac{na^2e}{\beta}\cos i\right)\dot{\Omega} + \left(\frac{na^2e}{\beta}\right)\dot{\omega} = \mathbf{F}\cdot\frac{\partial\mathbf{r}}{\partial e}. \tag{5.81}$$

Substituting Eq. 5.52 into Eq. 5.81 yields

$$(\cos i)\,\dot{\Omega} + \dot{\omega} = \frac{\beta}{na^2e}\left[-R\,(a\cos f) + S\,(a\sin f)\left(1 + \frac{r}{a\beta^2}\right)\right]. \tag{5.82}$$

Using the Lagrange brackets for the third element of $\mathbf{c}$, we find

$$\sum_{k=1}^{6}[i, c_k]\dot{c}_k = \mathbf{F}\cdot\frac{\partial\mathbf{r}}{\partial i}. \tag{5.83}$$

From Eqs. 5.10 and 5.11, we obtain

$$\left(na^2\beta\sin i\right)\dot{\Omega} = \mathbf{F}\cdot\frac{\partial\mathbf{r}}{\partial i}. \tag{5.84}$$

Substituting Eq. 5.75 into Eq. 5.84 yields

$$\left(na^2\beta\sin i\right)\dot{\Omega} = (r\sin u)\,W. \tag{5.85}$$

Using the Lagrange brackets for the fourth element of $\mathbf{c}$, we find

$$\sum_{k=1}^{6} [\Omega, c_k]\dot{c}_k = \mathbf{F} \cdot \frac{\partial \mathbf{r}}{\partial \Omega}. \tag{5.86}$$

From Eqs. 5.13 and 5.14, we obtain

$$-\left(\frac{a^2\beta}{3} \cos i\right) \dot{n} - \left(\frac{na^2e}{\beta} \cos i\right) \dot{e} - \left(na^2\beta \sin i\right) \frac{di}{dt} = \mathbf{F} \cdot \frac{\partial \mathbf{r}}{\partial \Omega}. \tag{5.87}$$

Substituting Eq. 5.68 into Eq. 5.87 yields

$$-\left(\frac{a^2\beta}{3} \cos i\right) \dot{n} - \left(\frac{na^2e}{\beta} \cos i\right) \dot{e} - \left(na^2\beta \sin i\right) \frac{di}{dt} = (r \cos i) \, S - (r \cos u \sin i) \, W. \tag{5.88}$$

Using the Lagrange brackets for the fifth element of **c**, we find

$$\sum_{k=1}^{6} [\omega, c_k]\dot{c}_k = \mathbf{F} \cdot \frac{\partial \mathbf{r}}{\partial \omega}. \tag{5.89}$$

From Eqs. 5.16 and 5.17, we obtain

$$-\left(\frac{a^2\beta}{3}\right) \dot{n} - \left(\frac{na^2e}{\beta}\right) \dot{e} = \mathbf{F} \cdot \frac{\partial \mathbf{r}}{\partial \omega}. \tag{5.90}$$

Substituting Eq. 5.72 into Eq. 5.90 yields

$$-\left(\frac{a^2\beta}{3}\right) \dot{n} - \left(\frac{na^2e}{\beta}\right) \dot{e} = (r) \, S. \tag{5.91}$$

Using the Lagrange brackets for the sixth element of **c**, we find

$$\sum_{k=1}^{6} [M_0, c_k]\dot{c}_k = \mathbf{F} \cdot \frac{\partial \mathbf{r}}{\partial M_0}. \tag{5.92}$$

From Eqs. 5.19 and 5.20, we obtain

$$-\frac{a^2}{3}\dot{n} = \mathbf{F} \cdot \frac{\partial \mathbf{r}}{\partial M_0}. \tag{5.93}$$

Substituting Eq. 5.56 into Eq. 5.93 yields

$$\dot{n} = -\left(\frac{3e}{a\beta} \sin f\right) R - \left(\frac{3\beta}{r}\right) S. \tag{5.94}$$

5.2.12 Solution for the Time Derivatives of the Orbital Elements

From Eq. 5.94, we have

$$\dot{n} = -\left(\frac{3e}{a\beta}\sin f\right) R - \left(\frac{3\beta}{r}\right) S. \tag{5.95}$$

Substituting Eq. 5.95 into Eq. 5.91 for $\dot{n}$ gives

$$\left(\frac{na^2 e}{\beta}\right)\dot{e} = -(r)\, S + (ae\sin f)\, R + \left(\frac{a^2\beta^2}{r}\right) S,$$

which simplifies to give

$$\dot{e} = \left(\frac{\beta}{na}\sin f\right) R + \frac{\beta}{na^2 e}\left(\frac{a^2\beta^2}{r} - r\right) S. \tag{5.96}$$

Substituting Eq. 5.91 into Eq. 5.88 gives

$$(r\cos i)\, S - \left(na^2\beta\sin i\right)\frac{di}{dt} = (r\cos i)\, S - (r\cos u\sin i)\, W. \tag{5.97}$$

Solving Eq. 5.97 for the inclination rate gives

$$\frac{di}{dt} = \frac{r\cos u}{na^2\beta} W. \tag{5.98}$$

Equation 5.85 can be rewritten as

$$\dot{\Omega} = \left(\frac{r\sin u}{na^2\beta\sin i}\right) W. \tag{5.99}$$

Substituting Eq. 5.99 into Eq. 5.82 yields

$$\left(\frac{na^2 e}{\beta}\right)\dot{\omega} = -R\,(a\cos f) + S\,(a\sin f)\left(1 + \frac{r}{a\beta^2}\right) - \left(\frac{na^2 e}{\beta}\cos i\right)\left(\frac{r\sin u}{na^2\beta\sin i}\right) W.$$

Solving for argument of periapsis rate gives

$$\dot{\omega} = -\left(\frac{\beta\cos f}{nae}\right) R + \left(\frac{\beta\sin f}{nae}\right)\left(1 + \frac{r}{a\beta^2}\right) S - \left(\frac{r\sin u\cos i}{na^2\beta\sin i}\right) W. \tag{5.100}$$

We rearrange Eq. 5.79 to obtain

$$\frac{a^2}{3}\dot{M}_0 = \frac{2}{na\beta}\left[R\left(1 - e\cos E\right) - \frac{3}{2}\frac{n}{\beta}t\left[(e\sin f)\,R + (1 + e\cos f)\,S\right]\right]$$

$$- \frac{a^2\beta}{3}\left[(\cos i)\,\dot{\Omega} + \dot{\omega}\right].$$

Substituting Eq. 5.82 for the last two terms on the right-hand side yields

$$\dot{M}_0 = \left(\frac{\beta^2}{nae}\cos f - \frac{2r}{na^2}\right)R - \frac{\beta^2}{nae}\sin f\left(1 + \frac{r}{a\beta^2}\right)S$$

$$+ \frac{3}{a\beta}t\left[(e\sin f)\,R + (1 + e\cos f)\,S\right]. \tag{5.101}$$

Finally, we substitute Eqs. 5.101 and 5.95 into the relation

$$\dot{M} = \dot{M}_0 + n + \dot{n}t$$

and obtain

$$\dot{M} = \left(\frac{\beta^2}{nae}\cos f - \frac{2r}{na^2}\right)R - \frac{\beta^2}{nae}\sin f\left(1 + \frac{r}{a\beta^2}\right)S$$

$$+ \frac{3}{a\beta}t\left[(e\sin f)\,R + (1 + e\cos f)\,S\right]$$

$$+ n - t\left[\left(\frac{3e}{a\beta}\sin f\right)R - \left(\frac{3\beta}{r}\right)S\right].$$

Simplification yields

$$\dot{M} = n + \left(\frac{\beta^2}{nae}\cos f - \frac{2r}{na^2}\right)R - \frac{\beta^2}{nae}\sin f\left(1 + \frac{r}{a\beta^2}\right)S. \tag{5.102}$$

In summary, we have

$$\dot{n} = -\left(\frac{3e}{a\beta}\sin f\right)R - \left(\frac{3\beta}{r}\right)S$$

$$\dot{e} = \left(\frac{\beta}{na}\sin f\right)R + \frac{\beta}{na^2e}\left(\frac{a^2\beta^2}{r} - r\right)S$$

$$\frac{di}{dt} = \frac{r\cos u}{na^2\beta}W$$

$$\dot{\Omega} = \left(\frac{r\sin u}{na^2\beta\sin i}\right)W \tag{5.103}$$

$$\dot{\omega} = -\left(\frac{\beta \cos f}{nae}\right) R + \left(\frac{\beta \sin f}{nae}\right)\left(1 + \frac{r}{a\beta^2}\right) S - \left(\frac{r \sin u \cos i}{na^2\beta \sin i}\right) W$$

$$\dot{M} = n + \left(\frac{\beta^2}{nae}\cos f - \frac{2r}{na^2}\right) R - \frac{\beta^2}{nae}\sin f \left(1 + \frac{r}{a\beta^2}\right) S.$$

5.3 Force System Case II: Normal–Tangential–Orthogonal (NTW)

The second coordinate system is is a satellite-centered coordinate system, NTW, and is sometimes referred to as the Frenet system. In this system, the primary axis, N, lies in the orbital plane, normal to the velocity vector. The T axis is tangential to the orbit, and the W axis is normal to the orbital plane (as in the RSW system). A depiction of the reference frame is given in Fig. 5.2.

We can use the NTW system to analyze the effects of atmospheric drag on the orbit because drag acts in the opposite direction of the velocity vector. Additionally, low-thrust propulsion systems often thrust along the instantaneous velocity vector.

Referring to Fig. 5.1, we can write

$$\begin{pmatrix} \mathbf{u}_P \\ \mathbf{u}_Q \end{pmatrix} = \begin{pmatrix} \cos f & -\sin f \\ \sin f & \cos f \end{pmatrix} \begin{pmatrix} \mathbf{u}_R \\ \mathbf{u}_S \end{pmatrix}. \tag{5.104}$$

In Chap. 4, we found Eqs. 4.49 and 4.50, which we repeat here:

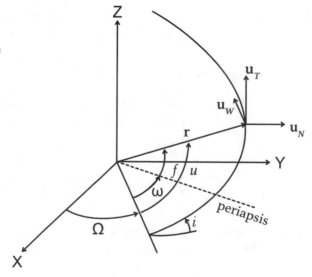

Fig. 5.2 The coordinate frame for Case II: normal–tangential–orthogonal system where $\mathbf{u}_W$ is orthogonal to the orbital plane

$$\dot{\xi} = -\frac{na}{\beta} \sin f \tag{5.105}$$

$$\dot{\eta} = \frac{na}{\beta} \left(\cos f + e \right) . \tag{5.106}$$

Then we have

$$\mathbf{v} = \dot{\xi} \mathbf{u}_P + \dot{\eta} \mathbf{u}_Q = -\frac{na}{\beta} \left[(\sin f) \, \mathbf{u}_P - (e + \cos f) \, \mathbf{u}_Q \right] , \tag{5.107}$$

so that the magnitude of the velocity is

$$v = \frac{na}{\beta} \sqrt{1 + 2e \cos f + e^2} = \frac{na}{\beta} \psi , \tag{5.108}$$

where we introduce

$$\psi = \sqrt{1 + 2e \cos f + e^2}$$

to simplify the notation.

Using Eqs. 5.107 and 5.108, we describe the unit velocity vector as

$$\mathbf{u}_T = \frac{\dot{\xi}}{\psi \frac{na}{\beta}} \mathbf{u}_P + \frac{\dot{\eta}}{\psi \frac{na}{\beta}} \mathbf{u}_Q = -\frac{1}{\psi} \left(\sin f \right) \mathbf{u}_P + \frac{1}{\psi} \left(e + \cos f \right) \mathbf{u}_Q . \tag{5.109}$$

It follows that

$$\begin{pmatrix} \mathbf{u}_N \\ \mathbf{u}_T \end{pmatrix} = \begin{pmatrix} \frac{e + \cos f}{\psi} & \frac{\sin f}{\psi} \\ -\frac{\sin f}{\psi} & \frac{e + \cos f}{\psi} \end{pmatrix} \begin{pmatrix} \mathbf{u}_P \\ \mathbf{u}_Q \end{pmatrix} . \tag{5.110}$$

Combining Eqs. 5.104 and 5.109, we can write

$$\begin{pmatrix} \mathbf{u}_R \\ \mathbf{u}_S \end{pmatrix} = \begin{pmatrix} \cos f & \sin f \\ -\sin f & \cos f \end{pmatrix} \begin{pmatrix} e + \cos f & -\sin f \\ \sin f & e + \cos f \end{pmatrix} \begin{pmatrix} \mathbf{u}_N \\ \mathbf{u}_T \end{pmatrix} \frac{1}{\psi} ,$$

so that

$$\begin{pmatrix} \mathbf{u}_R \\ \mathbf{u}_S \end{pmatrix} = \begin{pmatrix} 1 + e \cos f & e \sin f \\ -e \sin f & 1 + e \cos f \end{pmatrix} \begin{pmatrix} \mathbf{u}_N \\ \mathbf{u}_T \end{pmatrix} \frac{1}{\psi} . \tag{5.111}$$

It follows that the RS components of the force are related to the NT components of the force by

$$R = \frac{1}{\psi} (1 + e \cos f) N + \frac{1}{\psi} (e \sin f) T \tag{5.112}$$

$$S = -\frac{1}{\psi} (e \sin f) N + \frac{1}{\psi} (1 + e \cos f) T. \tag{5.113}$$

Substituting Eqs. 5.112 and 5.113 into the Lagrange planetary equations, Eq. 5.103, we get

$$\dot{n} = -\frac{3}{a\beta} \psi T$$

$$\dot{e} = \frac{1}{\psi} \frac{\beta}{na} \left(\frac{r}{a} \sin f \right) N + \frac{1}{\psi} \frac{2\beta}{na} (\cos f + e) T$$

$$\frac{di}{dt} = \frac{r \cos u}{na^2 \beta} W$$

$$\dot{\Omega} = \left(\frac{r \sin u}{na^2 \beta \sin i} \right) W \tag{5.114}$$

$$\dot{\omega} = \frac{\beta}{nae} \frac{1}{\psi} \left[2 (\sin f) T - \frac{r}{a\beta^2} \left(\cos f + 2e + e^2 \cos f \right) N \right] - \left(\frac{r \sin u \cos i}{na^2 \beta \sin i} \right) W$$

$$\dot{M} = n - \frac{1}{na^2} \frac{1}{\psi} 2 \sin f \left(re + \frac{a\beta^2}{e} \right) T + \frac{1}{na^2 e} \frac{1}{\psi} \cos f \left(r\beta^2 \right) N.$$

References

Battin, R. H. (1999). *An introduction to the mathematics and methods of astrodynamics* (Revised ed.). Reston: American Institute of Aeronautics and Astronautics, Inc.

Moulton, F. R. (1914). *An introduction to celestial mechanics* (2nd ed.). New York: The Macmillan Company.

Roy, A. E. (2005). *Orbital motion* (4th ed.). New York: Taylor & Francis Group.

Vallado, D. A. (2013). *Fundamentals of astrodynamics and applications* (4th ed.). El Segundo: Microcosm Press.

Chapter 6
Expansion of the Perturbation Function

We now consider an example scenario in which a body of interest is perturbed by a disturbing body, both of which are in orbit about a common central body (as depicted in Fig. 6.1).

In order to find the time rates of change of the orbital elements from Lagrange's planetary equations (Eqs. 5.27 or 5.37), the disturbing function, $\mathcal{R}$, is usually expressed as an infinite series in which the orbital elements appear either in the coefficients or in the arguments of trigonometric functions. In the figure, we use the following definitions:

$$m \equiv \text{perturbed mass}$$

$$m' \equiv \text{disturbing mass}$$

$$\mathbf{r} \equiv \text{position vector of m}$$

$$\mathbf{r}' \equiv \text{position vector of m'}$$

$$\rho \equiv \text{distance between disturbing body and disturbed body}$$

$$\phi \equiv \text{angle between disturbing body and disturbed body.}$$

In this example expansion, we assume the disturbing body is more distant than the disturbed body from the central body,

$$r' > r, \tag{6.1}$$

at all times, but the analysis that follows can be readily modified to accommodate the situation in which $r' < r$.

J. M. Longuski et al., *Introduction to Orbital Perturbations*, Space Technology
Library 40, https://doi.org/10.1007/978-3-030-89758-1_6

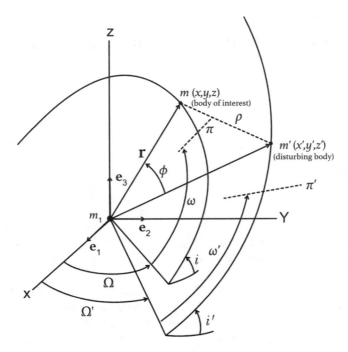

Fig. 6.1 The orbital geometry for the expansion of the perturbation function where m represents the mass of the body of interest (i.e., the perturbed body) and m' denotes the disturbing mass

The perturbation function in Eq. 3.8 can be written as

$$
\begin{aligned}
\mathcal{R} &= Gm' \left[\frac{1}{\rho} - \frac{xx' + yy' + zz'}{r'^3} \right] \\
&= Gm' \left[\left(r^2 + r'^2 - 2rr' \cos\phi \right)^{-1/2} - \frac{rr' \cos\phi}{r'^3} \right] \\
&= Gm' \left\{ \frac{1}{r'} \left[1 + \left(\frac{r}{r'} \right)^2 - 2 \left(\frac{r}{r'} \right) \cos\phi \right]^{-1/2} - \frac{r \cos\phi}{r'^2} \right\} \\
&= \frac{Gm'}{r'} \left\{ \left[1 + \left(\frac{r}{r'} \right)^2 - 2 \left(\frac{r}{r'} \right) \cos\phi \right]^{-1/2} - \left(\frac{r}{r'} \right) \cos\phi \right\},
\end{aligned}
\tag{6.2}
$$

where we have used the relation $\rho^2 = r^2 + r'^2 - 2rr' \cos\phi$ from the Law of Cosines and the relation $\mathbf{r} \cdot \mathbf{r}' = rr' \cos\phi$ for the dot product of the two position vectors. From the binomial theorem:

$$
(1 + w)^{-n} = 1 - nw + \frac{n(n+1)w^2}{2!} - \frac{n(n+1)(n+2)w^3}{3!} + \dots, \quad \left(w^2 < 1 \right).
\tag{6.3}
$$

Letting

$$w = \left(\frac{r}{r'}\right)^2 - 2\left(\frac{r}{r'}\right)\cos\phi \qquad (6.4)$$

and

$$n = \frac{1}{2}, \qquad (6.5)$$

we can expand the square bracket term of Eq. 6.2 as

$$\left[1 + \left(\frac{r}{r'}\right)^2 - 2\left(\frac{r}{r'}\right)\cos\phi\right]^{-1/2} = 1 - \frac{1}{2}\left(\frac{r}{r'}\right)^2 + \left(\frac{r}{r'}\right)\cos\phi$$

$$+ \frac{1}{2}\cdot\frac{1}{2}\cdot\frac{3}{2}\left[\left(\frac{r}{r'}\right)^2 - 2\left(\frac{r}{r'}\right)\cos\phi\right]^2$$

$$- \frac{1}{6}\cdot\frac{1}{2}\cdot\frac{3}{2}\cdot\frac{5}{2}\left[\left(\frac{r}{r'}\right)^2 - 2\left(\frac{r}{r'}\right)\cos\phi\right]^3$$

$$+ \frac{1}{24}\cdot\frac{1}{2}\cdot\frac{3}{2}\cdot\frac{5}{2}\cdot\frac{7}{2}\left[\left(\frac{r}{r'}\right)^2 - 2\left(\frac{r}{r'}\right)\cos\phi\right]^4$$

$$- \cdots \qquad (6.6)$$

From Eq. 6.6, we have

$$\left[\left(\frac{r}{r'}\right)^2 - 2\left(\frac{r}{r'}\right)\cos\phi\right]^2 = \left(\frac{r}{r'}\right)^4 - 4\left(\frac{r}{r'}\right)^3\cos\phi + 4\left(\frac{r}{r'}\right)^2\cos^2\phi, \qquad (6.7)$$

$$\left[\left(\frac{r}{r'}\right)^2 - 2\left(\frac{r}{r'}\right)\cos\phi\right]^3 = \left(\frac{r}{r'}\right)^6 - 4\left(\frac{r}{r'}\right)^5\cos\phi + 4\left(\frac{r}{r'}\right)^4\cos^2\phi$$

$$- 2\left(\frac{r}{r'}\right)^5\cos\phi + 8\left(\frac{r}{r'}\right)^4\cos^2\phi - 8\left(\frac{r}{r'}\right)^3\cos^3\phi$$

$$= \left(\frac{r}{r'}\right)^6 - 6\left(\frac{r}{r'}\right)^5\cos\phi + 12\left(\frac{r}{r'}\right)^4\cos^2\phi$$

$$- 8\left(\frac{r}{r'}\right)^3\cos^3\phi, \qquad (6.8)$$

$$\left[\left(\frac{r}{r'}\right)^2 - 2\left(\frac{r}{r'}\right)\cos\phi\right]^4 = \left(\frac{r}{r'}\right)^8 - 6\left(\frac{r}{r'}\right)^7\cos\phi + 12\left(\frac{r}{r'}\right)^6\cos^2\phi$$

$$- 8\left(\frac{r}{r'}\right)^5\cos^3\phi - 2\left(\frac{r}{r'}\right)^7\cos\phi + 12\left(\frac{r}{r'}\right)^6\cos^2\phi$$

$$- 24\left(\frac{r}{r'}\right)^5\cos^3\phi + 16\left(\frac{r}{r'}\right)^4\cos^4\phi$$

$$
= \left(\frac{r}{r'}\right)^8 - 8\left(\frac{r}{r'}\right)^7 \cos\phi + 24\left(\frac{r}{r'}\right)^6 \cos^2\phi
$$

$$
- 32\left(\frac{r}{r'}\right)^5 \cos^3\phi + 16\left(\frac{r}{r'}\right)^4 \cos^4\phi. \tag{6.9}
$$

Also, for the binomial coefficients in Eq. 6.6, we have

$$
\frac{1}{2} \cdot \frac{1}{2} \cdot \frac{3}{2} = \frac{3}{8},
$$

$$
-\frac{1}{6} \cdot \frac{1}{2} \cdot \frac{3}{2} \cdot \frac{5}{2} = -\frac{5}{16}, \tag{6.10}
$$

$$
\frac{1}{24} \cdot \frac{1}{2} \cdot \frac{3}{2} \cdot \frac{5}{2} \cdot \frac{7}{2} = \frac{35}{128}.
$$

Retaining terms to order $(r/r')^4$ in Eq. 6.6 (using Eqs. 6.7–6.9), we have

$$
\left[1 + \left(\frac{r}{r'}\right)^2 - 2\left(\frac{r}{r'}\right)\cos\phi\right]^{-1/2}
$$

$$
= 1 + \left(\frac{r}{r'}\right)\cos\phi + \left(\frac{r}{r'}\right)^2\left[-\frac{1}{2} + \frac{3}{8} \cdot 4\cos^2\phi\right]
$$

$$
+ \left(\frac{r}{r'}\right)^3\left[\frac{3}{8}(-4\cos\phi) - \frac{5}{16}(-8)\cos^3\phi\right]
$$

$$
+ \left(\frac{r}{r'}\right)^4\left[1 \cdot \frac{3}{8} - \frac{5}{16}(12)\cos^2\phi + \frac{35}{128}(16)\cos^4\phi\right], \tag{6.11}
$$

or

$$
\left[1 + \left(\frac{r}{r'}\right)^2 - 2\left(\frac{r}{r'}\right)\cos\phi\right]^{-1/2} = 1 + \left(\frac{r}{r'}\right)\cos\phi + \left(\frac{r}{r'}\right)^2\left[-\frac{1}{2} + \frac{3}{2}\cos^2\phi\right]
$$

$$
+ \left(\frac{r}{r'}\right)^3\left[-\frac{3}{2}\cos\phi + \frac{5}{2}\cos^3\phi\right]
$$

$$
+ \left(\frac{r}{r'}\right)^4\left[\frac{3}{8} - \frac{15}{4}\cos^2\phi + \frac{35}{8}\cos^4\phi\right] + \dots. \tag{6.12}
$$

The terms in square brackets are Legendre polynomials.

6.1 Legendre Polynomials of the First Kind

The groups of trigonometric functions in Eq. 6.12 are called the Legendre polynomials of the first kind, $P_n(\cos\phi)$:

$$P_0\,(\cos\phi) = 1$$
$$P_1\,(\cos\phi) = \cos\phi$$
$$P_2\,(\cos\phi) = \frac{1}{2}\left(3\cos^2\phi - 1\right) \tag{6.13}$$
$$P_3\,(\cos\phi) = \frac{1}{2}\left(5\cos^3\phi - 3\cos\phi\right)$$
$$P_4\,(\cos\phi) = \frac{1}{8}\left(35\cos^4\phi - 30\cos^2\phi + 3\right).$$

We make use of the trigonometric power relations:

$$\cos^2\phi = \frac{1}{2}(1 + \cos 2\phi)$$
$$\cos^3\phi = \frac{1}{4}(3\cos\phi + \cos 3\phi) \tag{6.14}$$
$$\cos^4\phi = \frac{1}{8}(3 + 4\cos 2\phi + \cos 4\phi).$$

Substituting Eq. 6.14 into Eq. 6.13, we obtain

$$\begin{aligned}
P_2 &= \frac{1}{2}\left(3\cos^2\phi - 1\right) \\
&= \frac{1}{2}\left(\frac{3}{2} + \frac{3}{2}\cos 2\phi - 1\right) \\
&= \frac{1}{2}\left(\frac{1}{2} + \frac{3}{2}\cos 2\phi\right) \\
&= \frac{1}{4}(1 + 3\cos 2\phi) \tag{6.15}
\end{aligned}$$

$$\begin{aligned}
P_3 &= \frac{1}{2}\left(5\cos^3\phi - 3\cos\phi\right) \\
&= \frac{1}{2}\left(\frac{15}{4}\cos\phi + \frac{5}{4}\cos 3\phi - 3\cos\phi\right) \\
&= \frac{1}{8}(5\cos 3\phi + 15\cos\phi - 12\cos\phi) \\
&= \frac{1}{8}(5\cos 3\phi + 3\cos\phi) \tag{6.16}
\end{aligned}$$

$$P_4 = \frac{1}{8} \left[\frac{35}{8} \left(3 + 4 \cos 2\phi + \cos 4\phi \right) - \frac{30}{2} \left(1 + \cos 2\phi \right) + 3 \right]$$

$$= \frac{1}{64} \left(105 + 140 \cos 2\phi + 35 \cos 4\phi - 120 - 120 \cos 2\phi + 24 \right)$$

$$= \frac{1}{64} \left(35 \cos 4\phi + 20 \cos 2\phi + 9 \right). \tag{6.17}$$

So the Legendre polynomials of the first kind may also be expressed as

$$P_0 \left(\cos \phi \right) = 1$$

$$P_1 \left(\cos \phi \right) = \cos \phi$$

$$P_2 \left(\cos \phi \right) = \frac{1}{4} \left(3 \cos 2\phi + 1 \right) \tag{6.18}$$

$$P_3 \left(\cos \phi \right) = \frac{1}{8} \left(5 \cos 3\phi + 3 \cos \phi \right)$$

$$P_4 \left(\cos \phi \right) = \frac{1}{64} \left(35 \cos 4\phi + 20 \cos 2\phi + 9 \right).$$

The Legendre coefficients are uniformly bounded:

$$\left| P_n \left(\cos \phi \right) \right| \leq 1 \qquad n = 0, 1, 2, \ldots, \tag{6.19}$$

so the series

$$\sum_{n=0}^{\infty} \left(\frac{r}{r'} \right)^n P_n \left(\cos \phi \right)$$

is convergent since we have chosen

$$\frac{r}{r'} < 1. \tag{6.20}$$

By combining Eq. 6.2 with Eqs. 6.12 and 6.18, we obtain

$$\mathcal{R} = \frac{Gm'}{r'} \left[1 + \left(\frac{r}{r'} \right)^2 P_2 + \left(\frac{r}{r'} \right)^3 P_3 + \left(\frac{r}{r'} \right)^4 P_4 + \ldots \right], \tag{6.21}$$

where we note that the P_1 coefficients cancel. To express $\mathcal{R}$ in terms of orbital elements requires an examination of $\left(r/r' \right)^n$ and $P_n \left(\cos \phi \right)$.

For illustration, we turn our attention to the term

$$\left(\frac{r}{r'} \right)^2 P_2 \left(\cos \phi \right),$$

and we will analyze the factors $\left(r/r' \right)^2$ and $P_2 \left(\cos \phi \right)$ separately.

6.1.1   The Factor $(r/r')^2$

It is known that (r/a) can be expanded into a series of trigonometric functions of the mean anomaly, M. Moulton (1914) gives

$$\frac{r}{a} = 1 - e \cos M - \frac{e^2}{2} (\cos 2M - 1) - \frac{e^3}{2! \, 2^2} (3 \cos 3M - 3 \cos M)$$

$$- \frac{e^4}{3! \, 2^3} \left(4^2 \cos 4M - 4 \cdot 2^2 \cos 2M \right)$$

$$- \frac{e^5}{4! \, 2^4} \left(5^3 \cos 5M - 5 \cdot 3^3 \cos 3M + 10 \cos M \right)$$

$$- \frac{e^6}{5! \, 2^5} \left(6^4 \cos 6M - 6 \cdot 4^4 \cos 4M + 15 \cdot 2^4 \cos 2M \right) - \dots . \qquad (6.22)$$

(See also Kovalevsky (1967) and Taff (1985).)

Taking terms from Eq. 6.22 to order e^2, we have

$$\frac{r}{a} = 1 - e \cos M - \frac{e^2}{2} (\cos 2M - 1) . \qquad (6.23)$$

To obtain a/r, we write (using the binomial theorem again)

$$\frac{a}{r} = \frac{1}{1 - e \cos M - \dfrac{e^2}{2} (\cos 2M - 1)}$$

$$= \frac{1}{1 + \epsilon} = 1 - \epsilon + \epsilon^2 + \dots$$

$$= 1 + e \cos M + \frac{e^2}{2} (\cos 2M - 1) + e^2 \cos^2 M + \dots$$

$$= 1 + e \cos M + \frac{e^2}{2} (\cos 2M - 1) + \frac{e^2}{2} (\cos 2M + 1) + \dots . \qquad (6.24)$$

We have (to order e^2)

$$\frac{a}{r} = 1 + e \cos M + e^2 \cos 2M . \qquad (6.25)$$

To find $\cos E$, we use

$$r = a \, (1 - e \cos E) \qquad (6.26)$$

$$\frac{r}{a} - 1 = - e \cos E \qquad (6.27)$$

$$\cos E = \frac{1}{e} \left(1 - \frac{r}{a} \right) . \qquad (6.28)$$

To expand $\cos E$ in terms of mean anomaly, M, to order e^2, we need (r/a) in Eq. 6.28 to order e^3:

$$\cos E = \frac{1}{e}\left[1 - 1 + e\cos M + \frac{e^2}{2}(\cos 2M - 1) + \frac{e^3}{8}(3\cos 3M - 3\cos M)\right]$$

$$= \cos M + \frac{e}{2}\cos 2M - \frac{e}{2} + \frac{3e^2}{8}\cos 3M - \frac{3e^2}{8}\cos M$$

$$= -\frac{1}{2}e + \left(1 - \frac{3}{8}e^2\right)\cos M + \frac{1}{2}e\cos 2M + \frac{3}{8}e^2\cos 3M. \qquad (6.29)$$

To find $\sin E$, we use Kepler's equation:

$$E - M = e\sin E, \qquad (6.30)$$

or

$$\sin E = \frac{1}{e}(E - M). \qquad (6.31)$$

Moulton (1914) gives the expansion for E in terms of M as follows:

$$E = M + e\sin M + \frac{e^2}{2}\sin 2M + \frac{e^3}{3! \, 2^2}\left(3^2\sin 3M - 3\sin M\right) + \ldots \qquad (6.32)$$

Substituting Eq. 6.32 into Eq. 6.31, we have

$$\sin E = \frac{1}{e}\left[M - M + e\sin M + \frac{e^2}{2}\sin 2M + \frac{1}{24}e^3(9\sin 3M - 3\sin M)\right],$$
$$\qquad (6.33)$$

so that

$$\sin E = \sin M + \frac{e}{2}\sin 2M + \frac{3e^2}{8}\sin 3M - \frac{1}{8}e^2\sin M, \qquad (6.34)$$

or

$$\sin E = \left(1 - \frac{1}{8}e^2\right)\sin M + \frac{1}{2}e\sin 2M + \frac{3}{8}e^2\sin 3M. \qquad (6.35)$$

From Eqs. 6.23 and 6.25, we can write

$$\frac{r}{r'} = \left(\frac{r}{a}\right)\left(\frac{a'}{r'}\right)\left(\frac{a}{a'}\right)$$

$$= \left(\frac{a}{a'}\right)\left[1 + \frac{1}{2}e^2 - e\cos M - \frac{1}{2}e^2\cos 2M\right]\left[1 + e'\cos M' + e'^2\cos 2M'\right].$$

$$(6.36)$$

Multiplying out the terms in Eq. 6.36 and squaring the result, we have (retaining terms only to e^2)

$$\left(\frac{r}{r'}\right)^2 = \left(\frac{a}{a'}\right)^2\left[1 + e'\cos M' + e'^2\cos 2M' + \frac{e^2}{2} - e\cos M\right.$$

$$\left. - ee'\cos M\cos M' - \frac{e^2}{2}\cos 2M\right]$$

$$\times\left[1 + e'\cos M' + e'^2\cos 2M' + \frac{e^2}{2} - e\cos M\right.$$

$$\left. - ee'\cos M\cos M' - \frac{e^2}{2}\cos 2M\right]$$

$$= \left(\frac{a}{a'}\right)^2 \times \left[1 + e'\cos M' + e'^2\cos 2M' + \frac{e^2}{2} - e\cos M - ee'\cos M\cos M'\right.$$

$$- \frac{e'^2}{2}\cos 2M + e'\cos M' + e'^2\cos^2 M' - e'e\cos M'\cos M + e'^2\cos 2M'$$

$$+ \frac{e^2}{2} - e\cos M - e'e\cos M'\cos M + e^2\cos^2 M - ee'\cos M\cos M'$$

$$\left. - \frac{e^2}{2}\cos 2M\right]$$

$$= \left(\frac{a}{a'}\right)^2\left[1 + e^2 + e^2\cos^2 M - 2e\cos M - e^2\cos 2M + e'^2\cos^2 M'\right.$$

$$\left. + 2e'\cos M' - 4ee'\cos M\cos M' + 2e'^2\cos 2M'\right].$$

$$(6.37)$$

Terms such as $\cos^2 M$ and $\cos M\cos M'$ can always be transformed by trigonometric identities into functions of the multiple angles or of sums and differences of angles:

$$\cos^2 M = \frac{1}{2}\left(1 + \cos 2M\right) \tag{6.38}$$

$$\cos M \cos M' = \frac{1}{2}\left[\cos\left(M + M'\right) + \cos\left(M - M'\right)\right]. \tag{6.39}$$

Equation 6.37 then becomes

$$\left(\frac{r}{r'}\right)^2 = \left(\frac{a}{a'}\right)^2\left[1 + e^2 + \frac{e^2}{2}\left(1 + \cos 2M\right) - 2e\cos M - e^2\cos 2M\right.$$

$$+ \frac{e'^2}{2}\left(1 + \cos 2M'\right) + 2e'\cos M'$$

$$\left. - 2ee'\left[\cos\left(M + M'\right) + \cos\left(M - M'\right)\right] + 2e'^2\cos 2M'\right]$$

$$= \left(\frac{a}{a'}\right)^2\left[1 + \frac{3}{2}e^2 + \frac{e'^2}{2} - 2e\cos M + 2e'\cos M' - \frac{e^2}{2}\cos 2M\right.$$

$$\left. + \frac{5}{2}e'^2\cos 2M' - 2ee'\cos\left(M + M'\right) - 2ee'\cos\left(M - M'\right)\right]. \tag{6.40}$$

We observe from Eq. 6.40 that $\left(r/r'\right)^2$ is the sum of terms of the form:

$$A_{pq}\cos\left(pM + qM'\right),$$

where p and q are positive or negative integers or zero, and the coefficients A_{pq} are functions of the elements a, a', e, and e':

$$p, q \equiv \text{integers} \quad (+, -, \text{ or } 0)$$

$$A_{pq} \equiv A_{pq}(a, a', e, e'). \tag{6.41}$$

6.1.2   The Factor $P_2(\cos\phi)$

In Eq. 6.21, we also consider the effect of the polynomial factor

$$P_2\left(\cos\phi\right) = -\frac{1}{2} + \frac{3}{2}\cos^2\phi. \tag{6.42}$$

Thus we seek the form of $\cos\phi$ in terms of the orbital elements. We recall from Eq. 4.4 and from the properties of the ellipse, the position vector in the perifocal coordinate system is

$$\mathbf{r} = \xi \mathbf{u}_P + \eta \mathbf{u}_Q = a \left[(\cos E - e) \, \mathbf{u}_P + \left(\sqrt{1 - e^2} \sin E \right) \mathbf{u}_Q \right], \qquad (6.43)$$

where $\mathbf{u}_P$ and $\mathbf{u}_Q$ are functions of the orientation elements, Ω, ω, and i as given by Eqs. 2.50 and 2.51. An equation similar to Eq. 6.43 relates $\mathbf{r}'$ to the appropriate unit vectors, $\mathbf{u}'_P$ and $\mathbf{u}'_Q$. Now we can write

$$\cos \phi = \frac{\mathbf{r} \cdot \mathbf{r}'}{rr'} = \frac{\left(\xi \mathbf{u}_P + \eta \mathbf{u}_Q \right) \cdot \left(\xi' \mathbf{u}'_P + \eta' \mathbf{u}'_Q \right)}{rr'}$$

$$= \frac{\xi \xi' \mathbf{u}_P \cdot \mathbf{u}'_P + \eta \xi' \mathbf{u}_Q \cdot \mathbf{u}'_P + \xi \eta' \mathbf{u}_P \cdot \mathbf{u}'_Q + \eta \eta' \mathbf{u}_Q \cdot \mathbf{u}'_Q}{rr'}. \qquad (6.44)$$

Let us consider a typical term such as $\xi \xi' \mathbf{u}_P \cdot \mathbf{u}'_P / (rr')$. Since

$$\xi = a \, (\cos E - e)$$

$$\xi' = a' \left(\cos E' - e' \right), \qquad (6.45)$$

and using the expression for $\cos E$ from Eq. 6.29:

$$\frac{\xi \xi'}{rr'} = \left(\frac{a}{r} \right) \left(\frac{a'}{r'} \right)$$

$$\times \left[-\frac{3}{2} e + \left(1 - \frac{3}{8} e^2 \right) \cos M + \frac{e}{2} \cos 2M + \frac{3}{8} e^2 \cos 3M \right]$$

$$\times \left[-\frac{3}{2} e' + \left(1 - \frac{3}{8} e'^2 \right) \cos M' + \frac{e'}{2} \cos 2M' + \frac{3}{8} e'^2 \cos 3M' \right]. \qquad (6.46)$$

A typical product of terms from Eq. 6.46 is of the form:

$$\cos pM \cos q M',$$

which can be converted into the form

$$\frac{1}{2} \left[\cos \left(pM + q M' \right) + \cos \left(pM - q M' \right) \right].$$

Furthermore, products of these typical terms by the series expansions for (a/r) and (a'/r') (see Eq. 6.25) can be reduced similarly. We conclude that the product $\xi \xi' / (rr')$ takes the form of (terms like):

$$B_{pq} \cos \left(pM + q M' \right),$$

where

$$p, q \equiv \text{integers} \quad (+, -, \text{ or } 0)$$

$$B_{pq} \equiv B_{pq}(a, a', e, e'). \tag{6.47}$$

6.1.3   The $(r/r')^2 P_2 (\cos \phi)$ Term

We have been considering (as an illustration) what terms are involved in the disturbing function, $\mathcal{R}$, as expressed in Eq. 6.21, when we expand the Legendre term $(r/r')^2 P_2 (\cos \phi)$ as a function of the mean anomalies, M and M'. In particular, we have been examining the factor $P_2 (\cos \phi)$. We must still consider the product of $\mathbf{u}_P \cdot \mathbf{u}'_P$ in Eq. 6.44, which can be written as

$$\mathbf{u}_P \cdot \mathbf{u}'_P = P_1 P'_1 + P_2 P'_2 + P_3 P'_3, \tag{6.48}$$

where

$$P_1 = \cos \Omega \cos \omega - \sin \Omega \sin \omega \cos i \tag{6.49}$$

$$P_2 = \sin \Omega \cos \omega + \cos \Omega \sin \omega \cos i \tag{6.50}$$

$$P_3 = \sin \omega \sin i, \tag{6.51}$$

from Eq. 2.51, with analogous expressions for P'_1, P'_2, and P'_3. For convenience, let us write

$$\cos i = 1 - 2 \sin^2 (i/2). \tag{6.52}$$

In many applications, the inclination, i, is a small quantity so that

$$\sin \left(\frac{i}{2} \right) \approx \frac{i}{2} = \gamma \tag{6.53}$$

$$\cos i \approx 1 - 2\gamma^2, \tag{6.54}$$

where we use gamma as the new inclination variable. Thus, for the small i approximation, P_1 becomes

$$P_1 \approx \cos \Omega \cos \omega - \sin \Omega \sin \omega + 2\gamma^2 \sin \Omega \sin \omega$$

$$\approx \cos (\Omega + \omega) + 2\gamma^2 \sin \Omega \sin \omega. \tag{6.55}$$

By using the trigonometric identity

$$2 \sin \omega \sin \Omega = \cos (\omega - \Omega) - \cos (\omega + \Omega), \tag{6.56}$$

we find

$$P_1 \approx \left(1 - \gamma^2\right) \cos\left(\Omega + \omega\right) + \gamma^2 \cos\left(\omega - \Omega\right), \tag{6.57}$$

and similarly,

$$P_1' \approx \left(1 - \gamma'^2\right) \cos\left(\Omega' + \omega'\right) + \gamma'^2 \cos\left(\omega' - \Omega'\right). \tag{6.58}$$

The other components of $\mathbf{u}_P$ and $\mathbf{u}_Q$ (see Eqs. 2.50 and 2.51) can be expressed similarly.

Here we are interested only in the *general form* of the product $\mathbf{u}_P \cdot \mathbf{u}_P'$ as typified by the term $P_1 P_1'$. From Eqs. 6.57 and 6.58, we see that the term $P_1 P_1'$ consists of terms of the form:

$$\cos\left(\Omega + \omega\right) \cos\left(\Omega' + \omega'\right),$$

which can be reduced to sums such as:

$$\frac{1}{2}\left[\cos\left(\Omega + \omega + \Omega' + \omega'\right) + \cos\left(\Omega + \omega - \Omega' - \omega'\right)\right].$$

Thus, all products arising from $\mathbf{u}_P \cdot \mathbf{u}_P'$ and the other scalar products in Eq. 6.44 $(\mathbf{u}_Q \cdot \mathbf{u}_P', \mathbf{u}_P \cdot \mathbf{u}_Q', \mathbf{u}_Q \cdot \mathbf{u}_Q')$ take the form

$$C_j \cos\left(j_1 \Omega + j_2 \Omega' + j_3 \omega + j_4 \omega'\right),$$

where the j_i $(i = 1, 2, 3, 4)$ are positive or negative integers or zero and the C_j are functions of the inclination variables γ and γ'.

From this outline of the term $\xi \xi' P_1 P_1' / (rr')$, we conclude that $\cos\phi$ can be expressed as a sum of cosine terms with arguments that are functions of M, M', Ω, Ω', ω, and ω' (see Eq. 6.44). Further, $\cos^2\phi$ (which appears in the Legendre polynomial $P_2\left(\cos\phi\right) = -\frac{1}{2} + \frac{3}{2}\cos^2\phi$) and higher powers of $\cos\phi$ can be expressed as products of such functions and ultimately reduced by trigonometric identities to sums of cosines of multiple angles.

6.2 Form of the Perturbing Function

In the final analysis, the perturbation function, $\mathcal{R}$, takes the form:

$$\mathcal{R} = Gm' \sum_p C_p\left(a, a', e, e', \gamma, \gamma'\right)$$

$$\times \cos\left(p_1 M + p_2 M' + p_3 \Omega + p_4 \Omega' + p_5 \omega + p_6 \omega'\right), \tag{6.59}$$

where the p_i ($i = 1, 2, 3, 4, 5, 6$) are integers (positive, negative, or zero). Equation 6.59 can now be used in Lagrange's planetary equations (Eqs. 5.27 or 5.37) to obtain the perturbations in the orbital elements.

Let

$$M = nt + \sigma$$
$$M' = n't + \sigma', \tag{6.60}$$

so that

$$p_1 M + p_2 M' = (p_1 n + p_2 n') t + p_1 \sigma + p_2 \sigma'. \tag{6.61}$$

Let the argument of the cosine in the perturbation function, $\mathcal{R}$, be denoted by

$$\theta = (p_1 n + p_2 n') t + p_1 \sigma + p_2 \sigma' + p_3 \Omega + p_4 \Omega' + p_5 \omega + p_6 \omega'. \tag{6.62}$$

We make the additional assumption that the orbital elements of the perturbing body, m', can be considered *constant*. That is, the body of interest is assumed to have no significant effect on the motion of the perturbing body. Thus, Eq. 6.62 can be written as

$$\theta = (p_1 n + p_2 n') t + p_1 \sigma + p_3 \Omega + p_5 \omega + \theta_0, \tag{6.63}$$

where θ_0 contains all the contributions due to combinations of p_2, p_4, and p_6 with σ', Ω', and ω'. Thus Eq. 6.59 can be written as

$$\mathcal{R} = Gm' \sum_p C_p \cos \left[(p_1 n + p_2 n') t + p_1 \sigma + p_3 \Omega + p_5 \omega + \theta_0 \right], \tag{6.64}$$

where the summation refers to all p_i ($i = 1, 2, 3, 4, 5, 6$). Taking partial derivatives of $\mathcal{R}$ with respect to the orbital elements, we obtain

$$\frac{\partial \mathcal{R}}{\partial \sigma} = -Gm' \sum_p C_p p_1 \sin \theta \qquad (p_1 \neq 0) \tag{6.65}$$

$$\frac{\partial \mathcal{R}}{\partial \sigma} = 0 \qquad (p_1 = 0) \tag{6.66}$$

$$\frac{\partial \mathcal{R}}{\partial \Omega} = -Gm' \sum_p C_p p_3 \sin \theta \tag{6.67}$$

$$\frac{\partial \mathcal{R}}{\partial \omega} = -Gm' \sum_p C_p p_5 \sin \theta \tag{6.68}$$

$$\frac{\partial \mathcal{R}}{\partial e} = Gm' \sum_p \frac{\partial C_p}{\partial e} \cos \theta \tag{6.69}$$

$$\frac{\partial \mathcal{R}}{\partial i} = Gm' \sum_p \frac{\partial C_p}{\partial \gamma} \frac{\partial \gamma}{\partial i} \cos \theta \tag{6.70}$$

$$\frac{\partial \mathcal{R}}{\partial i} = \frac{1}{2} Gm' \cos \left(\frac{i}{2} \right) \sum_p \frac{\partial C_p}{\partial \gamma} \cos \theta \tag{6.71}$$

$$\frac{\partial \mathcal{R}}{\partial a} = Gm' \sum_p \frac{\partial C_p}{\partial a} \cos \theta - Gm' \sum_p C_p \left(p_1 t \frac{\partial n}{\partial a} \right) \sin \theta, \tag{6.72}$$

where in Eq. 6.72 we use

$$\frac{\partial n}{\partial a} = -\frac{3}{2} \frac{n}{a}$$

from Eq. 4.27 to capture the dependence of n on a. Also, in Eq. 6.71, we have used the relation $\gamma = \sin(i/2)$ to evaluate the partial derivative

$$\frac{\partial \gamma}{\partial i} = \frac{1}{2} \cos \left(\frac{i}{2} \right),$$

which appears in Eq. 6.70.

Now let us consider (as an example) the $\dot{\Omega}$ equation of Lagrange's planetary equations (Eqs. 5.27 or 5.37):

$$\dot{\Omega} = \left(\frac{1}{a^2 n \sqrt{1 - e^2} \sin i} \right) \frac{\partial \mathcal{R}}{\partial i}. \tag{6.73}$$

Substituting Eq. 6.71 into Eq. 6.73, we obtain

$$\dot{\Omega} = \frac{Gm' \cos (i/2)}{2a^2 n \sqrt{1 - e^2} \sin i} \sum_p \frac{\partial C_p}{\partial \gamma} \cos \theta, \tag{6.74}$$

where

$$\theta = \left(p_1 n + p_2 n' \right) t + p_1 \sigma + p_3 \Omega + p_5 \varpi + \theta_0. \tag{6.75}$$

Usually the mass m' is small compared to that of the central body or the disturbing effect is small. Thus, we assume that the orbital elements on the right-hand side of Eq. 6.74 are constants—the osculating elements. We observe from Eqs. 6.74 and 6.75 that—if p_1 and p_2 are not both zero—the derivative, $\dot{\Omega}$, is a periodic function of time alone. That is, there is no secular change in Ω. On the other hand, if p_1 and p_2 are both zero, then $\dot{\Omega}$ is a constant, say A, and there is a secular change in Ω (if $A \neq 0$).

We may therefore consider $\dot{\Omega}$ to be separated into 2 pieces, as follows:

$$\dot{\Omega} = A + \sum_p B_p \cos\left[\left(p_1 n + p_2 n'\right) t + \theta_1\right], \qquad (6.76)$$

where

$$\theta_1 = p_1 \sigma_0 + p_3 \Omega_0 + p_5 \omega_0 + \theta_0 , \qquad (6.77)$$

and where the subscript zero denotes fixed elements. Integrating Eq. 6.76, we obtain

$$\Omega = \Omega_0 + At + \sum_p \frac{B_p}{(p_1 n + p_2 n')} \sin\left[\left(p_1 n + p_2 n'\right) t + \theta_1\right], \qquad (6.78)$$

where p_1 and p_2 are not zero simultaneously. The term At is known as a secular perturbation term. Of course, if

$$p_1 n + p_2 n' = 0 , \qquad (6.79)$$

we still have secular terms arising from Eq. 6.76. In the case of Eq. 6.79, we have a commensurability in the periods of the perturbed and perturbing bodies. That is, if P and P' are the periods, then

$$\frac{P'}{P} = \frac{n}{n'} = -\frac{p_2}{p_1}, \qquad (6.80)$$

where p_1 and p_2 are integers. (Such a scenario is highly improbable in the Solar System.) The nature of the periodic perturbations in an element such as Ω depends on the magnitude of B_p and on the sum $p_1 n + p_2 n'$. In the Solar System, B_p is not large.

6.3 Short-Period and Long-Period Inequalities

If $p_1 n + p_2 n'$ is large, then perturbations to the orbital element have small amplitudes and short periods (i.e., high frequencies). Such perturbations are referred to as short-period inequalities. On the other hand, if $p_1 n + p_2 n'$ is small, then perturbations to the element have large amplitudes and long periods (i.e., low frequencies). These perturbations are called long-period inequalities. Clearly, from Lagrange's planetary equations (Eqs. 5.27 or 5.37) and Eqs. 6.65–6.72, all of the elements except the semimajor axis, a, exhibit secular as well as periodic changes (to first order in m').

For the semimajor axis, however,

$$\dot{a} = \frac{2}{an}\frac{\partial \mathcal{R}}{\partial \sigma} = -\frac{2Gm'}{an}\sum_p C_p p_1 \sin\theta \qquad (p_1 \neq 0) \qquad (6.81)$$

$$\dot{a} = 0 \qquad\qquad\qquad (p_1 = 0). \qquad (6.82)$$

Integrating Eq. 6.81, we have

$$\delta(a) = \frac{2Gm'}{a_0 n}\sum_p C_p \left(\frac{p_1}{p_1 n + p_2 n'}\right)\cos\theta, \qquad (6.83)$$

where

$$a = a_0 + \delta(a). \qquad (6.84)$$

We see that the length of the semimajor axis oscillates about the mean value of a_0 with period:

$$P = \frac{2\pi}{p_1 n + p_2 n'}. \qquad (6.85)$$

We have outlined the analysis of the perturbations for the elements Ω and a; a similar approach can be used with the other orbital elements. The results we have obtained are only to the first power in m'. In principle, we can take the revised elements (such as $\Omega = \Omega_0 + At+$ periodic terms) and re-substitute them into Lagrange's planetary equations to obtain a second approximation (with new periodic terms in t^2 and m'^2). Thus, a higher-order solution is possible.

6.4 Stability

We have shown that the semimajor axis exhibits no secular change in the first-order theory. If it did, then the orbit would expand or contract indefinitely, and the orbit would be unstable. In another situation, if the eccentricity were to increase secularly, a close approach with another planet could occur and disrupt the system. In 1776, Lagrange showed that when all powers of e are included to first order in m' that the semimajor axis undergoes no secular change. In 1809, Poisson obtained the same result to second order in m'; however, in 1885, Haretu showed that a secular term exists for a third-order theory in m' (Moulton, 1914).

References

Kovalevsky, J. (1967). *Introduction to celestial mechanics*. Netherlands: Springer.
Moulton, F. R. (1914). *An introduction to celestial mechanics* (2nd ed.). New York: The Macmillan Company.
Taff, L. G. (1985). *Celestial mechanics, a computational guide for the practitioner*. New York: John Wiley & Sons.

Chapter 7
The Gravitational Potential

7.1 Description of the Gravitational Potential

We consider the gravitational potential due to a planet that is not a uniform sphere. We let P be the location of a unit mass at a distance r from C (the center of mass) of a bounded distribution of matter having a total mass M. Let dm be a differential element of the mass distribution, located at point Q, a distance ρ from C. This situation is depicted in Fig. 7.1.

Then the gravitational potential function at P due to the differential clement, dm, at point Q is

$$dU = \frac{G\,dm}{\left(\rho^2 + r^2 - 2r\rho \cos\theta\right)^{1/2}}, \tag{7.1}$$

where we note that the distance from Q to P is given by the trigonometric law of cosines

$$\left|\mathbf{r}^{QP}\right| = \left(\rho^2 + r^2 - 2r\rho \cos\theta\right)^{1/2}. \tag{7.2}$$

The total potential function at P is

$$U = G \int_M \frac{dm}{\left(\rho^2 + r^2 - 2r\rho \cos\theta\right)^{1/2}}. \tag{7.3}$$

We note that r is held constant in the integration over the entire mass of the spheroid.

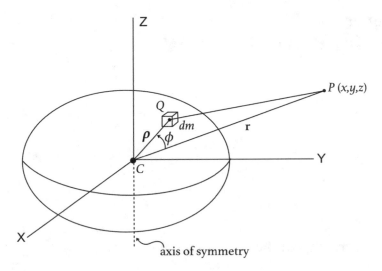

Fig. 7.1 Potential at a fixed exterior point, P, due to an irregular mass distribution; $\mathbf{r}$ is held constant during integration

Expanding the denominator of Eq. 7.3, we have (see Eq. 6.12)

$$\left(\rho^2 + r^2 - 2r\rho\cos\theta\right)^{-1/2} = \frac{1}{r}\left[1 - \frac{2\rho}{r}\cos\theta + \left(\frac{\rho}{r}\right)^2\right]^{-1/2}$$

$$= \frac{1}{r}\left[1 + \left(\frac{\rho}{r}\right)\cos\theta + \left(\frac{\rho}{r}\right)^2\left(-\frac{1}{2} + \frac{3}{2}\cos^2\theta\right) + \ldots\right]$$

(7.4)

in which powers greater than 2 in ρ/r are not needed in the following analysis. The expression in its more complete form, however, is necessary when precise studies of the shape of the Earth (for instance) are being carried out.

For a general expansion we would use the Legendre polynomials:

$$\left(\rho^2 + r^2 - 2r\rho\cos\theta\right)^{-1/2} = \frac{1}{r}\sum_{n=1}^{\infty}\left(\frac{\rho}{r}\right)^n P_n(\cos\theta).$$

(7.5)

(Refer to Eq. 6.18 for a list of the Legendre coefficients, $P_n(\cos\theta)$.) The function on the left-hand side of Eq. 7.5 is called the generating function of the polynomials. The polynomials are sometimes called "zonal harmonics." Substituting Eq. 7.4 into Eq. 7.3, we obtain

$$U = \frac{G}{r}\int_M\left[1 + \left(\frac{\rho}{r}\right)\cos\theta + \left(\frac{\rho}{r}\right)^2\left(-\frac{1}{2} + \frac{3}{2}\cos^2\theta\right)\right]dm,$$

(7.6)

and after expanding, we have

$$U = \frac{G}{r} \int_M dm + \frac{G}{r^2} \int_M \rho \cos\theta \, dm + \frac{G}{2r^3} \int_M \rho^2 \left(3\cos^2\theta - 1\right) dm. \quad (7.7)$$

For convenience, let us write the potential function as

$$U = U_0 + U_1 + U_2 + \ldots , \quad (7.8)$$

where

$$U_0 = \frac{G}{r} \int P_0 dm$$

$$U_1 = \frac{G}{r} \int \frac{\rho}{r} P_1 dm \quad (7.9)$$

$$U_2 = \frac{G}{r} \int \left(\frac{\rho}{r}\right)^2 P_2 dm$$

$$\vdots$$

Let us evaluate the integrals in Eq. 7.9. For U_0, we have

$$U_0 = \frac{G}{r} \int dm = \frac{GM}{r}, \quad (7.10)$$

which is the potential of the point mass m due to a point mass M located at C. To evaluate U_1, we use body-fixed unit vectors to specify the locations of $P(x, y, z)$ and $Q(x, y, z)$ with respect to the central body:

$$\mathbf{r}^{CP} = \mathbf{r} = x\mathbf{b}_1 + y\mathbf{b}_2 + z\mathbf{b}_3 \quad (7.11)$$

$$\mathbf{r}^{CQ} = \rho = \xi\mathbf{b}_1 + \eta\mathbf{b}_2 + \zeta\mathbf{b}_3, \quad (7.12)$$

so that

$$U_1 = \frac{G}{r} \int \frac{\rho}{r} \cos\theta \, dm = \frac{G}{r} \int \frac{\rho}{r} \frac{\mathbf{r} \cdot \rho}{|\mathbf{r}||\rho|} dm, \quad (7.13)$$

where

$$\frac{\mathbf{r} \cdot \rho}{|\mathbf{r}|} = \frac{x\xi + y\eta + z\zeta}{r}. \quad (7.14)$$

Using Eq. 7.14 in Eq. 7.13 provides

$$U_1 = \frac{G}{r^3} \left(z \int_M \xi \, dm + y \int_M \eta \, dm + z \int_M \zeta \, dm\right). \quad (7.15)$$

However, since the origin of the body frame is located at the center of mass, we have

$$\int_M \xi \, dm = \int_M \eta \, dm = \int_M \zeta \, dm = 0 \,, \tag{7.16}$$

by the definition of the center of mass. Thus we find

$$U_1 = 0. \tag{7.17}$$

Next we evaluate U_2:

$$\begin{aligned} U_2 &= \frac{G}{r} \int \frac{1}{2} \left(3 \cos^2 \theta - 1 \right) \frac{\rho^2}{r^2} dm \\ &= \frac{G}{2r^3} \int \left(3 \cos^2 \theta - 1 \right) \rho^2 dm. \end{aligned} \tag{7.18}$$

The instantaneous moment of inertia of the spheroid about the line CP, I, is

$$I = \int_M \rho^2 \sin^2 \theta \, dm. \tag{7.19}$$

Substituting $\cos^2 \theta = 1 - \sin^2 \theta$ into Eq. 7.18 provides

$$\begin{aligned} U_2 &= \frac{G}{2r^3} \int \left(3 - 3 \sin^2 \theta - 1 \right) \rho^2 dm \\ &= \frac{G}{2r^3} \left[-3I + 2 \int \rho^2 dm \right], \end{aligned} \tag{7.20}$$

where we have used Eq. 7.19 in Eq. 7.20. For the integral in Eq. 7.20, we have

$$2 \int \rho^2 dm = 2 \int \left(\xi^2 + \eta^2 + \zeta^2 \right) dm. \tag{7.21}$$

Let us define the moments of inertia as follows:

$$\begin{aligned} I_1 &= \int \left(\eta^2 + \zeta^2 \right) dm \\ I_2 &= \int \left(\xi^2 + \zeta^2 \right) dm \\ I_3 &= \int \left(\xi^2 + \eta^2 \right) dm. \end{aligned} \tag{7.22}$$

We see that, using Eq. 7.22 in Eq. 7.21, we have

$$2 \int \rho^2 dm = I_1 + I_2 + I_3. \tag{7.23}$$

Substituting Eq. 7.23 into Eq. 7.20 provides

$$U_2 = \frac{G}{2r^3} (I_1 + I_2 + I_3 - 3I).\tag{7.24}$$

Thus, from Eqs. 7.8, 7.10, 7.17, and 7.24, we have the potential function given by MacCullagh's formula:

$$U = \frac{GM}{r} \left[1 + \frac{1}{2Mr^2} (I_1 + I_2 + I_3 - 3I) \right].\tag{7.25}$$

For an *oblate spheroid*, we have the property:

$$I_1 = I_2 \neq I_3,\tag{7.26}$$

and we can write

$$U = \frac{GM}{r} \left[1 + \frac{1}{2Mr^2} (2I_e + I_3 - 3I) \right],\tag{7.27}$$

where I_e is the moment of inertia about any axis in the equatorial plane and I_3 is the moment of inertia about the axis of rotation. We note that while I_e and I_3 are constant quantities, I varies with the position of P in space. We also note that, in the case of a sphere, the parenthetical term in Eq. 7.25 is zero.

7.2 Zonal Harmonics

We recall the gravity potential function from Eq. 7.8

$$U = U_0 + U_1 + U_2 + \ldots,\tag{7.28}$$

where we have from Eq. 7.10

$$U_0 = \frac{GM}{r}\tag{7.29}$$

and from Eq. 7.17

$$U_1 = 0,\tag{7.30}$$

due to the choice of the center of mass as the origin. For U_2, we have from Eq. 7.18:

$$U_2 = \frac{G}{2r^3} \int \left(3\cos^2\theta - 1 \right) \rho^2 dm.\tag{7.31}$$

The last term in the integral is given by Eq. 7.23 as

$$\int \rho^2 dm = \frac{1}{2}(I_1 + I_2 + I_3)$$

$$= \int \xi^2 dm + \int \eta^2 dm + \int \zeta^2 dm, \tag{7.32}$$

where I_1, I_2, and I_3 are the moments of inertia about the $\mathbf{b}_1$, $\mathbf{b}_2$, and $\mathbf{b}_3$ body axes, respectively.

Now we examine the first term in the integral in Eq. 7.31, which we denote as U_2':

$$U_2' = \frac{G}{2r^3} \int 3\rho^2 \cos^2 \theta \, dm. \tag{7.33}$$

We can make use of the expression for $\cos \theta$ in Eqs. 7.13 and 7.14:

$$\cos \theta = \frac{\mathbf{r} \cdot \boldsymbol{\rho}}{r\rho} = \frac{x\xi + y\eta + z\zeta}{r\rho}. \tag{7.34}$$

Substituting Eq. 7.34 into Eq. 7.33 provides

$$U_2' = \frac{3}{2}\frac{G}{r^5}\left[x^2 \int \xi^2 dm + y^2 \int \eta^2 dm + z^2 \int \zeta^2 dm\right]$$

$$+ \frac{3G}{r^5}\left[xy \int \xi\eta dm + yz \int \eta\zeta dm + xz \int \xi\zeta dm\right]. \tag{7.35}$$

We recall the definitions, Eq. 7.22:

$$I_1 = \int \left(\eta^2 + \zeta^2\right) dm \tag{7.36}$$

$$I_2 = \int \left(\xi^2 + \zeta^2\right) dm \tag{7.37}$$

$$I_3 = \int \left(\xi^2 + \eta^2\right) dm. \tag{7.38}$$

Also, let us define the products of inertia:

$$I_{xy} = -\int \xi\eta dm \tag{7.39}$$

$$I_{xz} = -\int \xi\zeta dm \tag{7.40}$$

$$I_{yz} = -\int \eta \zeta \, dm. \tag{7.41}$$

Assuming the use of principal axes, we have

$$I_{xy} = I_{xz} = I_{yz} = 0, \tag{7.42}$$

and thus the second square bracket in Eq. 7.35 vanishes.

We note from Eqs. 7.36–7.38 that

$$I_1 + I_2 + I_3 = 2 \int \left(\xi^2 + \eta^2 + \zeta^2 \right) dm$$

$$= 2 \int \xi^2 dm + 2 I_1, \tag{7.43}$$

so we can write

$$\int \xi^2 dm = \frac{1}{2} (I_1 + I_2 + I_3) - I_1 \tag{7.44}$$

$$\int \eta^2 dm = \frac{1}{2} (I_1 + I_2 + I_3) - I_2 \tag{7.45}$$

$$\int \zeta^2 dm = \frac{1}{2} (I_1 + I_2 + I_3) - I_3. \tag{7.46}$$

Using Eqs. 7.42 and 7.44–7.46 in Eq. 7.35, we find

$$U_2' = \frac{3}{2} \frac{G}{r^5} \left\{ x^2 \left[\frac{1}{2} (I_1 + I_2 + I_3) - I_1 \right] + y^2 \left[\frac{1}{2} (I_1 + I_2 + I_3) - I_2 \right] \right.$$

$$\left. + z^2 \left[\frac{1}{2} (I_1 + I_2 + I_3) - I_3 \right] \right\}. \tag{7.47}$$

From Eqs. 7.31–7.33 and Eq. 7.47, we have for U_2:

$$U_2 = \frac{G}{r^3} \left\{ \left(\frac{3x^2}{2r^2} - \frac{1}{2} \right) \left[\frac{1}{2} (I_1 + I_2 + I_3) - I_1 \right] \right.$$

$$+ \left(\frac{3y^2}{2r^2} - \frac{1}{2} \right) \left[\frac{1}{2} (I_1 + I_2 + I_3) - I_2 \right]$$

$$\left. + \left(\frac{3z^2}{2r^2} - \frac{1}{2} \right) \left[\frac{1}{2} (I_1 + I_2 + I_3) - I_3 \right] \right\}. \tag{7.48}$$

We use spherical coordinates, as shown in Fig. 7.2.

Fig. 7.2 Spherical coordinate system where λ is the longitude and ϕ is the latitude

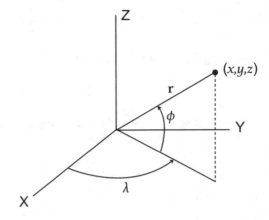

$$x = r \cos\phi \cos\lambda \qquad (7.49)$$

$$y = r \cos\phi \sin\lambda \qquad (7.50)$$

$$z = r \sin\phi, \qquad (7.51)$$

where we note that here λ is the longitude angle. Substituting Eqs. 7.49–7.51 into Eq. 7.48, we obtain

$$U_2 = \frac{G}{r^3}\left\{\left[\frac{I_1 + I_2 + I_3}{2} - I_1\right]\left(\frac{3\cos^2\phi\cos^2\lambda}{2} - \frac{1}{2}\right)\right.$$
$$+ \left[\frac{I_1 + I_2 + I_3}{2} - I_2\right]\left(\frac{3\cos^2\phi\sin^2\lambda}{2} - \frac{1}{2}\right)$$
$$\left. + \left[\frac{I_1 + I_2 + I_3}{2} - I_3\right]\left(\frac{3\sin^2\phi}{2} - \frac{1}{2}\right)\right\}. \qquad (7.52)$$

We can arrange Eq. 7.52 as follows:

$$U_2 = \frac{G}{r^3}\left\{\left[\frac{I_2 - I_1}{2}\right]\left(\frac{3\cos^2\phi\cos^2\lambda}{2} - \frac{1}{2}\right) + \frac{I_3}{2}\left(\frac{3}{2}\cos^2\phi\cos^2\lambda - \frac{1}{2}\right)\right.$$
$$+ \left[\frac{I_1 - I_2}{2}\right]\left(\frac{3\cos^2\phi\sin^2\lambda}{2} - \frac{1}{2}\right) + \frac{I_3}{2}\left(\frac{3}{2}\cos^2\phi\sin^2\lambda - \frac{1}{2}\right)$$
$$\left. + \left[I_3 - \frac{I_1 + I_2}{2}\right]\left(\frac{1}{2} - \frac{3\sin^2\phi}{2}\right) + \frac{I_3}{2}\left(\frac{3}{2}\sin^2\phi - \frac{1}{2}\right)\right\}$$
$$= \frac{G}{r^3}\left\{\left[\frac{I_2 - I_1}{2}\right]\left[\frac{3\cos^2\phi}{2}\left(\cos^2\lambda - \sin^2\lambda\right)\right] + \frac{I_3}{2}\left(\frac{3}{2}\cos^2\phi - \frac{3}{2}\right)\right.$$

$$+ \left[I_3 - \frac{I_1 + I_2}{2} \right] \left(\frac{1}{2} - \frac{3 \sin^2 \phi}{2} \right) + \frac{I_3}{2} \left(\frac{3}{2} \sin^2 \phi \right) \Bigg\}. \tag{7.53}$$

Further simplifying and noting that

$$\cos^2 \lambda - \sin^2 \lambda = \cos 2\lambda, \tag{7.54}$$

Eq. 7.53 becomes

$$U_2 = \frac{G}{r^3} \left[\left(I_3 - \frac{I_1 + I_2}{2} \right) \left(\frac{1}{2} - \frac{3}{2} \sin^2 \phi \right) - \frac{3}{4} (I_1 - I_2) \cos^2 \phi \cos 2\lambda \right]. \tag{7.55}$$

The same expression for U_2 is given by Roy (2005).

We recall the general form for the total gravitational potential function at the point P:

$$U = G \int_M \frac{dm}{\left(\rho^2 + r^2 - 2r\rho \cos \lambda \right)^{1/2}}. \tag{7.56}$$

As shown by Roy (2005), Eq. 7.56 can be written as

$$U = U_0 + U_1 + U_2 + U_3 + \ldots \tag{7.57}$$

where

$$U_0 = \frac{G}{r} \int P_0 dm \tag{7.58}$$

$$U_1 = \frac{G}{r} \int \frac{\rho}{r} P_1 dm \tag{7.59}$$

$$U_2 = \frac{G}{r} \int \left(\frac{\rho}{r} \right)^2 P_2 dm \tag{7.60}$$

$$U_3 = \frac{G}{r} \int \left(\frac{\rho}{r} \right)^3 P_3 dm \tag{7.61}$$

$$\vdots$$

where the P_i are the Legendre polynomials

$$P_0 = 1 \tag{7.62}$$

$$P_1 = q \tag{7.63}$$

$$P_2 = \frac{1}{2} \left(3q^2 - 1 \right) \tag{7.64}$$

$$P_3 = \frac{1}{2}\left(5q^3 - 3q\right) \qquad (7.65)$$

$$\vdots$$

and where

$$q = \cos \lambda. \qquad (7.66)$$

In Eq. 7.10, we have

$$U_0 = \frac{GM}{r}, \qquad (7.67)$$

and for the origin placed at the center of mass, we have from Eq. 7.17

$$U_1 = 0. \qquad (7.68)$$

For the further assumption of principal axes, we have from Eq. 7.55

$$U_2 = \frac{G}{r^3}\left[\left(I_3 - \frac{I_1 + I_2}{2}\right)\left(\frac{1}{2} - \frac{3}{2}\sin^2\phi\right) - \frac{3}{4}(I_1 - I_2)\cos^2\phi\cos 2\lambda\right].$$
$$(7.69)$$

We now consider the U_3 term in Eq. 7.61:

$$U_3 = \frac{G}{2r^4}\int\left(5q^3 - 3q\right)\rho^3 dm. \qquad (7.70)$$

According to Roy (2005), Eq. 7.70 leads to a complicated expression containing integrals of the form

$$\int \xi^a \eta^b \zeta^c dm,$$

where a, b, and c are positive integers that satisfy the relation

$$a + b + c = 3. \qquad (7.71)$$

If the body is symmetrical about all three coordinate planes (e.g., a triaxial homogeneous ellipsoid with three distinct axes), all the integrals vanish so that U_3 is zero. Indeed, all the odds U_i vanish in this case so that

$$U_3 = U_5 = U_7 = \ldots = 0. \qquad (7.72)$$

Satellite studies have established that the Earth deviates slightly from Eq. 7.72, being slightly pear shaped (as in Fig. 7.3) so that

Fig. 7.3 Rotationally
symmetric body with a slight
pear shape and where the
origin is at the center of mass

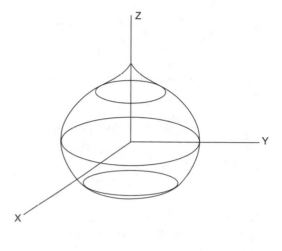

$$U_3 \approx 0. \tag{7.73}$$

If the body has rotational symmetry about the z axis, then

$$I_1 = I_2 = I_e \tag{7.74}$$

and our expression for U_2 in Eq. 7.69 reduces to

$$U_2 = \frac{G}{r^3} (I_3 - I_e) \left(\frac{1}{2} - \frac{3}{2} \sin^2 \phi \right). \tag{7.75}$$

We note that in Eq. 7.75, we are not assuming that the body is symmetrical with respect to the equator: it may be pear shaped as in the figure. We recall, however, that the center of mass must be at the intersection of the xy plane and the axis of rotational symmetry.

From McCuskey (1963), the Earth's potential function at a distance r from the center of mass of the planet may be approximated by

$$U = \frac{GM_e}{r} \left[1 + J \left(\frac{r_e}{r} \right)^2 \left(\frac{1}{3} - \sin^2 \phi \right) + H \left(\frac{r_e}{r} \right)^3 \left(\frac{3}{5} - \sin^2 \phi \right) \sin \phi \right.$$
$$\left. + K \left(\frac{r_e}{r} \right)^4 \left(\frac{1}{10} - \sin^2 \phi + \frac{7}{6} \sin^4 \phi \right) + \ldots \right], \tag{7.76}$$

where M_e is the mass of the Earth, r_e is the equatorial radius of the Earth, and the constants J, H, K are called the coefficients of the second, third, and fourth harmonics of the Earth's gravitational potential.

If we assume that the Earth is a spheroid (i.e., a rotationally symmetric ellipsoid), then its potential may be written as a series of spherical harmonics of the form:

$$U = \frac{GM_e}{r}\left[1 - \sum_{n=2}^{\infty} J_n \left(\frac{r_e}{r}\right)^n P_n (\sin\phi)\right],\tag{7.77}$$

where $P_n(\sin\phi)$ is the Legendre polynomial. The first three of these polynomials are

$$P_0(\sin\phi) = 1 \tag{7.78}$$

$$P_1(\sin\phi) = \sin\phi \tag{7.79}$$

$$P_2(\sin\phi) = \frac{1}{2}\left(3\sin^2\phi - 1\right), \tag{7.80}$$

where the origin is the center of mass. The general result was first obtained by Laplace. From Eqs. 7.76 and 7.77, we have

$$J = \frac{3}{2}J_2 \tag{7.81}$$

$$H = \frac{5}{2}J_3 \tag{7.82}$$

$$K = -\frac{15}{4}J_4. \tag{7.83}$$

The terms involving J and K (i.e., J_2 and J_4) are due to the Earth's oblateness. The term containing H (J_3) arises because of asymmetry with respect to the equatorial plane (that is, the "pear shape").

We recall that the angle ϕ denotes the latitude of the satellite, which is related to the orbital elements i and ω and the true anomaly f via the Law of Sines in spherical trigonometry

$$\sin\phi = \sin i \sin u = \sin i \sin(\omega + f), \tag{7.84}$$

as depicted in Fig. 7.4.

We substitute the expression for $\sin\phi$ from Eq. 7.84 into the Legendre polynomials in Eq. 7.76:

$$\frac{1}{3} - \sin^2\phi = \frac{1}{3} - \sin^2 i \sin^2 u. \tag{7.85}$$

Using the trigonometric identity

$$\sin^2 u = \frac{1}{2}(1 - \cos 2u), \tag{7.86}$$

we obtain

Fig. 7.4 Relationship between satellite latitude and orbital elements

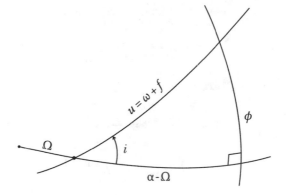

$$\frac{1}{3} - \sin^2 \phi = \frac{1}{3} - \frac{\sin^2 i}{2} + \frac{1}{2} \sin^2 i \cos 2u. \tag{7.87}$$

Next we have

$$\left(\frac{3}{5} - \sin^2 \phi\right) \sin \phi = \frac{3}{5} \sin i \sin u - \sin^3 i \sin^3 u. \tag{7.88}$$

Using the trigonometric identity

$$\sin^3 u = \frac{3}{4} \sin u - \frac{1}{4} \sin 3u \tag{7.89}$$

gives

$$\left(\frac{3}{5} - \sin^2 \phi\right) \sin \phi = \frac{3}{5} \sin i \sin u - \sin^3 i \left(\frac{3}{4} \sin u - \frac{1}{4} \sin 3u\right)$$

$$= \sin u \left(\frac{3}{5} \sin i - \frac{3}{4} \sin^3 i\right) + \frac{1}{4} \sin^3 i \sin 3u$$

$$= \sin u \sin i \left(\frac{3}{5} - \frac{3}{4} \sin^2 i\right) + \frac{1}{4} \sin^3 i \sin 3u$$

$$= \frac{3}{5} \left[\sin u \sin i \left(1 - \frac{5}{4} \sin^2 i\right) + \frac{5}{12} \sin^3 i \sin 3u\right]$$

$$= \frac{3}{5} \left[\left(1 - \frac{5}{4} \sin^2 i\right) \sin u + \frac{5}{12} \sin^2 i \sin 3u\right] \sin i. \tag{7.90}$$

For the last parenthetical term in Eq. 7.76, we have

$$\frac{1}{10} - \sin^2 \phi + \frac{7}{6} \sin^4 \phi = \frac{1}{10} - \sin^2 i \sin^2 u + \frac{7}{6} \sin^4 i \sin^4 u. \tag{7.91}$$

Making use of the trigonometric identity

$$\sin^4 u = \frac{3}{8} - \frac{1}{2}\cos 2u + \frac{1}{8}\cos 4u \tag{7.92}$$

and Eq. 7.86, we obtain

$$\frac{1}{10} - \sin^2\phi + \frac{7}{6}\sin^4\phi = \frac{1}{10} - \frac{1}{2}\sin^2 i + \frac{1}{2}\sin^2 i\cos 2u$$

$$+ \frac{7}{6}\sin^4 i\left(\frac{3}{8} - \frac{1}{2}\cos 2u + \frac{1}{8}\cos 4u\right)$$

$$= \frac{1}{10} - \frac{1}{2}\sin^2 i + \frac{21}{48}\sin^4 i$$

$$+ \frac{1}{2}\sin^2 i\cos 2u - \frac{7}{12}\sin^4 i\cos 2u + \frac{7}{48}\sin^4 i\cos 4u$$

$$= \frac{1}{10} - \frac{1}{2}\sin^2 i + \frac{21}{48}\sin^4 i$$

$$+ \left(\frac{1}{2} - \frac{7}{12}\sin^i\right)\left(\sin^2 i\cos 2u\right) + \frac{7}{48}\sin^4 i\cos 4u$$

$$= \frac{7}{6}\left[\frac{3}{35} - \frac{3}{7}\sin^2 i + \frac{3}{8}\sin^4 i\right.$$

$$\left. + \left(\frac{3}{7} - \frac{1}{2}\sin^2 i\right)\sin^2 i\cos 2u + \frac{1}{8}\sin^4 i\cos 4u\right]. \tag{7.93}$$

The disturbing function, $\mathcal{R}$, is deduced from Eq. 7.76 after substituting the results from Eqs. 7.85, 7.90, and 7.93, where we recall that the disturbing function is given by subtracting the two-body term from the total gravitational potential function:

$$\mathcal{R} = U - \frac{GM_e}{r} \tag{7.94}$$

so that

$$\mathcal{R} = \mu\left\{\frac{Jr_e^2}{a^3}\left(\frac{a}{r}\right)^3\left[\frac{1}{3} - \frac{1}{2}\sin^2 i + \frac{1}{2}\sin^2 i\cos 2u\right]\right.$$

$$+ \frac{Hr_e^3}{a^4}\left(\frac{a}{r}\right)^4\frac{3}{5}\left[\left(1 - \frac{5}{4}\sin^2 i\right)\sin u + \frac{5}{12}\sin^2 i\sin 3u\right]\sin i$$

$$+ \frac{Kr_e^4}{a^5}\left(\frac{a}{r}\right)^5\frac{7}{6}\left[\frac{3}{35} - \frac{3}{7}\sin^2 i + \frac{3}{8}\sin^4 i + \left(\frac{3}{7} - \frac{1}{2}\sin^2 i\right)\sin^2 i\cos 2u\right.$$

$$+\frac{1}{8}\sin^4 i \cos 4u \Big]\Big\}.$$ (7.95)

7.3   An Alternate Description of the Gravitational Potential

The following discussion provides an overview of the development of the most general formulation of the gravitational potential. We substantially follow Fitzpatrick (1970) and Kaula (1966) (also published by Dover in 2000), where the reader can find further details.

The gravitational potential V must satisfy Laplace's equation

$$\nabla^2 V = \frac{\partial}{\partial r}\left(r^2\frac{\partial V}{\partial r}\right) + \frac{1}{\cos\phi}\frac{\partial}{\partial\phi}\left(\cos\phi\frac{\partial V}{\partial\phi}\right) + \frac{1}{\cos^2\phi}\frac{\partial^2 V}{\partial\lambda^2} = 0.$$ (7.96)

We seek a solution $V(r, \phi, \lambda)$ of Eq. 7.96. A common approach to solving such a partial differential equation is to assume the solution takes the form

$$V = R(r)\,\Phi(\phi)\,\Lambda(\lambda),$$ (7.97)

which allows use of the separation of variables technique.

We substitute Eq. 7.97 into Eq. 7.96 and divide by $R\Phi\Lambda$ to get

$$\frac{1}{R}\frac{d}{dr}\left(r^2\frac{dR}{dr}\right) + \frac{1}{\Phi\cos\phi}\frac{d}{d\phi}\left(\cos\phi\frac{d\Phi}{d\phi}\right) + \frac{1}{\Lambda\cos^2\phi}\frac{d^2\Lambda}{d\lambda^2} = 0,$$ (7.98)

where the terms are now functions of a single variable and thus become total derivatives rather than partial derivatives.

Since the first term is the only function of r, it must be a constant. Therefore

$$\frac{1}{R}\frac{d}{dr}\left(r^2\frac{dR}{dr}\right) = \frac{1}{R}r^2\frac{d^2R}{dr^2} + \frac{1}{R}2r\frac{dR}{dr} = c$$

$$r^2\frac{d^2R}{dr^2} + 2r\frac{dR}{dr} - cR = 0.$$ (7.99)

Since our practical problem of interest is representing a general potential based on an inverse square law force, we postulate that R is of the form r^l. Substituting this form into Eq. 7.99, we obtain

$$r^2 l\,(l-1)\,r^{l-2} + 2rl r^{l-1} - c r^l = 0$$

$$l\,(l-1) + 2l - c = 0$$

$$l\,(l+1) = c.$$ (7.100)

Thus, we have

$$F(R) = r^2 \frac{d^2 R}{dr^2} + 2r \frac{dR}{dr} - l(l+1)R.$$

This expression suggests two possible solutions, which we now demonstrate. Let $R = r^l$, so that

$$F\left(r^l\right) = r^2 l(l-1)r^{l-2} + 2rlr^{l-1} - l(l+1)r^l$$

$$= l(l-1)r^l + 2lr^l - l(l+1)r^l$$

$$= r^l \left(l^2 - l + 2l - l^2 - l\right) = 0.$$

Now we let $R = r^{-l-1}$ and obtain

$$F\left(r^{-l-1}\right) = r^2(-l-1)(-l-2)r^{-l-3} + 2r(-l-1)r^{-l-2} - l(l+1)r^{-l-1}$$

$$= (-l-1)(-l-2)r^{-l-1} + 2(-l-1)r^{-l-1} - l(l+1)r^{-l-1}$$

$$= r^{-l-1}\left(l^2 + 3l + 2 - 2l - 2 - l^2 - l\right) = 0.$$

Thus, the two possible solutions are r^l and r^{-l-1}. Since we expect the potential to approach zero as r increases, we reject the first solution. Hence, we take the second solution

$$R = r^{-(l+1)}. \tag{7.101}$$

We substitute Eq. 7.101 into Eq. 7.98 and multiply by $\cos^2\phi$ to obtain

$$\cos^2\phi \frac{1}{R}\frac{d}{dr}\left[r^2(-l-1)r^{-l-2}\right] + \frac{\cos\phi}{\Phi}\frac{d}{d\phi}\left(\cos\phi\frac{d\Phi}{d\phi}\right) + \frac{1}{\Lambda}\frac{d^2\Lambda}{d\lambda^2} = 0$$

$$l(l+1)\cos^2\phi + \frac{\cos\phi}{\Phi}\frac{d}{d\phi}\left(\cos\phi\frac{d\Phi}{d\phi}\right) + \frac{1}{\Lambda}\frac{d^2\Lambda}{d\lambda^2} = 0. \tag{7.102}$$

We can now separate variables in Eq. 7.102 to write

$$l(l+1)\cos^2\phi + \frac{\cos\phi}{\Phi}\frac{d}{d\phi}\left(\cos\phi\frac{d\Phi}{d\phi}\right) = -\frac{1}{\Lambda}\frac{d^2\Lambda}{d\lambda^2} = m^2,$$

where the separation constant m^2, m being an integer, has been chosen to lead to solutions for Λ that have period 2π. We can write the separated equations in the forms

$$\Lambda'' + m^2 \Lambda = 0 \tag{7.103}$$

and

$$\cos \phi \frac{d}{d\phi} \left(\cos \phi \frac{d\Phi}{d\phi} \right) + l\,(l+1)\,\Phi \cos^2 \phi = m^2 \Phi$$

$$\cos \phi \left(- \sin \phi \frac{d\Phi}{d\phi} + \cos \phi \frac{d^2\Phi}{d\phi^2} \right) + l\,(l+1)\,\Phi \cos^2 \phi = m^2 \Phi$$

$$\cos^2 \phi\, \Phi'' - \sin \phi \cos \phi\, \Phi' + \left[l\,(l+1) \cos^2 \phi - m^2 \right] \Phi = 0. \tag{7.104}$$

The general solution of Eq. 7.103 is

$$\Lambda = C_1 \cos (m\lambda) + C_2 \sin (m\lambda). \tag{7.105}$$

Equation 7.104 is known as Legendre's associated equation, and the solution Φ corresponding to a particular pair of subscripts l and m is called the Legendre associated function $P_{lm} (\sin \phi)$. The specific formula is

$$P_{lm} (\sin \phi) = \cos^m \phi \sum_{w=0}^{k} T_{lmw} \sin^{l-m-2w} \phi, \tag{7.106}$$

where k is the integer part of $(l - m)/2$ and

$$T_{lmw} = \frac{(-1)^w\,(2l - 2w)!}{2^l\,w!\,(l - w)!\,(l - m - 2w)!}. \tag{7.107}$$

Equations 7.101, 7.105, and 7.106 provide a particular solution for Laplace's equation (Eq. 7.96), that is

$$V_{lm} = r^{-(l+1)} \left[C_1 \cos (m\lambda) + C_2 \sin (m\lambda) \right] P_{lm} (\sin \phi). \tag{7.108}$$

Since any linear combination of solutions of Eq. 7.96 also is a solution, we take the most general formulation of the solution to be

$$V = \sum_{l=0}^{\infty} \sum_{m=0}^{l} \frac{1}{r^{l+1}} P_{lm} (\sin \phi) \left[C_{lm} \cos m\lambda + S_{lm} \sin m\lambda \right]. \tag{7.109}$$

7.4 Gravitational Potential in Terms of Keplerian Elements

Following Kaula (1966), we can convert the spherical harmonic potential (given by
Eq. 7.109) into Keplerian elements. We make use of some trigonometric identities

$$\cos mx = \Re\exp(mjx) = \Re\left(\cos x + j\sin x\right)^m$$

$$= \Re\sum_{s=0}^{m}\binom{m}{s}j^s\cos^{m-s}x\sin^s x, \qquad (7.110)$$

where $\Re$ indicates the real part, $j = \sqrt{-1}$, and $\binom{m}{s}$ is the binomial coefficient:

$$\binom{m}{s} = \frac{m!}{s!(m-s)!}, \qquad (7.111)$$

$$\sin mx = \Re\left[-j\exp(mjx)\right] = \Re\left[-j\left(\cos x + j\sin x\right)^m\right]$$

$$= \Re\sum_{s=0}^{m}\binom{m}{s}j^{s-1}\cos^{m-s}x\sin^s x, \qquad (7.112)$$

$$\sin^a x\cos^b x = \left[-\frac{j}{2}\left(e^{jx}-e^{-jx}\right)\right]^a\left[\frac{1}{2}\left(e^{jx}+e^{-jx}\right)\right]^b$$

$$= \frac{(-1)^a j^a}{2^a}\sum_{c=0}^{a}\binom{a}{c}e^{(a-c)jx}(-1)^c e^{-cjx}\cdot\frac{1}{2^b}\sum_{d=0}^{b}\binom{b}{d}\cdot e^{(b-d)jx}e^{-djx}$$

$$= \frac{(-1)^a j^a}{2^{a+b}}\sum_{c=0}^{a}\sum_{d=0}^{b}\binom{a}{c}\binom{b}{d}e^{(a+b-2c-2d)jx}(-1)^c$$

$$= \frac{(-1)^a j^a}{2^{a+b}}\sum_{c=0}^{a}\sum_{d=0}^{b}\binom{a}{c}\binom{b}{d}(-1)^c$$

$$\times\left[\cos\left(a+b-2c-2d\right)x + j\sin\left(a+b-2c-2d\right)x\right], \qquad (7.113)$$

$$\cos a\cos b = \frac{1}{2}\cos\left(a+b\right) + \frac{1}{2}\cos\left(a-b\right)$$

$$\sin a\sin b = -\frac{1}{2}\cos\left(a+b\right) + \frac{1}{2}\cos\left(a-b\right)$$

$$\sin a \cos b = \frac{1}{2} \sin (a + b) + \frac{1}{2} \sin (a - b) \tag{7.114}$$

$$\cos a \sin b = \frac{1}{2} \sin (a + b) - \frac{1}{2} \sin (a - b).$$

Let a particular term of Eq. 7.109 be

$$V_{lm} = \frac{\mu r_e^l}{r^{l+1}} P_{lm} (\sin \phi) (C_{lm} \cos m\lambda + S_{lm} \sin m\lambda). \tag{7.115}$$

We have made the C_{lm}, S_{lm} non-dimensional by applying the factor μr_e^l, where r_e is the equatorial radius of the Earth. We then substitute $[m (\alpha - \Omega) + m (\Omega - \theta)]$ for $m\lambda$, where α is the right ascension of the satellite and θ is the Greenwich Sidereal Time:

$$\cos m\lambda = \cos m (\alpha - \Omega) \cos m (\Omega - \theta) - \sin m (\alpha - \Omega) \sin m (\Omega - \theta)$$

$$\sin m\lambda = \sin m (\alpha - \Omega) \cos m (\Omega - \theta) + \cos m (\alpha - \Omega) \sin m (\Omega - \theta). \tag{7.116}$$

Referring to Fig. 7.4, in the spherical triangle formed by the orbit, the equator, and the satellite meridian, we have

$$\cos (\omega + f) = \cos (\alpha - \Omega) \cos \phi + \sin (\alpha - \Omega) \sin \phi \cos (\pi/2)$$

$$\cos \phi = \cos (\omega + f) \cos (\alpha - \Omega) + \sin (\omega + f) \sin (\alpha - \Omega) \cos i,$$

so that

$$\cos (\alpha - \Omega) = \frac{\cos (\omega + f)}{\cos \phi}$$

$$\sin (\alpha - \Omega) = \frac{\sin (\omega + f) \cos i}{\cos \phi} \tag{7.117}$$

and

$$\sin \phi = \sin i \sin (\omega + f). \tag{7.118}$$

If we apply Eqs. 7.110 and 7.112 to the $(\alpha - \Omega)$ functions in Eq. 7.116 and substitute Eq. 7.117, we obtain

$$\cos m\lambda = \Re \sum_{s=0}^{m} \binom{m}{s} j^s \frac{\cos^{m-s} (\omega + f) \sin^s (\omega + f) \cos^s i}{\cos^m \phi}$$

$$\times [\cos m (\Omega - \theta) + j \sin m (\Omega - \theta)]$$

$$\sin m\lambda = \Re \sum_{s=0}^{m} \binom{m}{s} j^s \frac{\cos^{m-s}(\omega + f) \sin^s (\omega + f) \cos^s i}{\cos^m \phi} \tag{7.119}$$

$$\times \left[\sin m (\Omega - \theta) - j \cos m (\Omega - \theta) \right].$$

If we substitute Eq. 7.118 for $\sin \phi$ in Eq. 7.106 and then substitute Eqs. 7.106 and 7.119 into Eq. 7.115, by canceling out the $\cos^m \phi$ terms, we have

$$V_{lm} = \frac{\mu r_e^l}{r^{l+1}} \sum_{w=0}^{k} T_{lmw} \sin^{l-m-2w} i$$

$$\times \Re \left[(C_{lm} - j S_{lm}) \cos m (\Omega - \theta) + (S_{lm} + j C_{lm}) \sin m (\Omega - \theta) \right]$$

$$\times \sum_{s=0}^{m} \binom{m}{s} j^s \sin^{l-m-2w+s}(\omega + f) \cos^{m-s}(\omega + f) \cos^s i, \tag{7.120}$$

where k is the integer part of $(l - m)/2$.

Applying Eq. 7.113 in Eq. 7.120, with $a = l - m - 2w + s$, and $b = m - s$, we get

$$V_{lm} = \frac{\mu r_e^l}{r^{l+1}} \sum_{w=0}^{k} T_{lmw} \sin^{l-m-2w} i$$

$$\times \Re \left[(C_{lm} - j S_{lm}) \cos m (\Omega - \theta) + (S_{lm} + j C_{lm}) \sin m (\Omega - \theta) \right]$$

$$\times \sum_{s=0}^{m} \binom{m}{s} j^s \cos^s i \frac{(-j)^{l-m-2w+s}}{2^{l-2w}} \sum_{c=0}^{l-m-2w+s} \sum_{d=0}^{m-s} \binom{l-m-2w+s}{c} \binom{m-s}{d}$$

$$\times (-1)^c \left[\cos(l - 2w - 2c - 2d)(\omega + f) + j \sin(l - 2w - 2c - 2d)(\omega + f) \right]. \tag{7.121}$$

We apply Eq. 7.114 to the products of $(\Omega - \theta)$ and $(\omega + f)$ trigonometric functions in Eq. 7.121 and drop any term with an odd power of j as a coefficient (since V_{lm} is real, any such term has another term canceling it out), to obtain

$$V_{lm} = \frac{\mu r_e^l}{r^{l+1}} \sum_{w=0}^{k} T_{lmw} \sin^{l-m-2w} i \, (-1)^{k+w} \sum_{s=0}^{m} \binom{m}{s} \frac{\cos^s i}{2^{l-2w}}$$

$$\times \sum_{c=0}^{l-m-2w+s} \sum_{d=0}^{m-s} \binom{l-m-2w+s}{c} \binom{m-s}{d} (-1)^c$$

$$\times \left\{ \left[\begin{array}{c} C_{lm} \\ -S_{lm} \end{array} \right]_{l-m \text{ odd}}^{l-m \text{ even}} \cos \left[(l - 2w - 2c - 2d)(\omega + f) + m (\Omega - \theta) \right] \right.$$

$$\left. + \left[\begin{array}{c} S_{lm} \\ C_{lm} \end{array} \right]_{l-m \text{ odd}}^{l-m \text{ even}} \sin \left[(l - 2w - 2c - 2d)(\omega + f) + m (\Omega - \theta) \right] \right\}.$$

$$(7.122)$$

Next, we transform Eq. 7.122 so that terms of the same argument $[(l - 2p)(\omega + f) + m (\Omega - \theta)]$ are collected together. Substituting p for $(w + c + d)$ necessitates the elimination of one subscript from the factors. We opt to use $p - w - c$ in place of d. The limits of the d summation place limits on the possible values of c, which arc those that make the binomial coefficients nonzero. In addition, $w \le p$.

The expression for V_{lm} becomes

$$V_{lm} = \frac{\mu r_e^l}{r^{l+1}} \sum_{p=0}^{l} F_{lmp}(i) \left\{ \left[\begin{array}{c} C_{lm} \\ -S_{lm} \end{array} \right]_{l-m \text{ odd}}^{l-m \text{ even}} \cos \left[(l - 2p)(\omega + f) + m (\Omega - \theta) \right] \right.$$

$$\left. + \left[\begin{array}{c} S_{lm} \\ C_{lm} \end{array} \right]_{l-m \text{ odd}}^{l-m \text{ even}} \sin \left[(l - 2p)(\omega + f) + m (\Omega - \theta) \right] \right\}, \qquad (7.123)$$

where, substituting from Eq. 7.107

$$T_{lmw} = \frac{(-1)^w (2l - 2w)!}{2^l w! (l - w)! (l - m - 2w)!},$$

we have

$$F_{lmp}(i) = \sum_w \frac{(2l - 2w)!}{w! (l - w)! (l - m - 2w)! 2^{2l-2w}} \sin^{l-m-2w} i$$

$$\times \sum_{s=0}^{m} \binom{m}{s} \cos^s i \sum_c \binom{l - m - 2w + s}{c} \binom{m - s}{p - w - c} (-1)^{c-k}.$$

$$(7.124)$$

Here, k is the integer part of $(l - m)/2$, w is summed from 0 to the lesser of p or k, and c is summed over all values for which the binomial coefficients are nonzero.

The final transformation necessary to obtain a disturbing function in terms of Keplerian orbital elements is to replace r and f in Eq. 7.123 by a, M, and e. We make the replacement

$$\frac{1}{r^{l+1}} \begin{bmatrix} \cos \\ \sin \end{bmatrix} [(l - 2p)(\omega + f) + m(\Omega - \theta)]$$

$$= \frac{1}{a^{l+1}} \sum_q G_{lpq}(e) \begin{bmatrix} \cos \\ \sin \end{bmatrix} [(l - 2p)\omega + (l - 2p + q)M + m(\Omega - \theta)].$$

The development of $G_{lpq}(e)$ is very complicated, and Kaula (1966) quotes the result of one solution, published by Tisserand in 1889:

$$G_{lpq}(e) = (-1)^{|q|}\left(1 + \Gamma^2\right)^l \Gamma^{|q|} \sum_{k=0}^{\infty} P_{lpqk} Q_{lpqk} \Gamma^{2k}, \tag{7.125}$$

where

$$\Gamma = \frac{e}{1 + \sqrt{1 - e^2}}$$

$$P_{lpqk} = \sum_{r=0}^{\kappa} \binom{2p' - 2l}{\kappa - r} \frac{(-1)^r}{r!} \left(\frac{(l - 2p' + q')e}{2\Gamma}\right)^r, \tag{7.126}$$

$$\kappa = k + q', \quad q' > 0; \quad \kappa = k, \quad q' < 0;$$

and

$$Q_{lpqk} = \sum_{r=0}^{\kappa} \binom{-2p'}{\kappa - r} \frac{1}{r!} \left(\frac{(l - 2p' + q')e}{2\Gamma}\right)^r, \tag{7.127}$$

$$\kappa = k, \quad q' > 0; \quad \kappa = k - q', \quad q' < 0;$$

$$p' = p, \quad q' = q \text{ for } p \leq l/2; \quad p' = l - p, \quad q' = -q \text{ for } p > l/2.$$

The final result for the transformation of V_{lm} in spherical coordinates (Eq. 7.115) into orbital elements is expressed by

$$V_{lm} = \frac{\mu r_e^l}{a^{l+1}} \sum_{p=0}^{l} F_{lmp}(i) \sum_{q=-\infty}^{\infty} G_{lpq}(e) S_{lmpq}(\omega, M, \Omega, \theta), \tag{7.128}$$

where

$$
S_{lmpq} = \begin{bmatrix} C_{lm} \\ -S_{lm} \end{bmatrix}_{\substack{l-m \text{ even} \\ l-m \text{ odd}}} \cos\left[(l-2p)\,\omega + (l-2p+q)\,M + m\,(\Omega - \theta)\right]
$$
$$
+ \begin{bmatrix} S_{lm} \\ C_{lm} \end{bmatrix}_{\substack{l-m \text{ even} \\ l-m \text{ odd}}} \sin\left[(l-2p)\,\omega + (l-2p+q)\,M + m\,(\Omega - \theta)\right].
$$

$$(7.129)$$

The total spherical harmonic potential is given by the double summation in Eq. 7.109. A particular term V_{lm} of the total spherical harmonic potential, as shown in Eq. 7.128, yields insight into effects of particular harmonics on the long-term motion of an orbit. Chao and Hoots (2018) list tables for the inclination functions, F, and eccentricity functions, G, which are reproduced here for ease of reference. The subscript indices, l, m, p, and q, are integers that identify the terms in the F and G functions for a particular harmonic term (l, m).

The series of a particular G term is identified by the integer indexes of the subscript, l, p, and q, or $G_{lpq}(e)$. The first index, l, is determined by the particular pair of gravity harmonic coefficients, C_{lm} and S_{lm}; the second index, p, can be a positive integer or zero, and p is less than or equal to l. The third index, q, can be negative or positive, and its magnitude determines the power of eccentricity of the first term of that infinite series. Symmetry exists in most of the terms, and some terms are represented in closed-form functions of eccentricity, including the following:

$$
G_{210} = \left(1 - e^2\right)^{-3/2}
$$
$$
G_{31-1} = G_{321} = e\left(1 - e^2\right)^{-5/2}
$$
$$
G_{41-2} = G_{432} = \left(3e^2/4\right)\left(1 - e^2\right)^{-7/2} \qquad (7.130)
$$
$$
G_{420} = \left(1 + 3e^2/2\right)\left(1 - e^2\right)^{-7/2}.
$$

Table 7.1 provides values for the eccentricity function, $G_{lpq}(e)$, with lpq up to 664 and eccentricity up to e^4, from Chao and Hoots (2018).

Table 7.2 provides values for the inclination function, $F_{lmp}(i)$, with lmp up to 555, from Chao and Hoots (2018).

Table 7.1 Eccentricity functions

l	p	q	l	p	q	$G_{lpq}(e)$
2	0	-2	2	2	2	0
2	0	-1	2	2	1	$-e/2 + e^3/16 + \dots$
2	0	0	2	2	0	$1 - 5e^2/2 + 13e^4/16 + \dots$
2	0	1	2	2	-1	$7e/2 - 123e^3/16 + \dots$
2	0	2	2	2	-2	$17e^2/2 - 115e^4/6 + \dots$
2	1	-2	2	1	2	$9e^2/4 + 7e^4/4 + \dots$
2	1	-1	2	1	1	$3e/2 + 27e^3/16 + \dots$
			2	1	0	$(1 - e^2)^{-3/2}$
3	0	-2	3	3	2	$e^2/8 + e^4/48 + \dots$
3	0	-1	3	3	1	$-e + 5e^3/4 + \dots$
3	0	0	3	3	0	$1 - 6e^2 + 423e^4/64 + \dots$
3	0	1	3	3	-1	$5e - 22e^3 + \dots$
3	0	2	3	3	-2	$127e^2/8 - 3065e^4/48 + \dots$
3	1	-2	3	2	2	$11e^2/8 + 49e^4/16 + \dots$
3	1	-1	3	2	1	$e(1 - e^2)^{-5/2}$
3	1	0	3	2	0	$1 + 2e^2 + 239e^4/64 + \dots$
3	1	1	3	2	-1	$3e + 11e^3/4 + \dots$
3	1	2	3	2	-2	$53e^2/8 + 39e^4/16 + \dots$
4	0	-2	4	4	2	$e^2/2 - e^4/3 + \dots$
4	0	-1	4	4	1	$-3e/2 + 75e^3/16 + \dots$
4	0	0	4	4	0	$1 - 11e^2 + 199e^4/8 + \dots$
4	0	1	4	4	-1	$13e/2 - 765e^3/16 + \dots$
4	0	2	4	4	-2	$51e^2/2 - 321e^4/2 + \dots$
4	0	3	4	4	-3	$3751e^3/48 + \dots$
4	0	4	4	4	-4	$4943e^4/24 + \dots$
4	1	-4	4	3	4	$67e^4/48 + \dots$
4	1	-3	4	3	3	$49e^3/48 + \dots$
4	1	-2	4	3	2	$(3e^2/4)(1 - e^2)^{-7/2}$
4	1	-1	4	3	1	$e/2 + 33e^3/16 + \dots$
4	1	0	4	3	0	$1 + e^2 + 65e^4/16 + \dots$
4	1	1	4	3	-1	$9e/2 - 3e^3/16 + \dots$
4	1	2	4	3	-2	$53e^2/4 - 179e^4/24 + \dots$
4	1	3	4	3	-3	$1541e^3/48 + \dots$
4	1	4	4	3	-4	$555e^4/8 + \dots$
4	2	-4	4	2	4	$745e^4/48 + \dots$
4	2	-3	4	2	3	$145e^3/16 + \dots$
4	2	-2	4	2	2	$5e^2 + 155e^4/12 + \dots$
4	2	-1	4	2	1	$5e/2 + 135e^3/16 + \dots$
			4	2	0	$(1 + 3e^2/2)(1 - e^2)^{-7/2}$
5	5	4	5	0	-4	$e^4/384 + \dots$
5	5	3	5	0	-3	$-e^3/6 + \dots$

(continued)

Table 7.1 (continued)

l	p	q	l	p	q	$G_{lpq}(e)$
5	5	2	5	0	−2	$9e^2/8 - 9e^4/4 + \ldots$
5	5	1	5	0	−1	$-2e + 23e^3/2 + \ldots$
5	5	0	5	0	0	$1 - 35e^2/2 + 4255e^4/64 + \ldots$
5	5	−1	5	0	1	$8e - 177e^3/2 + \ldots$
5	5	−2	5	0	2	$299e^2/8 - 4067e^4/12 + \ldots$
5	5	−3	5	0	3	$799e^3/6 + \ldots$
5	5	−4	5	0	4	$51,287e^4/128 + \ldots$
5	4	4	5	1	−4	$87e^4/128 + \ldots$
5	4	3	5	1	−3	$e^3/2 + \ldots$
5	4	2	5	1	−2	$3e^2/8 + 7e^4/4 + \ldots$
5	4	1	5	1	−1	$3e^3/2 + \ldots$
5	4	0	5	1	0	$1 - 3e^2/2 + 303e^4/64 + \ldots$
5	4	−1	5	1	1	$6e - 21e^3/2 + \ldots$
5	4	−2	5	1	2	$177e^2/8 - 177e^4/4 + \ldots$
5	4	−3	5	1	3	$129e^3/2 + \ldots$
5	4	−4	5	1	4	$20,875e^4/128 + \ldots$
5	3	4	5	2	−4	$1291e^4/128 + \ldots$
5	3	3	5	2	−3	$37e^3/6 + \ldots$
5	3	2	5	2	−2	$29e^2/8 + 193e^4/12 + \ldots$
5	3	1	5	2	−1	$2e + 21e^3/2 + \ldots$
5	3	0	5	2	0	$1 + 13e^2/2 + 1399e^4/64 + \ldots$
5	3	−1	5	2	1	$4e + 29e^3/2 + \ldots$
5	3	−2	5	2	2	$87e^2/8 + 107e^4/4 + \ldots$
5	3	−3	5	2	3	$149e^3/6 + \ldots$
5	3	−4	5	2	4	$19,669e^4/384 + \ldots$
6	6	4	6	0	−4	$e^4/24 + \ldots$
6	6	3	6	0	−3	$-9e^3/16 + \ldots$
6	6	2	6	0	−2	$2e^2 - 23e^4/3 + \ldots$
6	6	1	6	0	−1	$-5e/2 + 365e^3/16 + \ldots$
6	6	0	6	0	0	$1 - 51e^2/2 + 2331e^4/16 + \ldots$
6	6	−1	6	0	1	$19e/2 - 2359e^3/16 + \ldots$
6	6	−2	6	0	2	$103e^2/2 - 1907e^4/3 + \ldots$
6	6	−3	6	0	3	$3347e^3/16 + \ldots$
6	6	−4	6	0	4	$33,965e^4/48 + \ldots$
6	5	4	6	1	−4	$5e^4/16 + \ldots$
6	5	3	6	1	−3	$11e^3/48 + \ldots$
6	5	2	6	1	−2	$e^2/4 + e^4 + \ldots$
6	5	1	6	1	−1	$-e/2 + 31e^3/16 + \ldots$

(continued)

Table 7.1 (continued)

l	p	q	l	p	q	$G_{lpq}(e)$
6	5	0	6	1	0	$1 - 11e^2/2 + 11e^4 + \ldots$
6	5	-1	6	1	1	$15e/2 - 505e^3/16 + \ldots$
6	5	-2	6	1	2	$133e^2/4 - 131e^4 + \ldots$
6	5	-3	6	1	3	$5443e^3/48 + \ldots$
6	5	-4	6	1	4	$5259e^4/16 + \ldots$
6	4	4	6	2	-4	$101e^4/16 + \ldots$
6	4	3	6	2	-3	$193e^3/48 + \ldots$
6	4	2	6	2	-2	$5e^2/2 + 15e^4 + \ldots$
6	4	1	6	2	-1	$3e/2 + 161e^3/16 + \ldots$
6	4	0	6	2	0	$1 + 13e^2/2 + 419e^4/16 + \ldots$
6	4	-1	6	2	1	$11e/2 + 277e^3/16 + \ldots$
6	4	-2	6	2	2	$19e^2 + 35e^4 + \ldots$
6	4	-3	6	2	3	$2525e^3/48 + \ldots$
6	4	-4	6	2	4	$255e^4/2 + \ldots$
6	3	4	6	3	-4	$889e^4/24 + \ldots$
6	3	3	6	3	-3	$301e^3/16 + \ldots$
6	3	2	6	3	-2	$35e^2/4 + 133e^4/3 + \ldots$
6	3	1	6	3	-1	$7e/2 + 371e^3/16 + \ldots$
6	3	0	6	3	0	$1 + 21e^2/2 + 189e^4/4 + \ldots$

Table 7.2 Inclination functions

l	m	p	$F_{lmp}(i)$; $S = \sin i, C = \cos i$
2	0	0	$-0.375S^2$
2	0	1	$0.75S^2 - 0.5$
2	0	2	$-0.375S^2$
2	1	0	$0.75S(1 + C)$
2	1	1	$-1.5SC$
2	1	2	$-0.75S(1 - C)$
2	2	0	$0.75(1 + C)^2$
2	2	1	$1.5S^2$
2	2	2	$0.75(1 - C)^2$
3	0	0	$-0.3125S^3$
3	0	1	$0.9375S^3 - 0.75S$
3	0	2	$-0.9375S^3 + 0.75S$
3	0	3	$0.3125S^3$
3	1	0	$-0.9375S^2(1 + C)$
3	1	1	$0.9375S^2(1 + 3C) - 0.75(1 + C)$
3	1	2	$0.9375S^2(1 - 3C) - 0.75(1 - C)$
3	1	3	$-0.9375S^2(1 - C)$
3	2	0	$1.875S(1 + C)^2$
3	2	1	$1.875S(1 - 2C - 3C^2)$
3	2	2	$-1.875S(1 + 2C - 3C^2)$
3	2	3	$-1.875S(1 - C)^2$
3	3	0	$1.875(1 + C)^3$
3	3	1	$5.625S^2(1 + C)$
3	3	2	$5.625S^2(1 - C)$
3	3	3	$1.875(1 - C)^3$
4	0	0	$0.27344S^4$
4	0	1	$-1.09375S^4 + 0.9375S^2$
4	0	2	$1.64063S^4 - 1.875S^2 + 0.375$
4	0	3	$-1.09375S^4 + 0.9375S^2$
4	0	4	$0.27344S^4$
4	1	0	$-1.09375S^3(1 + C)$
4	1	1	$2.1875S^3(1 + 2C) - 1.875(1 + C)S$
4	1	2	$C(3.75S - 6.5625S^3)$
4	1	3	$-2.1875S^3(1 - 2C) + 1.875(1 - C)S$
4	1	4	$1.09375S^3(1 - C)$
4	2	0	$-3.28125S^2(1 + C)^2$
4	2	1	$13.125S^2C(1 + C) - 1.875(1 + C)^2$
4	2	2	$6.5625S^2(1 - 3C^2) - 3.75S^2$
4	2	3	$-13.125S^2C(1 - C) - 1.875(1 - C)^2$
4	2	4	$-3.28125S^2(1 - C)^2$
4	3	0	$6.5625S(1 + C)^3$
4	3	1	$13.125S(1 - 3C^2 - 2C^3)$
4	3	2	$-39.375S^3C$

(continued)

Table 7.2 (continued)

l	m	p	$F_{lmp}(i)$; $S = \sin i, C = \cos i$
4	3	3	$-13.125S(1 - 3C^2 + 2C^3)$
4	3	4	$-6.5625S(1 - C)^3$
4	4	0	$6.5625(1 + C)^4$
4	4	1	$26.25S^2(1 + C)^2$
4	4	2	$39.375S^4$
4	4	3	$26.25S^2(1 - C)^2$
4	4	4	$6.5625(1 - C)^4$
5	0	0	$0.24609S^5$
5	0	1	$1.09375S^3 - 1.23047S^5$
5	0	2	$0.93750S - 3.28125S^3 + 2.46094S^5$
5	0	3	$-0.93750S + 3.28125S^3 - 2.46094S^5$
5	0	4	$-1.09375S^3 + 1.23047S^5$
5	0	5	$-0.24609S^5$
5	1	0	$1.23047S^4(1 + C)$
5	1	1	$3.28125S^2(1 + C) - 6.15234S^4(0.6 + C)$
5	1	2	$0.93750(1 + C) - 3.28125S^2(1 + 3C) + 2.46094S^4(1 + 5C)$
5	1	3	$0.93750(1 - C) - 3.28125S^2(1 - 3C) + 2.46094S^4(1 - 5C)$
5	1	4	$3.28125S^2(1 - C) - 6.15234S^4(0.6 - C)$
5	1	5	$1.23047S^4(1 - C)$
5	2	0	$-4.92188S^3(1 + C)^2$
5	2	1	$-6.56250S(1 + C)^2 + 4.92188S^3(1 + 6C + 5C^2)$
5	2	2	$-6.56250S(1 - 2C - 3C^2) + 9.84375S^3(1 - 2C - 5C^2)$
5	2	3	$6.56250S(1 + 2C - 3C^2) - 9.84375S^3(1 + 2C - 5C^2)$
5	2	4	$6.56250S(1 - C)^2 - 4.92188S^3(1 - 6C + 5C^2)$
5	2	5	$4.92188S^3(1 - C)^2$
5	3	0	$-14.76563S^2(1 + C)^3$
5	3	1	$-6.56250(1 + C)^3 - 14.76563S^2(1 - 3C - 9C^2 - 5C^3)$
5	3	2	$-19.68750(1 + C - C^2 - C^3) + 29.53125S^2(1 + 3C - 3C^2 - 5C^3)$
5	3	3	$-19.68750(1 - C - C^2 + C^3) + 29.53125S^2(1 - 3C - 3C^2 + 5C^3)$
5	3	4	$-6.56250(1 - C)^3 - 14.76563S^2(1 + 3C - 9C^2 + 5C^3)$
5	3	5	$-14.76563S^2(1 - C)^3$
5	4	0	$29.53125S(1 + C)^4$
5	4	1	$88.59375S(1 + 1.33333C - 2C^2 - 4C^3 - 1.66666C^4)$
5	4	2	$59.0625S(1 - 4C - 6C^2 + 4C^3 + C^4)$
5	4	3	$-59.0625S(1 + 4C - 6C^2 - 4C^3 + 5C^4)$
5	4	4	$-88.59375S(1 - 1.33333C - 2C^2 + 4C^3 - 1.66666C^4)$
5	4	5	$-29.53125S(1 - C)^4$
5	5	0	$29.53125(1 + C)^5$
5	5	1	$147.65625(1 + 3C + 2C^2 - 2C^3 - 3C^4 - C^5)$
5	5	2	$295.31251(1 + C - 2C^2 - 2C^3 + C^4 + C^5)$
5	5	3	$295.31251(1 - C - 2C^2 + 2C^3 + 3C^4 - C^5)$
5	5	4	$147.65625(1 - 3C + 2C^2 + 2C^3 - 3C^4 + C^5)$
5	5	5	$29.53125(1 - C)^5$

References

Chao, C.-C., & Hoots, F. R. (2018). *Applied orbit perturbations and maintenance* (2nd ed.). El Segundo: The Aerospace Press.

Fitzpatrick, P. M. (1970). *Principles of celestial mechanics*. New York: Academic Press.

Kaula, W. M. (1966). *Theory of satellite geodesy*. Waltham: Blaisdell Publishing Company.

McCuskey, S. W. (1963). *Introduction to celestial mechanics*. Reading: Addison-Wesley Publishing Company, Inc.

Roy, A. E. (2005). *Orbital motion* (4th ed.). New York: Taylor & Francis Group.

Chapter 8
The Generalized Method of Averaging and Application

In this chapter, we bring together multiple elements to enable solving an important orbital perturbation problem—the motion of a satellite due to the oblate shape of the Earth. In Chap. 7, we developed a model of the Earth gravitational potential including the representation of the first three zonal harmonics. From Eq. 7.95, we select the first term, which contains the effect due to the second zonal harmonic

$$R = \mu \frac{3 J_2 r_e^2}{2 a^3} \left(\frac{a}{r}\right)^3 \left[\frac{1}{3} - \frac{1}{2} \sin^2 i + \frac{1}{2} \sin^2 i \cos 2u\right], \tag{8.1}$$

where we have replaced J with $3 J_2/2$ as the more conventional notation.

In Chap. 5, we developed the Lagrange planetary equations in terms of a perturbing force arising from a geopotential. Using our geopotential model in the perturbation equations brings us to a system of six first-order nonlinear differential equations. In the following section, we describe a general process that can be used to transform that system to a simpler form that may allow finding an approximate solution. The general process is often referred to as the method of averaging (MOA).

8.1 The Concept of Averaging

The MOA is sometimes misunderstood and misused, leading to incorrect or incomplete results. One can certainly invoke a term called averaging, but then one also must invent a *consistent methodology for applying it*. Some practitioners assume that one can simply compute an average of the right-hand side of the Lagrange equations and that gives them the secular rate. As we demonstrate below, we can get four different answers for the "average" value of r and argue that each of them represents the "average."

Case 1: The "average" with respect to time

$$\langle r \rangle = \frac{1}{P} \int_0^P r \, dt$$

$$= \frac{1}{P} \int_0^P r \left(\frac{r}{na} dE \right) = \frac{1}{2\pi} \frac{1}{a} \int_0^{2\pi} a^2 (1 - e \cos E)^2 dE$$

$$= \frac{1}{2\pi} a \int_0^{2\pi} \left(1 - 2e \cos E + \frac{1}{2} e^2 + \frac{1}{2} e^2 \cos 2E \right) dE$$

$$= a \left(1 + \frac{1}{2} e^2 \right).$$

Case 2: The "average" with respect to true anomaly

$$\langle r \rangle = \frac{1}{2\pi} \int_0^{2\pi} r \, df$$

$$= \frac{1}{2\pi} \int_0^{2\pi} r \frac{a\beta}{r} dE = \frac{1}{2\pi} \int_0^{2\pi} a\beta \, dE$$

$$= a\beta.$$

Case 3: The "average" with respect to eccentric anomaly

$$\langle r \rangle = \frac{1}{2\pi} \int_0^{2\pi} r \, dE$$

$$= \frac{1}{2\pi} \int_0^{2\pi} a (1 - e \cos E) \, dE = \frac{1}{2\pi} a [E - e \sin E]_0^{2\pi}$$

$$= a.$$

Case 4: The "average" with respect to mean anomaly

$$\langle r \rangle = \frac{1}{2\pi} \int_0^{2\pi} r \, dM$$

$$= \frac{1}{2\pi} \int_0^{2\pi} r \, (1 - e \cos E) \, dE = \frac{1}{2\pi} a \int_0^{2\pi} (1 - e \cos E)^2 dE$$

$$= \frac{1}{2\pi} a \left[E - 2e \sin E + \frac{1}{2} e^2 E + \frac{1}{4} e^2 \sin 2E \right]_0^{2\pi}$$

$$= a \left(1 + \frac{1}{2} e^2 \right).$$

Which is the correct "average"?

- Average with respect to time: $\langle r \rangle = a \left(1 + \frac{1}{2} e^2 \right)$
- Average with respect to true anomaly: $\langle r \rangle = a\beta$
- Average with respect to eccentric anomaly: $\langle r \rangle = a$
- Average with respect to mean anomaly: $\langle r \rangle = a \left(1 + \frac{1}{2} e^2 \right)$

The point of this illustration is that one must have a well-defined methodology to know what is being averaged, how to do it, and the justification for doing it. To answer these questions, we begin by forgetting for a moment about the term "method of averaging." Instead, we undertake an effort to find a *transformation of variables* that can simplify our set of first-order differential equations.

When we talk about perturbations, we are referring to small changes to the basic two-body motion. We often use the parameter ε to denote a term that is small relative to the two-body force. In the development of a solution for a perturbed two-body orbit, we frequently introduce a series expansion that may lead to terms of higher powers in the small parameter ε.

Subsequently, we must make a judgment concerning the maximum power of the small parameter for which the solution will be carried. The solution usually contains both periodic terms and secular terms. The periodic terms are bounded, but the effect of the secular terms will increase as prediction time increases. For this reason, it is customary to carry the solution development to one higher power of the small parameter for the secular terms than for the periodic terms. We follow that convention in this book.

Consider the following system of first-order differential equations where ε is a small parameter:

$$\dot{x} = \varepsilon \sin y \tag{8.2}$$

$$\dot{y} = x + \varepsilon \cos y. \tag{8.3}$$

Working through the details of this example will extend for a few pages. Therefore, it will be helpful if we provide a roadmap of the steps we are about to undertake. We begin by introducing a transformation of variables that is periodic in the new variables. For now we will not discuss how we conceived of this transformation, but rather we focus on what the transformation accomplishes. We will literally substitute the transformation into the differential equations (Eqs. 8.2 and 8.3). After the substitution, we will do a significant amount of algebra to arrive at equations for the time rate of change of the new transformed variables. After all that algebra, we will see that the differential equations for the new variables are trivial to solve. This example demonstrates that it is possible to find a transformation of variables that makes a given differential equation system simpler to solve. But what about all the algebra that we had to endure? And, by the way, where did that transformation come from? Well, the good news is that, after we finish the example, we will provide a recipe for both the transformation and the transformed differential equations. So, let us get on with working through this example.

We now introduce a transformation of variables from x, y to ξ, χ as follows:

$$x = \xi - \varepsilon \frac{1}{\xi} \cos \chi \tag{8.4}$$

$$y = \chi - \varepsilon \frac{1}{\xi^2} \sin \chi + \varepsilon \frac{1}{\xi} \sin \chi. \tag{8.5}$$

We want to determine the new differential equations resulting from this transformation. Applying the transformation in Eqs. 8.4 and 8.5 to our original differential equations 8.2 and 8.3, we have

$$\dot{x} = \varepsilon \sin y = \varepsilon \sin \left(\chi - \varepsilon \frac{1}{\xi^2} \sin \chi + \varepsilon \frac{1}{\xi} \sin \chi \right)$$

$$= \varepsilon \sin \chi \cos \left(\varepsilon \frac{1}{\xi^2} \sin \chi - \varepsilon \frac{1}{\xi} \sin \chi \right) - \varepsilon \cos \chi \sin \left(\varepsilon \frac{1}{\xi^2} \sin \chi - \varepsilon \frac{1}{\xi} \sin \chi \right).$$

Since ε is a small parameter, we can use the small angle approximation to write

$$\dot{x} \approx \varepsilon \sin \chi - \varepsilon^2 \frac{1}{\xi^2} \cos \chi \sin \chi + \varepsilon^2 \frac{1}{\xi} \cos \chi \sin \chi, \tag{8.6}$$

where we have dropped terms of order $O\left(\varepsilon^3\right)$. Similarly

$$\dot{y} = x + \varepsilon \cos y = \xi - \varepsilon \frac{1}{\xi} \cos \chi + \varepsilon \cos \left(\chi - \varepsilon \frac{1}{\xi^2} \sin \chi + \varepsilon \frac{1}{\xi} \sin \chi \right)$$

$$= \xi - \varepsilon \frac{1}{\xi} \cos \chi + \varepsilon \cos \chi \cos \left(\varepsilon \frac{1}{\xi^2} \sin \chi - \varepsilon \frac{1}{\xi} \sin \chi \right)$$

$$+ \varepsilon \sin \chi \sin \left(\varepsilon \frac{1}{\xi^2} \sin \chi - \varepsilon \frac{1}{\xi} \sin \chi \right)$$

$$\dot{y} \approx \xi - \varepsilon \frac{1}{\xi} \cos \chi + \varepsilon \cos \chi + \varepsilon^2 \frac{1}{\xi^2} \sin^2 \chi - \varepsilon^2 \frac{1}{\xi} \sin^2 \chi, \tag{8.7}$$

where we have dropped terms of order $O\left(\varepsilon^3\right)$.

Now we consider the transformation itself. By differentiating the transformation Eqs. 8.4 and 8.5, we get

$$\dot{x} = \dot{\xi} + \varepsilon\dot{\xi}\,\frac{1}{\xi^2}\cos\chi + \varepsilon\dot{\chi}\,\frac{1}{\xi}\sin\chi$$

$$\dot{y} = \dot{\chi} + \varepsilon\dot{\xi}\,\frac{2}{\xi^3}\sin\chi - \varepsilon\dot{\chi}\,\frac{1}{\xi^2}\cos\chi - \varepsilon\dot{\xi}\,\frac{1}{\xi^2}\sin\chi + \varepsilon\dot{\chi}\,\frac{1}{\xi}\cos\chi$$

or

$$\dot{x} = A\dot{\xi} + B\dot{\chi} \tag{8.8}$$

$$\dot{y} = C\dot{\xi} + D\dot{\chi}, \tag{8.9}$$

where we have introduced the shorthand notation

$$A = \left(1 + \varepsilon\frac{1}{\xi^2}\cos\chi\right)$$

$$B = \left(\varepsilon\frac{1}{\xi}\sin\chi\right)$$

$$C = \left(\varepsilon\frac{2}{\xi^3}\sin\chi - \varepsilon\frac{1}{\xi^2}\sin\chi\right) \tag{8.10}$$

$$D = \left(1 - \varepsilon\frac{1}{\xi^2}\cos\chi + \varepsilon\frac{1}{\xi}\cos\chi\right).$$

To solve Eqs. 8.8 and 8.9 for $\dot{\xi}$ and $\dot{\chi}$, we combine Eqs. 8.8 and 8.9 into the linear combinations

$$D\dot{x} - B\dot{y} = AD\dot{\xi} + BD\dot{\chi} - BC\dot{\xi} - BD\dot{\chi} = (AD - BC)\dot{\xi}$$

and

$$C\dot{x} - A\dot{y} = AC\dot{\xi} + BC\dot{\chi} - AC\dot{\xi} - AD\dot{\chi} = -(AD - BC)\dot{\chi}.$$

Rearranging, we have

$$\dot{\xi} = \frac{1}{(AD - BC)}(D\dot{x} - B\dot{y}) \tag{8.11}$$

and

$$\dot{\chi} = \frac{1}{(AD - BC)}(-C\dot{x} + A\dot{y}). \tag{8.12}$$

Using the definitions introduced in Eq. 8.10, we find

$$
AD - BC = \left(1 + \varepsilon \frac{1}{\xi^2} \cos \chi\right)\left(1 - \varepsilon \frac{1}{\xi^2} \cos \chi + \varepsilon \frac{1}{\xi} \cos \chi\right)
$$

$$
- \left(\varepsilon \frac{1}{\xi} \sin \chi\right)\left(\varepsilon \frac{2}{\xi^3} \sin \chi - \varepsilon \frac{1}{\xi^2} \sin \chi\right)
$$

$$
= 1 - \varepsilon \frac{1}{\xi^2} \cos \chi + \varepsilon \frac{1}{\xi} \cos \chi + \varepsilon \frac{1}{\xi^2} \cos \chi - \varepsilon^2 \frac{1}{\xi^4} \cos^2 \chi + \varepsilon^2 \frac{1}{\xi^3} \cos^2 \chi
$$

$$
- \varepsilon^2 \frac{2}{\xi^4} \sin^2 \chi + \varepsilon^2 \frac{1}{\xi^3} \sin^2 \chi
$$

$$
= 1 + \varepsilon \frac{1}{\xi} \cos \chi - \varepsilon^2 \frac{1}{\xi^4} - \varepsilon^2 \frac{1}{\xi^4} \sin^2 \chi + \varepsilon^2 \frac{1}{\xi^3}   \tag{8.13}
$$

and

$$
(D\dot{x} - B\dot{y}) = \left(1 - \varepsilon \frac{1}{\xi^2} \cos \chi + \varepsilon \frac{1}{\xi} \cos \chi\right)\dot{x} - \left(\varepsilon \frac{1}{\xi} \sin \chi\right)\dot{y}.   \tag{8.14}
$$

Using Eqs. 8.6 and 8.7 in Eq. 8.14, we find

$$
(D\dot{x} - B\dot{y}) = \left(1 - \varepsilon \frac{1}{\xi^2} \cos \chi + \varepsilon \frac{1}{\xi} \cos \chi\right)
$$

$$
\times \left(\varepsilon \sin \chi - \varepsilon^2 \frac{1}{\xi^2} \sin \chi \cos \chi + \varepsilon^2 \frac{1}{\xi} \sin \chi \cos \chi\right)
$$

$$
- \left(\varepsilon \frac{1}{\xi} \sin \chi\right)\left(\xi - \varepsilon \frac{1}{\xi} \cos \chi + \varepsilon \cos \chi + \varepsilon^2 \frac{1}{\xi^2} \sin^2 \chi - \varepsilon^2 \frac{1}{\xi} \sin^2 \chi\right)
$$

$$
= \varepsilon \sin \chi - \varepsilon^2 \frac{1}{\xi^2} \sin \chi \cos \chi + \varepsilon^2 \frac{1}{\xi} \sin \chi \cos \chi - \varepsilon^2 \frac{1}{\xi^2} \sin \chi \cos \chi
$$

$$
+ \varepsilon^2 \frac{1}{\xi} \sin \chi \cos \chi - \varepsilon \sin \chi + \varepsilon^2 \frac{1}{\xi^2} \sin \chi \cos \chi - \varepsilon^2 \frac{1}{\xi} \sin \chi \cos \chi
$$

$$
= - \varepsilon^2 \frac{1}{\xi^2} \sin \chi \cos \chi + \varepsilon^2 \frac{1}{\xi} \sin \chi \cos \chi,   \tag{8.15}
$$

where we have dropped terms of order $O\left(\varepsilon^3\right)$.

From the definitions in Eq. 8.10

$$
(-C\dot{x} + A\dot{y}) = - \left(\varepsilon \frac{2}{\xi^3} \sin \chi - \varepsilon \frac{1}{\xi^2} \sin \chi\right)\dot{x} + \left(1 + \varepsilon \frac{1}{\xi^2} \cos \chi\right)\dot{y}.   \tag{8.16}
$$

Using Eqs. 8.6 and 8.7 in Eq. 8.16, we find

$$(-C\dot{x} + A\dot{y}) = \left(\varepsilon \frac{2}{\xi^3} \sin \chi - \varepsilon \frac{1}{\xi^2} \sin \chi \right)$$

$$\times \left(-\varepsilon \sin \chi + \varepsilon^2 \frac{1}{\xi^2} \cos \chi \sin \chi - \varepsilon^2 \frac{1}{\xi} \cos \chi \sin \chi \right)$$

$$+ \left(1 + \varepsilon \frac{1}{\xi^2} \cos \chi \right)$$

$$\times \left(\xi - \varepsilon \frac{1}{\xi} \cos \chi + \varepsilon \cos \chi + \varepsilon^2 \frac{1}{\xi^2} \sin^2 \chi - \varepsilon^2 \frac{1}{\xi} \sin^2 \chi \right)$$

$$= -\varepsilon^2 \frac{2}{\xi^3} \sin^2 \chi + \varepsilon^2 \frac{1}{\xi^2} \sin^2 \chi + \xi - \varepsilon \frac{1}{\xi} \cos \chi + \varepsilon \cos \chi$$

$$+ \varepsilon^2 \frac{1}{\xi^2} \sin^2 \chi - \varepsilon^2 \frac{1}{\xi} \sin^2 \chi + \varepsilon \frac{1}{\xi} \cos \chi - \varepsilon^2 \frac{1}{\xi^3} \cos^2 \chi + \varepsilon^2 \frac{1}{\xi^2} \cos^2 \chi$$

$$= \xi + \varepsilon \cos \chi - \varepsilon^2 \frac{1}{\xi} \sin^2 \chi + \varepsilon^2 \frac{1}{\xi^2}$$

$$+ \varepsilon^2 \frac{1}{\xi^2} \sin^2 \chi - \varepsilon^2 \frac{1}{\xi^3} - \varepsilon^2 \frac{1}{\xi^3} \sin^2 \chi. \tag{8.17}$$

Returning to Eq. 8.11 and substituting Eqs. 8.13 and 8.15, we obtain

$$\dot{\xi} = \frac{1}{(AD - BC)} (D\dot{x} - B\dot{y})$$

$$= \frac{1}{\left(1 + \varepsilon \frac{1}{\xi} \cos \chi - \varepsilon^2 \frac{1}{\xi^4} - \varepsilon^2 \frac{1}{\xi^4} \sin^2 \chi + \varepsilon^2 \frac{1}{\xi^3} \right)}$$

$$\times \left(-\varepsilon^2 \frac{1}{\xi^2} \sin \chi \cos \chi + \varepsilon^2 \frac{1}{\xi} \sin \chi \cos \chi \right)$$

$$= 0, \tag{8.18}$$

where we have dropped terms of order $O\left(\varepsilon^2\right)$.

Substituting Eqs. 8.13 and 8.17 into Eq. 8.12, we obtain

$$\dot{\chi} = \frac{1}{(AD - BC)} (-C\dot{x} + A\dot{y})$$

$$= \frac{1}{\left(1 + \varepsilon \frac{1}{\xi} \cos \chi - \varepsilon^2 \frac{1}{\xi^4} - \varepsilon^2 \frac{1}{\xi^4} \sin^2 \chi + \varepsilon^2 \frac{1}{\xi^3} \right)}$$

$$\times \left[\xi + \varepsilon \cos \chi - \varepsilon^2 \frac{1}{\xi} \sin^2 \chi + \varepsilon^2 \frac{1}{\xi^2} + \varepsilon^2 \frac{1}{\xi^2} \sin^2 \chi - \varepsilon^2 \frac{1}{\xi^3} - \varepsilon^2 \frac{1}{\xi^3} \sin^2 \chi \right]$$

$$= \xi + \varepsilon \cos \chi - \varepsilon \cos \chi$$

$$= \xi, \tag{8.19}$$

where we have dropped terms of order $O\left(\varepsilon^2\right)$.

Thus, we have shown that we can start with the differential equations, Eqs. 8.2 and 8.3

$$\dot{x} = \varepsilon \sin y$$

$$\dot{y} = x + \varepsilon \cos y$$

and apply the transformation of variables in Eqs. 8.4 and 8.5

$$x = \xi - \varepsilon \frac{1}{\xi} \cos \chi$$

$$y = \chi - \varepsilon \frac{1}{\xi^2} \sin \chi + \varepsilon \frac{1}{\xi} \sin \chi$$

to arrive at the transformed differential equations, Eqs. 8.18 and 8.19:

$$\dot{\xi} = 0$$

$$\dot{\chi} = \xi.$$

It is now straightforward to determine the solution of the system of differential equations. Equations 8.18 and 8.19 are easily integrated to give

$$\xi = \xi_0$$

$$\chi = \chi_0 + \xi_0 t.$$

Having found a solution for the transformed variables, we can apply the transformation (Eqs. 8.4 and 8.5) to determine the solution in terms of the original variables

$$x = \xi_0 - \varepsilon \frac{1}{\xi_0} \cos \left(\chi_0 + \xi_0 t \right) \tag{8.20}$$

$$y = \left(\chi_0 + \xi_0 t \right) - \varepsilon \frac{1}{\xi_0^2} \sin \left(\chi_0 + \xi_0 t \right) + \varepsilon \frac{1}{\xi_0} \sin \left(\chi_0 + \xi_0 t \right). \tag{8.21}$$

Thus, we have demonstrated a transformation of variables that admits a ready solution of the transformed nonlinear system of differential equations. We now turn our attention to a methodology that allows us to create a transformation of variables that is guaranteed to simplify the transformed system of differential equations. Furthermore, the transformed differential equations can be developed through explicit formulas.

8.2 The Generalized Method of Averaging

The method of averaging (MOA) in orbit perturbation theories played a vital role in the development of several semianalytical orbit-propagation computer tools for satellite mission design and analysis over the past several decades. The widely used Draper Semianalytical Satellite Theory Standalone Orbit Propagator was developed by Cefola et al. of the Draper Laboratory (Cefola et al., 2014) utilizing the singly averaged equations in equinoctial elements. The recursive formulation allows efficient computation with good accuracy. The formulation of this method is well documented by Danielson et al. (1994). Other long-term orbit propagators applying the method of averaging include LOP (Long-term Orbit Predictor), developed by Kwok of the Jet Propulsion Laboratory (Kwok, 1986); SALT (Semi-Analytic Liu Theory), developed by Liu of Air Force Space Command (Liu and Alford, 1980); and GEOSYN and LIFETIME, developed by Chao et al. of The Aerospace Corporation (Chao and Hoots, 2018). These tools have been used by orbit analysts at various organizations for studying long-term orbit perturbations and station keeping.

To assure a consistent understanding of the MOA, it is beneficial to carefully present the mathematically rigorous basis for the method as well as a clear recipe for its application. The following development closely follows that of Morrison (1966). Briefly, the MOA applies to any system of first-order differential equations in $(\mathbf{x}, \mathbf{y})$ with the following properties:

(1) The differential equations are periodic functions of the variables $\mathbf{y}$, whose primary rate of change is a function only of the variables $\mathbf{x}$.
(2) All other rates of change of the variables are proportional to some small parameter ε.

This means that the rate of change of the $\mathbf{y}$ variables is a factor of $1/\varepsilon$ larger than that of the $\mathbf{x}$ variables. Hence, the $\mathbf{y}$ variables are called fast variables, while the $\mathbf{x}$ variables are called slow variables. Thus, we can assume that the slow $\mathbf{x}$ variables can be treated as constants when performing an integral with respect to the fast $\mathbf{y}$ variables. This property allows us to find a periodic transformation of variables in which the transformed differential equations are free of the $\mathbf{y}$ variables and therefore are less complex. Furthermore, the transformed differential equations can be integrated using a much larger step size (since they contain no fast-varying quantities) and may even yield an analytical solution. Once a solution is obtained, the periodics from the transformation can be included to obtain the complete solution. Note that the variables are not required to be canonical nor are the forces required to be conservative as they are in many other perturbation methods.

The MOA, like the von Zeipel method (von Zeipel, 1916) and the Lie series method (Steinberg, 1984), is a method for finding a *transformation of variables that simplifies the differential equations.* The word "averaging" simply refers to the means by which we determine the transformed differential equations. A similar process is used by von Zeipel and Lie.

We consider the system of first-order differential equations

$$\frac{dx_i}{dt} = \varepsilon f_i\,(\mathbf{x}, \mathbf{y}) \qquad\qquad i = 1, \ldots, n \qquad\qquad (8.22)$$

$$\frac{dy_\alpha}{dt} = \omega_\alpha\,(\mathbf{x}) + \varepsilon u_\alpha\,(\mathbf{x}, \mathbf{y}) \qquad \alpha = 1, \ldots, m, \qquad\qquad (8.23)$$

where f_i and u_α are periodic with period 2π in each component of $\mathbf{y}$, ε is a small parameter, and

$$f_i\,(\mathbf{x}, \mathbf{y}) = f_i^{(1)}\,(\mathbf{x}, \mathbf{y}) + \varepsilon f_i^{(2)}\,(\mathbf{x}, \mathbf{y}) + \cdots$$

$$u_\alpha\,(\mathbf{x}, \mathbf{y}) = u_\alpha^{(1)}\,(\mathbf{x}, \mathbf{y}) + \varepsilon u_\alpha^{(2)}\,(\mathbf{x}, \mathbf{y}) + \cdots,$$

where the superscript indicates the order of the term. We assume that $\omega_\alpha\,(\mathbf{x}) \neq 0$ leading to a rapid rotation of the $\mathbf{y}$ variables in comparison to the slower rate of change (proportional to ε) of the $\mathbf{x}$ variables.

We assume asymptotic expansions of the form

$$x_i = \xi_i + \varepsilon \eta_i^{(1)}\,(\boldsymbol{\xi}, \boldsymbol{\chi}) + \varepsilon^2 \eta_i^{(2)}\,(\boldsymbol{\xi}, \boldsymbol{\chi}) + \cdots \qquad\qquad (8.24)$$

$$y_\alpha = \chi_\alpha + \varepsilon \phi_\alpha^{(1)}\,(\boldsymbol{\xi}, \boldsymbol{\chi}) + \varepsilon^2 \phi_\alpha^{(2)}\,(\boldsymbol{\xi}, \boldsymbol{\chi}) + \cdots, \qquad\qquad (8.25)$$

where $\eta_i^{(j)}\,(\boldsymbol{\xi}, \boldsymbol{\chi})$ and $\phi_\alpha^{(j)}\,(\boldsymbol{\xi}, \boldsymbol{\chi})$ have period 2π and are unknown functions of $\boldsymbol{\xi}$ and $\boldsymbol{\chi}$ to be determined in such a way as to effect a simplification in the transformed dynamical system. It is desired that the transformed differential equations have the form

$$\frac{d\xi_i}{dt} = \varepsilon M_i^{(1)}\,(\boldsymbol{\xi}) + \varepsilon^2 M_i^{(2)}\,(\boldsymbol{\xi}) + \cdots \qquad\qquad (8.26)$$

$$\frac{d\chi_\alpha}{dt} = \omega_\alpha\,(\boldsymbol{\xi}) + \varepsilon \Omega_\alpha^{(1)}\,(\boldsymbol{\xi}) + \varepsilon^2 \Omega_\alpha^{(1)}\,(\boldsymbol{\xi}) + \cdots. \qquad\qquad (8.27)$$

We note that the symbols $M_i^{(j)}$ and $\Omega_\alpha^{(j)}$ should not be confused with mean anomaly and right ascension (or longitude) of the ascending node.

We refer to Eqs. 8.24 and 8.25 as the transformation of variables from $(\mathbf{x}, \mathbf{y})$ to $(\boldsymbol{\xi}, \boldsymbol{\chi})$ and refer to Eqs. 8.26 and 8.27 as the transformed, or averaged, equations since the right-hand sides depend only on $\boldsymbol{\xi}$. By substituting the transformation into the original differential equations and expanding in the small parameter ε, Morrison (1966) shows that we must have

$$M_i^{(1)} + \omega_\beta \frac{\partial \eta_i^{(1)}}{\partial \chi_\beta} = f_i^{(1)}(\xi, \chi) \tag{8.28}$$

$$\Omega_\alpha^{(1)} + \omega_\beta \frac{\partial \phi_\alpha^{(1)}}{\partial \chi_\beta} = \eta_j^{(1)} \frac{\partial \omega_\alpha}{\partial \xi_j} + u_\alpha^{(1)}(\xi, \chi). \tag{8.29}$$

Here and in all subsequent equations, the Einstein summation convention applies. The convention states that when an index occurs more than once in the same expression, the expression is implicitly summed over all possible values of the index.

We introduce the notation $\langle\,\rangle$ to indicate the average of a function, and the average operation is defined by

$$\langle f(\xi, \chi) \rangle_\chi = \frac{1}{(2\pi)^m} \int_0^{2\pi} \cdots \int_0^{2\pi} f(\xi, \chi)\, d\chi_1 \cdots d\chi_m. \tag{8.30}$$

The subscript χ denotes that the integration is performed for each of the rapidly changing χ_α variables.

Applying Eq. 8.30 to Eq. 8.28, it follows that

$$\left\langle M_i^{(1)} \right\rangle_\chi + \left\langle \omega_\beta \frac{\partial \eta_i^{(1)}}{\partial \chi_\beta} \right\rangle_\chi = \left\langle f_i^{(1)}(\xi, \chi) \right\rangle_\chi, \tag{8.31}$$

and because $\eta_i^{(1)}$ is periodic in χ

$$M_i^{(1)} = \left\langle f_i^{(1)}(\xi, \chi) \right\rangle_\chi \tag{8.32}$$

$$\omega_\beta \frac{\partial \eta_i^{(1)}}{\partial \chi_\beta} = f_i^{(1)}(\xi, \chi) - \left\langle f_i^{(1)}(\xi, \chi) \right\rangle_\chi. \tag{8.33}$$

Because this methodology computes average rates for the orbital elements over one period of the satellite motion, the resulting orbital elements are often referred to as "mean" elements because the transformation contains the periodic variation. When the transformation is applied to the mean elements, the results are osculating elements.

For the fast variable equation Eq. 8.29, we have

$$\left\langle \Omega_\alpha^{(1)} \right\rangle_\chi + \left\langle \omega_\beta \frac{\partial \phi_\alpha^{(1)}}{\partial \chi_\beta} \right\rangle_\chi = \left\langle \eta_j^{(1)} \frac{\partial \omega_\alpha}{\partial \xi_j} \right\rangle_\chi + \left\langle u_\alpha^{(1)}(\xi, \chi) \right\rangle_\chi. \tag{8.34}$$

Since $\phi_\alpha^{(1)}$ is periodic in χ, it follows that

$$\Omega_\alpha^{(1)} = \frac{\partial \omega_\alpha}{\partial \xi_j} \left\langle \eta_j^{(1)} \right\rangle_\chi + \left\langle u_\alpha^{(1)} (\xi, \chi) \right\rangle_\chi . \tag{8.35}$$

Then we have

$$\omega_\beta \frac{\partial \phi_\alpha^{(1)}}{\partial \chi_\beta} = \frac{\partial \omega_\alpha}{\partial \xi_j} \eta_j^{(1)} - \frac{\partial \omega_\alpha}{\partial \xi_j} \left\langle \eta_j^{(1)} \right\rangle_\chi + u_\alpha^{(1)} (\xi, \chi) - \left\langle u_\alpha^{(1)} (\xi, \chi) \right\rangle_\chi . \tag{8.36}$$

Thus, the first-order terms of the transformed differential equations are given by Eqs. 8.32 and 8.35, and the first-order periodic transformation is given by simply integrating Eqs. 8.33 and 8.36. It should be noted that, for a given order, we must first compute the transformed slow variable terms followed by the transformation of the slow variables. Then we must compute the transformed fast variable terms followed by the transformation of the fast variables. To obtain second-order effects, one must apply this process to the pair of equations

$$M_i^{(2)} + M_j^{(1)} \frac{\partial \eta_i^{(1)}}{\partial \xi_j} + \Omega_\beta^{(1)} \frac{\partial \eta_i^{(1)}}{\partial \chi_\beta} + \omega_\beta \frac{\partial \eta_i^{(2)}}{\partial \chi_\beta} = \eta_j^{(1)} \frac{\partial f_i^{(1)}}{\partial \xi_j} + \phi_\beta^{(1)} \frac{\partial f_i^{(1)}}{\partial \chi_\beta} + f_i^{(2)} (\xi, \chi) \tag{8.37}$$

$$\Omega_\alpha^{(2)} + M_j^{(1)} \frac{\partial \phi_\alpha^{(1)}}{\partial \xi_j} + \Omega_\beta^{(1)} \frac{\partial \phi_\alpha^{(1)}}{\partial \chi_\beta} + \omega_\beta \frac{\partial \phi_\alpha^{(2)}}{\partial \chi_\beta} = \eta_j^{(2)} \frac{\partial \omega_\alpha}{\partial \xi_j} + \frac{1}{2} \eta_j^{(1)} \eta_k^{(1)} \frac{\partial^2 \omega_\alpha}{\partial \xi_j \partial \xi_k}$$

$$+ \eta_j^{(1)} \frac{\partial u_\alpha^{(1)}}{\partial \xi_j} + \varphi_\beta^{(1)} \frac{\partial u_\alpha^{(1)}}{\partial \chi_\beta}$$

$$+ u_\alpha^{(2)} (\xi, \chi) . \tag{8.38}$$

The solution of Eqs. 8.33 and 8.36 provides an explicit transformation from the mean state to the osculating state. If variables need to be transformed from osculating to mean, the calculation can be done by a numerical iteration on Eqs. 8.33 and 8.36.

8.3   The Motion of a Satellite About an Oblate Planet

This development closely follows the paper by Liu and Alford (1980). The Lagrange planetary equations for a perturbing potential are given by Eq. 5.37

$$\dot{n} = -\frac{3}{a^2} \frac{\partial \mathcal{R}}{\partial M}$$

$$\dot{e} = \left(\frac{\beta^2}{na^2 e} \right) \frac{\partial \mathcal{R}}{\partial M} - \left(\frac{\beta}{na^2 e} \right) \frac{\partial \mathcal{R}}{\partial \omega}$$

$$\frac{di}{dt} = \left(\frac{\cos i}{na^2 \beta \sin i} \right) \frac{\partial R}{\partial \omega} - \left(\frac{1}{na^2 \beta \sin i} \right) \frac{\partial R}{\partial \Omega}$$

$$\dot{\Omega} = \left(\frac{1}{na^2 \beta \sin i} \right) \frac{\partial R}{\partial i} \tag{8.39}$$

$$\dot{\omega} = \frac{\beta}{na^2 e} \frac{\partial R}{\partial e} - \left(\frac{\cos i}{na^2 \beta \sin i} \right) \frac{\partial R}{\partial i}$$

$$\dot{M} = n - \left(\frac{\beta^2}{na^2 e} \right) \frac{\partial R}{\partial e} + \left(\frac{3}{a^2} \right) \left(\frac{\partial R}{\partial n} \right)_M .$$

Our perturbing potential R is of size $J_2 \approx 10^{-3}$. We take this as our small parameter ε. Thus, our perturbation problem can be characterized as

$$\dot{n} = O\,(\varepsilon)$$

$$\dot{e} = O\,(\varepsilon)$$

$$\frac{di}{dt} = O\,(\varepsilon)$$

$$\dot{\Omega} = O\,(\varepsilon)$$

$$\dot{\omega} = O\,(\varepsilon)$$

$$\dot{M} = n + O\,(\varepsilon).$$

This system of equations is of the general form for application of the method of averaging with a single fast variable M and the slow variables n, e, i, Ω, ω. Furthermore, we select an integration constant of the mean motion periodic such that the periodic has zero average value. Given that approach, the transformed rates of both the slow and fast variables are just the average of the right-hand sides of Eq. 8.39. Thus, the transformed differential equations for the slow and fast variables are

$$\langle \dot{n} \rangle^{(1)} = -\frac{3}{a^2} \left\langle \frac{\partial R}{\partial M} \right\rangle$$

$$\langle \dot{e} \rangle^{(1)} = \left(\frac{\beta^2}{na^2 e} \right) \left\langle \frac{\partial R}{\partial M} \right\rangle - \left(\frac{\beta}{na^2 e} \right) \left\langle \frac{\partial R}{\partial \omega} \right\rangle$$

$$\left\langle \frac{di}{dt} \right\rangle^{(1)} = \left(\frac{\cos i}{na^2 \beta \sin i} \right) \left\langle \frac{\partial R}{\partial \omega} \right\rangle - \left(\frac{1}{na^2 \beta \sin i} \right) \left\langle \frac{\partial R}{\partial \Omega} \right\rangle$$

$$\langle \dot{\Omega} \rangle^{(1)} = \left(\frac{1}{na^2 \beta \sin i} \right) \left\langle \frac{\partial R}{\partial i} \right\rangle \tag{8.40}$$

$$\langle \dot{\omega} \rangle^{(1)} = \frac{\beta}{na^2 e} \left\langle \frac{\partial \mathcal{R}}{\partial e} \right\rangle - \left(\frac{\cos i}{na^2 \beta \sin i} \right) \left\langle \frac{\partial \mathcal{R}}{\partial i} \right\rangle$$

$$\langle \dot{M} \rangle^{(1)} = - \left(\frac{\beta^2}{na^2 e} \right) \left\langle \frac{\partial \mathcal{R}}{\partial e} \right\rangle + \left(\frac{3}{a^2} \right) \left\langle \left(\frac{\partial \mathcal{R}}{\partial n} \right)_M \right\rangle,$$

where all quantities on the right-hand side are mean elements.

From Eq. 8.1, our perturbing potential is

$$\mathcal{R} = \mu \frac{3 J_2 r_e^2}{2a^3} \left(\frac{a}{r} \right)^3 \left[\frac{1}{3} - \frac{1}{2} \sin^2 i + \frac{1}{2} \sin^2 i \cos 2u \right],$$

which can be simplified to

$$\mathcal{R} = n^2 a^3 \frac{3}{2} \frac{J_2 r_e^2}{a^3} \left(\frac{a}{r} \right)^3 \frac{1}{6} \left[2 - 3\sin^2 i + 3\sin^2 i \cos 2u \right]$$

$$= \frac{1}{4} n^2 J_2 r_e^2 \left(\frac{a}{r} \right)^3 \left(2 - 3\sin^2 i + 3\sin^2 i \cos 2u \right). \tag{8.41}$$

We need the integral

$$\int \mathcal{R} \, dM = \frac{1}{4} n^2 J_2 r_e^2 \int \left(\frac{a}{r} \right)^3 \left(2 - 3\sin^2 i + 3\sin^2 i \cos 2u \right) dM. \tag{8.42}$$

Now from Eq. 4.47, we see that

$$\frac{df}{dt} = \frac{na^2 \beta}{r^2}$$

$$n \, dt = dM = \frac{r^2}{a^2 \beta} df,$$

so

$$\int \left(\frac{a}{r} \right)^3 dM = \int \left(\frac{a}{r} \right)^3 \frac{r^2}{a^2 \beta} df = \int \frac{a}{r} \frac{1}{\beta} df = \frac{1}{\beta^3} \int (1 + e \cos f) \, df.$$

Then we have

$$\int \left(\frac{a}{r} \right)^3 dM = \frac{1}{\beta^3} (f + e \sin f) \tag{8.43}$$

and

$$\int \left(\frac{a}{r} \right)^3 (\cos 2u) \, dM = \frac{1}{\beta^3} \int (1 + e \cos f) (\cos 2u) \, df$$

$$= \frac{1}{2\beta^3} \sin 2u + \frac{e}{6\beta^3} \sin(3f + 2\omega) + \frac{e}{2\beta^3} \sin(f + 2\omega).$$

$$(8.44)$$

Using Eqs. 8.43 and 8.44 in Eq. 8.42 gives

$$\int \mathcal{R} dM = \frac{1}{4} \frac{n^2 J_2 r_e^2}{\beta^3} \left(2 - 3\sin^2 i\right)(f + e \sin f)$$

$$+ \frac{1}{8} \frac{n^2 J_2 r_e^2}{\beta^3} \left(3\sin^2 i\right) \left[\sin 2u + \frac{e}{3} \sin(3f + 2\omega) + e \sin(f + 2\omega)\right].$$

$$(8.45)$$

We can also conclude, in evaluating the definite integral over the interval from 0 to 2π, that

$$\langle \mathcal{R} \rangle = \frac{1}{2\pi} \int_0^{2\pi} \mathcal{R} dM = \frac{1}{4} \frac{n^2 J_2 r_e^2}{\beta^3} \left(2 - 3\sin^2 i\right),$$

$$(8.46)$$

and we also have

$$\int [\mathcal{R} - \langle \mathcal{R} \rangle] dM = \frac{1}{4} \frac{n^2 J_2 r_e^2}{\beta^3} \left(2 - 3\sin^2 i\right)(f - M + e \sin f)$$

$$+ \frac{1}{8} \frac{n^2 J_2 r_e^2}{\beta^3} \left(3\sin^2 i\right)$$

$$\times \left[\sin 2u + \frac{e}{3} \sin(3f + 2\omega) + e \sin(f + 2\omega)\right]. \quad (8.47)$$

Now from the secular equations, Eqs. 8.40 and 8.41, we need the following quantities:

$$\left\langle \frac{\partial \mathcal{R}}{\partial M} \right\rangle \quad \left\langle \frac{\partial \mathcal{R}}{\partial \omega} \right\rangle \quad \left\langle \frac{\partial \mathcal{R}}{\partial \Omega} \right\rangle \quad \left\langle \frac{\partial \mathcal{R}}{\partial i} \right\rangle \quad \left\langle \frac{\partial \mathcal{R}}{\partial e} \right\rangle \quad \left\langle \left(\frac{\partial \mathcal{R}}{\partial n}\right)_M \right\rangle. \quad (8.48)$$

The first item can be computed as

$$\left\langle \frac{\partial \mathcal{R}}{\partial M} \right\rangle = \frac{1}{2\pi} \int_0^{2\pi} \frac{\partial \mathcal{R}}{\partial M} dM = \frac{1}{2\pi} \int_0^{2\pi} d\mathcal{R} = \frac{1}{2\pi} \mathcal{R} \Big|_0^{2\pi}.$$

Using Eq. 8.41, we see

$$\left\langle \frac{\partial \mathcal{R}}{\partial M} \right\rangle = \frac{1}{2\pi} \mathcal{R} \Big|_0^{2\pi} = \frac{1}{2\pi} \frac{1}{4} n^2 J_2 r_e^2 \left(\frac{a}{r}\right)^3 \left(2 - 3\sin^2 i + 3\sin^2 i \, \cos 2u\right) \Big|_0^{2\pi} = 0.$$

(8.49)

For all the other partial derivatives in Eq. 8.48, we can use the fact that the partial can be moved outside the integral sign since all quantities except M are constants. Let Λ represent any orbital element other than the mean anomaly. Then

$$\left\langle \frac{\partial \mathcal{R}}{\partial \Lambda} \right\rangle = \frac{1}{2\pi} \int_0^{2\pi} \frac{\partial \mathcal{R}}{\partial \Lambda} dM = \frac{\partial}{\partial \Lambda} \frac{1}{2\pi} \int_0^{2\pi} \mathcal{R} \, dM = \frac{\partial}{\partial \Lambda} \langle \mathcal{R} \rangle. \tag{8.50}$$

We can now compute the items listed in Eq. 8.48 using Eqs. 8.46, 8.49, and 8.50.

$$\left\langle \frac{\partial \mathcal{R}}{\partial M} \right\rangle = 0$$

$$\left\langle \frac{\partial \mathcal{R}}{\partial \omega} \right\rangle = \frac{\partial}{\partial \omega} \langle \mathcal{R} \rangle = \frac{\partial}{\partial \omega} \left[\frac{1}{4} \frac{n^2 J_2 r_e^2}{\beta^3} \left(2 - 3\sin^2 i\right) \right] = 0$$

$$\left\langle \frac{\partial \mathcal{R}}{\partial \Omega} \right\rangle = \frac{\partial}{\partial \Omega} \langle \mathcal{R} \rangle = \frac{\partial}{\partial \Omega} \left[\frac{1}{4} \frac{n^2 J_2 r_e^2}{\beta^3} \left(2 - 3\sin^2 i\right) \right] = 0$$

$$\left\langle \frac{\partial \mathcal{R}}{\partial i} \right\rangle = \frac{\partial}{\partial i} \langle \mathcal{R} \rangle = \frac{\partial}{\partial i} \left[\frac{1}{4} \frac{n^2 J_2 r_e^2}{\beta^3} \left(2 - 3\sin^2 i\right) \right] = -\frac{3}{2} \frac{n^2 J_2 r_e^2}{\beta^3} \left(\sin i \, \cos i\right)$$

(8.51)

$$\left\langle \frac{\partial \mathcal{R}}{\partial e} \right\rangle = \frac{\partial}{\partial e} \langle \mathcal{R} \rangle = \frac{\partial}{\partial e} \left[\frac{1}{4} \frac{n^2 J_2 r_e^2}{\beta^3} \left(2 - 3\sin^2 i\right) \right] = \frac{3}{4} \frac{n^2 e J_2 r_e^2}{\beta^5} \left(2 - 3\sin^2 i\right)$$

$$\left\langle \frac{\partial \mathcal{R}}{\partial n} \right\rangle = \frac{\partial}{\partial n} \langle \mathcal{R} \rangle = \frac{\partial}{\partial n} \left[\frac{1}{4} \frac{n^2 J_2 r_e^2}{\beta^3} \left(2 - 3\sin^2 i\right) \right] = \frac{1}{2} \frac{n J_2 r_e^2}{\beta^3} \left(2 - 3\sin^2 i\right).$$

We substitute Eq. 8.51 into Eqs. 8.40 and 8.41 to get

$$\langle \dot{n} \rangle^{(1)} = 0$$

$$\langle \dot{e} \rangle^{(1)} = 0$$

$$\left\langle \frac{di}{dt} \right\rangle^{(1)} = 0$$

$$\langle \dot{\Omega} \rangle^{(1)} = \left(\frac{1}{na^2 \beta \sin i} \right) \left[-\frac{3}{2} \frac{n^2 J_2 r_e^2}{\beta^3} \left(\sin i \, \cos i\right) \right]$$

$$\langle \dot{\Omega} \rangle^{(1)} = -\frac{3}{2} \frac{n J_2 r_e^2}{a^2 \beta^4} \cos i$$

$$\langle \dot{\omega} \rangle^{(1)} = \frac{\beta}{na^2 e} \left[-\frac{3}{4} \frac{n^2 e J_2 r_e^2}{\beta^5} \left(2 - 3\sin^2 i \right) \right]$$
$$- \left(\frac{\cos i}{na^2 \beta \sin i} \right) \left[-\frac{3}{2} \frac{n^2 J_2 r_e^2}{\beta^3} \left(\sin i \cos i \right) \right]$$

$$\langle \dot{\omega} \rangle^{(1)} = \frac{3}{4} \frac{n J_2 r_e^2}{a^2 \beta^4} \left(-1 + 5\cos^2 i \right)$$

$$\langle \dot{M} \rangle^{(1)} = - \left(\frac{\beta^2}{na^2 e} \right) \left[\frac{3}{4} \frac{n^2 e J_2 r_e^2}{\beta^5} \left(2 - 3\sin^2 i \right) \right] + \left(\frac{3}{a^2} \right) \left[\frac{1}{2} \frac{n J_2 r_e^2}{\beta^3} \left(2 - 3\sin^2 i \right) \right]$$

$$\langle \dot{M} \rangle^{(1)} = \frac{3}{4} \frac{n J_2 r_e^2}{a^2 \beta^3} \left(-1 + 3\cos^2 i \right).$$

In summary, we have the transformed differential equations

$$\langle \dot{n} \rangle^{(1)} = 0$$

$$\langle \dot{e} \rangle^{(1)} = 0$$

$$\left\langle \frac{di}{dt} \right\rangle^{(1)} = 0$$

$$\langle \dot{\Omega} \rangle^{(1)} = -\frac{3}{2} \frac{n J_2 r_e^2}{a^2 \beta^4} \cos i \tag{8.52}$$

$$\langle \dot{\omega} \rangle^{(1)} = \frac{3}{4} \frac{n J_2 r_e^2}{a^2 \beta^4} \left(-1 + 5\cos^2 i \right)$$

$$\langle \dot{M} \rangle^{(1)} = \frac{3}{4} \frac{n J_2 r_e^2}{a^2 \beta^3} \left(-1 + 3\cos^2 i \right).$$

The transformed differential Eqs. 8.52 are free of the fast variable, M, to first order in the small parameter J_2. The method of averaging requires that the transformation be purely periodic in the satellite anomaly angle. We refer to these transformation periodics as "short period" because they have the same period as the satellite motion around the Earth. This process may seem a little inverted in that we have found the transformed equations (Eqs. 8.52), but we do not yet know the transformation that produced them. Rest assured that we can determine those equations. They will be determined by integration of explicit constraint equations provided by the method of averaging.

The periodics for the slow variables are given by the solution of Eq. 8.33

$$\omega_\beta \frac{\partial \eta_i^{(1)}}{\partial \chi_\beta} = f_i^{(1)} (\xi, \chi) - M_i^{(1)} (\xi).$$

For the problem we are considering, there is only one fast variable. The solution of the partial differential equation Eq. 8.33 is

$$\eta_i^{(1)} = \frac{1}{\omega_\beta} \int \left[f_i^{(1)}(\xi, \chi) - M_i^{(1)}(\xi) \right] d\chi + c_i(\xi), \tag{8.53}$$

where $c_i(\xi)$ is an arbitrary constant of integration. It can be chosen to be zero for convenience, or it can be chosen such that the resulting transformed equation has some desirable property such as being canonical. For our purpose, it is useful to select $c_i(\xi)$ so that the periodic has zero mean value. This choice gives

$$c_i(\xi) = -\frac{1}{\omega_\beta} \left\langle f_i^{(1)}(\xi, \chi) - M_i^{(1)}(\xi) \right\rangle.$$

Now we apply Eq. 8.53 to each of our five orbital elements that are slow variables

$$\delta n_{SP} = -\frac{3}{na^2} \int \left[\frac{\partial \mathcal{R}}{\partial M} - \left\langle \frac{\partial \mathcal{R}}{\partial M} \right\rangle \right] dM + c_n^{(1)}(\xi)$$

$$\delta e_{SP} = \left(\frac{\beta^2}{n^2 a^2 e} \right) \int \left[\frac{\partial \mathcal{R}}{\partial M} - \left\langle \frac{\partial \mathcal{R}}{\partial M} \right\rangle \right] dM - \left(\frac{\beta}{n^2 a^2 e} \right) \int \left[\frac{\partial \mathcal{R}}{\partial \omega} - \left\langle \frac{\partial \mathcal{R}}{\partial \omega} \right\rangle \right] dM$$
$$+ c_e^{(1)}(\xi)$$

$$\delta i_{SP} = \left(\frac{\cos i}{n^2 a^2 \beta \sin i} \right) \int \left[\frac{\partial \mathcal{R}}{\partial \omega} - \left\langle \frac{\partial \mathcal{R}}{\partial \omega} \right\rangle \right] dM$$
$$- \left(\frac{1}{na^2 \beta \sin i} \right) \int \left[\frac{\partial \mathcal{R}}{\partial \Omega} - \left\langle \frac{\partial \mathcal{R}}{\partial \Omega} \right\rangle \right] dM + c_i^{(1)}(\xi) \tag{8.54}$$

$$\delta \Omega_{SP} = \left(\frac{1}{n^2 a^2 \beta \sin i} \right) \int \left[\frac{\partial \mathcal{R}}{\partial i} - \left\langle \frac{\partial \mathcal{R}}{\partial i} \right\rangle \right] dM + c_\Omega^{(1)}(\xi)$$

$$\delta \omega_{SP} = \frac{\beta}{n^2 a^2 e} \int \left[\frac{\partial \mathcal{R}}{\partial e} - \left\langle \frac{\partial \mathcal{R}}{\partial e} \right\rangle \right] dM - \left(\frac{\cos i}{n^2 a^2 \beta \sin i} \right) \int \left[\frac{\partial \mathcal{R}}{\partial i} - \left\langle \frac{\partial \mathcal{R}}{\partial i} \right\rangle \right] dM$$
$$+ c_\omega^{(1)}(\xi),$$

where we introduce the notation δx_{SP} to represent the short-period periodic transformation of the variable x.

The first-order transformation equations need the following integrals:

$$\int \left[\frac{\partial \mathcal{R}}{\partial M} - \left\langle \frac{\partial \mathcal{R}}{\partial M} \right\rangle \right] dM \quad \int \left[\frac{\partial \mathcal{R}}{\partial \omega} - \left\langle \frac{\partial \mathcal{R}}{\partial \omega} \right\rangle \right] dM \quad \int \left[\frac{\partial \mathcal{R}}{\partial \Omega} - \left\langle \frac{\partial \mathcal{R}}{\partial \Omega} \right\rangle \right] dM$$

$$\int \left[\frac{\partial \mathcal{R}}{\partial i} - \left\langle \frac{\partial \mathcal{R}}{\partial i} \right\rangle \right] dM \quad \int \left[\frac{\partial \mathcal{R}}{\partial e} - \left\langle \frac{\partial \mathcal{R}}{\partial e} \right\rangle \right] dM. \tag{8.55}$$

Now all the orbital elements on the right-hand sides are held constant except for the mean anomaly. Thus, any partial derivative with respect to any classical element except the mean anomaly can be moved outside the integral. First, we consider terms involving partial derivatives with respect to mean anomaly.

Then, using the result of Eq. 8.49, we obtain

$$\int \left[\frac{\partial \mathcal{R}}{\partial M} - \left\langle \frac{\partial \mathcal{R}}{\partial M} \right\rangle \right] dM = \int \frac{\partial \mathcal{R}}{\partial M} dM - \left\langle \frac{\partial \mathcal{R}}{\partial M} \right\rangle \int dM = \mathcal{R}. \qquad (8.56)$$

Using Eq. 8.56 in the first equation of Eqs. 8.54, we find

$$\delta n_{SP} = -\frac{3}{na^2} \mathcal{R} + c_n^{(1)}(\xi) .$$

Since δn_{SP} is to be purely periodic, we require that

$$\langle \delta n_{SP} \rangle = -\frac{3}{na^2} \langle \mathcal{R} \rangle + c_n^{(1)}(\xi) = 0.$$

Then we have

$$c_n^{(1)}(\xi) = \frac{3}{na^2} \langle \mathcal{R} \rangle$$

and

$$\delta n_{SP} = -\frac{3}{na^2} \mathcal{R} + \frac{3}{na^2} \langle \mathcal{R} \rangle = -\frac{3}{na^2} [\mathcal{R} - \langle \mathcal{R} \rangle]$$
$$= -\frac{3}{2} \frac{n J_2 r_e^2}{a^2} \left[\left(\frac{a}{r} \right)^3 \left(1 - \frac{3}{2} \sin^2 i + \frac{3}{2} \sin^2 i \cos 2u \right) - \frac{1}{\beta^3} \left(1 - \frac{3}{2} \sin^2 i \right) \right].$$
$$(8.57)$$

For all the other partial derivatives, we can use the fact that the partial operator can be moved outside the integral sign since all quantities except M are constants:

$$\int \left[\frac{\partial \mathcal{R}}{\partial \Lambda} - \left\langle \frac{\partial \mathcal{R}}{\partial \Lambda} \right\rangle \right] dM = \frac{\partial}{\partial \Lambda} \int [\mathcal{R} - \langle \mathcal{R} \rangle] dM.$$

From Eq. 8.47, we have found that

$$\int [\mathcal{R} - \langle \mathcal{R} \rangle] dM = \frac{1}{4} \frac{n^2 J_2 r_e^2}{\beta^3} \left(2 - 3 \sin^2 i \right) (f - M + e \sin f) + \frac{1}{8} \frac{n^2 J_2 r_e^2}{\beta^3} \left(3 \sin^2 i \right)$$
$$\times \left[\sin 2u + \frac{e}{3} \sin (3f + 2\omega) + e \sin (f + 2\omega) \right]. \qquad (8.58)$$

We can see by inspection that this function has zero mean value. So we set all other constants of integration to zero.

We need to calculate various partial derivatives of Eq. 8.58 for use in Eq. 8.54. Thus we obtain

$$\frac{\partial}{\partial \omega} \int [\mathcal{R} - \langle \mathcal{R} \rangle] dM = \frac{1}{8} \frac{n^2 J_2 r_e^2}{\beta^3} \left(3\sin^2 i \right)$$

$$\times \frac{\partial}{\partial \omega} \left[\sin 2u + \frac{e}{3} \sin (3f + 2\omega) + e \sin (f + 2\omega) \right]$$

$$\frac{\partial}{\partial \omega} \int [\mathcal{R} - \langle \mathcal{R} \rangle] dM = \frac{1}{8} \frac{n^2 J_2 r_e^2}{\beta^3} \left(3\sin^2 i \right)$$

$$\times \left[2\cos 2u + \frac{2e}{3} \cos (3f + 2\omega) + 2e \cos (f + 2\omega) \right]$$

$$\tag{8.59}$$

$$\frac{\partial}{\partial \Omega} \int [\mathcal{R} - \langle \mathcal{R} \rangle] dM = 0 \tag{8.60}$$

$$\frac{\partial}{\partial i} \int [\mathcal{R} - \langle \mathcal{R} \rangle] dM = \frac{1}{4} \frac{n^2 J_2 r_e^2}{\beta^3} \frac{\partial}{\partial i} \left(2 - 3\sin^2 i \right) (f - M + e \sin f)$$

$$+ \frac{1}{8} \frac{n^2 J_2 r_e^2}{\beta^3} \frac{\partial}{\partial i} \left(3\sin^2 i \right)$$

$$\times \left[\sin 2u + \frac{e}{3} \sin (3f + 2\omega) + e \sin (f + 2\omega) \right]$$

$$= -\frac{3}{2} \frac{n^2 J_2 r_e^2}{\beta^3} (\sin i \cos i) (f - M + e \sin f)$$

$$+ \frac{3}{4} \frac{n^2 J_2 r_e^2}{\beta^3} (\sin i \cos i)$$

$$\times \left[\sin 2u + \frac{e}{3} \sin (3f + 2\omega) + e \sin (f + 2\omega) \right].$$

$$\tag{8.61}$$

When we developed the Lagrange planetary equations, we selected the set of independent variables n, e, i, Ω, ω, and M. Thus, any auxiliary angle variable such as the true anomaly or eccentric anomaly will have a dependence on eccentricity, so we must be particularly careful when we take partial derivatives. Taking the partial derivative of Eq. 2.76 with respect to eccentricity gives

$$-\sin f \frac{\partial f}{\partial e} = \frac{(1 - e \cos E) \left(-\sin E \frac{\partial E}{\partial e} - 1 \right) - (\cos E - e) \left(e \sin E \frac{\partial E}{\partial e} - \cos E \right)}{(1 - e \cos E)^2},$$

and using Eq. 4.33 for $\partial E / \partial e$ finally yields

$$\frac{\partial f}{\partial e} = \frac{1}{\beta^2} \sin f \, (2 + e \cos f).$$ (8.62)

Now we compute the partial with respect to eccentricity

$$\frac{\partial}{\partial e} \int [\mathcal{R} - \langle \mathcal{R} \rangle] dM = \frac{3}{4} \frac{n^2 e J_2 r_e^2}{\beta^5} \left(2 - 3\sin^2 i \right) (f - M + e \sin f)$$

$$+ \frac{1}{4} \frac{n^2 J_2 r_e^2}{\beta^3} \left(2 - 3\sin^2 i \right) (\sin f)$$

$$+ \frac{1}{4} \frac{n^2 J_2 r_e^2}{\beta^3} \left(2 - 3\sin^2 i \right) (1 + e \cos f) \frac{\partial f}{\partial e}$$

$$+ \frac{3}{8} \frac{n^2 e J_2 r_e^2}{\beta^5} \left(3\sin^2 i \right)$$

$$\times \left[\sin 2u + \frac{e}{3} \sin (3f + 2\omega) + e \sin (f + 2\omega) \right]$$

$$+ \frac{1}{8} \frac{n^2 J_2 r_e^2}{\beta^3} \left(3\sin^2 i \right) \left[\frac{1}{3} \sin (3f + 2\omega) + \sin (f + 2\omega) \right]$$

$$+ \frac{1}{8} \frac{n^2 J_2 r_e^2}{\beta^3} \left(3\sin^2 i \right)$$

$$\times \left[2 \cos 2u + e \cos (3f + 2\omega) + e \cos (f + 2\omega) \right] \frac{\partial f}{\partial e}.$$

Using Eq. 8.62 and considerable algebra, we find

$$\frac{\partial}{\partial e} \int [\mathcal{R} - \langle \mathcal{R} \rangle] dM = \frac{3}{4} \frac{n^2 e J_2 r_e^2}{\beta^5} \left(2 - 3\sin^2 i \right) (f - M + e \sin f)$$

$$+ \frac{3}{2} \frac{n^2 J_2 r_e^2}{\beta^5} \left(1 - \frac{3}{2} \sin^2 i \right)$$

$$\times \left[\left(1 - \frac{1}{4} e^2 \right) \sin f + \frac{1}{2} e \sin 2f + \frac{1}{12} e^2 \sin 3f \right]$$

$$+ \frac{3}{8} \frac{n^2 J_2 r_e^2}{\beta^5} \left(\sin^2 i \right) \left[\sin (f + 2\omega) \left(-1 + \frac{7}{4} e^2 \right) \right.$$

$$+ 3e \sin (2f + 2\omega) + \left(\frac{7}{3} + \frac{11}{12} e^2 \right) \sin (3f + 2\omega)$$

$$+ \frac{3}{2} e \sin (4f + 2\omega) + \frac{1}{4} e^2 \sin (5f + 2\omega)$$

$$- \frac{1}{4}e^2 \sin (2\omega - f) - \frac{3}{2}e \sin 2\omega \bigg]. \tag{8.63}$$

We have completed the calculation of all the integrals listed in Eq. 8.55.

We now return to the computation of the short-period periodics given by Eq. 8.54, which we compute one by one. Using Eqs. 8.56 and 8.59 in the second of Eqs. 8.54, we find

$$\delta e_{SP} = \left(\frac{\beta^2}{n^2 a^2 e} \right)$$

$$\times \left[\frac{1}{2} n^2 J_2 r_e^2 \left(\frac{a}{r} \right)^3 \left(1 - \frac{3}{2} \sin^2 i + \frac{3}{2} \sin^2 i \cos 2u \right) - \frac{1}{2} \frac{n^2 J_2 r_e^2}{\beta^3} \left(1 - \frac{3}{2} \sin^2 i \right) \right]$$

$$- \left(\frac{\beta}{n^2 a^2 e} \right) \frac{1}{8} \frac{n^2 J_2 r_e^2}{\beta^3} \left(3 \sin^2 i \right)$$

$$\times \left[2 \cos 2u + \frac{2e}{3} \cos (3f + 2\omega) + 2e \cos (f + 2\omega) \right],$$

which can be simplified to obtain

$$\delta e_{SP} = \frac{1}{2} \frac{J_2 r_e^2}{a^2 \beta^4 e} \left(1 - \frac{3}{2} \sin^2 i \right) \left(1 + \frac{3e^2}{2} - \beta^3 \right)$$

$$+ \frac{1}{2} \frac{J_2 r_e^2}{a^2 \beta^4} \left(1 - \frac{3}{2} \sin^2 i \right) \left[3 \left(1 + \frac{e^2}{4} \right) \cos f + \frac{3e}{2} \cos 2f + \frac{e^2}{4} \cos 3f \right]$$

$$+ \frac{1}{8} \frac{J_2 r_e^2}{a^2 \beta^4} \left(3 \sin^2 i \right) \left[\left(1 + \frac{11e^2}{4} \right) \cos (f + 2\omega) + \frac{e^2}{4} \cos (-f + 2\omega) \right.$$

$$+ 5e \cos (2f + 2\omega) + \frac{1}{3} \left(7 + \frac{17e^2}{4} \right) \cos (3f + 2\omega) + \frac{3e}{2} \cos (4f + 2\omega)$$

$$+ \frac{e^2}{4} \cos (5f + 2\omega) + \frac{3e}{2} \cos 2\omega \bigg]. \tag{8.64}$$

Using Eqs. 8.59 and 8.60 in the third of Eqs. 8.54, we find

$$\delta i_{SP} = \left(\frac{\cos i}{n^2 a^2 \beta \sin i} \right) \frac{1}{8} \frac{n^2 J_2 r_e^2}{\beta^3} \left(3 \sin^2 i \right)$$

$$\times \left[2 \cos 2u + \frac{2e}{3} \cos (3f + 2\omega) + 2e \cos (f + 2\omega) \right],$$

which can be simplified to obtain

$$\delta i_{SP} = \frac{3}{8}\frac{J_2 r_e^2}{a^2 \beta^4}(\sin 2i)\left[\cos 2u + \frac{e}{3}\cos(3f + 2\omega) + e\cos(f + 2\omega)\right]. \quad (8.65)$$

Using Eq. 8.61 in the fourth of Eqs. 8.54, we find

$$\delta\Omega_{SP} = -\left(\frac{1}{n^2 a^2 \beta \sin i}\right)\frac{3}{2}\frac{n^2 J_2 r_e^2}{\beta^3}(\sin i \cos i)(f - M + e\sin f)$$

$$+ \left(\frac{1}{n^2 a^2 \beta \sin i}\right)\frac{3}{4}\frac{n^2 J_2 r_e^2}{\beta^3}(\sin i \cos i)$$

$$\times \left[\sin 2u + \frac{e}{3}\sin(3f + 2\omega) + e\sin(f + 2\omega)\right],$$

which can be simplified to obtain

$$\delta\Omega_{SP} = -\frac{3}{2}\frac{J_2 r_e^2}{a^2 \beta^4}(\cos i)$$

$$\times \left[(f - M + e\sin f) - \frac{1}{2}\sin 2u - \frac{e}{6}\sin(3f + 2\omega) - \frac{e}{2}\sin(f + 2\omega)\right]. \quad (8.66)$$

Using Eqs. 8.61 and 8.63 in the fifth of Eqs. 8.54, we find

$$\delta\omega_{SP} = \frac{\beta}{n^2 a^2 e}\frac{3}{4}\frac{n^2 e J_2 r_e^2}{\beta^5}\left(2 - 3\sin^2 i\right)(f - M + e\sin f)$$

$$+ \frac{\beta}{n^2 a^2 e}\frac{3}{2}\frac{n^2 J_2 r_e^2}{\beta^5}\left(1 - \frac{3}{2}\sin^2 i\right)$$

$$\times \left[\left(1 - \frac{1}{4}e^2\right)\sin f + \frac{1}{2}e\sin 2f + \frac{1}{12}e^2\sin 3f\right]$$

$$+ \frac{\beta}{n^2 a^2 e}\frac{3}{8}\frac{n^2 J_2 r_e^2}{\beta^5}\left(\sin^2 i\right)\left[\sin(f + 2\omega)\left(-1 + \frac{7}{4}e^2\right) + 3e\sin(2f + 2\omega)\right.$$

$$+ \left(\frac{7}{3} + \frac{11}{12}e^2\right)\sin(3f + 2\omega) + \frac{3}{2}e\sin(4f + 2\omega) + \frac{1}{4}e^2\sin(5f + 2\omega)$$

$$\left. - \frac{1}{4}e^2\sin(2\omega - f) - \frac{3}{2}e\sin 2\omega\right]$$

$$+ \left(\frac{\cos i}{n^2 a^2 \beta \sin i}\right)\frac{3}{2}\frac{n^2 J_2 r_e^2}{\beta^3}(\sin i \cos i)(f - M + e\sin f)$$

$$- \left(\frac{\cos i}{n^2 a^2 \beta \sin i}\right)\frac{3}{4}\frac{n^2 J_2 r_e^2}{\beta^3}(\sin i \cos i)$$

$$\times \left[\sin 2u + \frac{e}{3} \sin (3f + 2\omega) + e \sin (f + 2\omega) \right],$$

which can be simplified to obtain

$$\delta\omega_{SP} = \frac{3}{4} \frac{J_2 r_e^2}{a^2 \beta^4} \left(4 - 5\sin^2 i \right) (f - M + e \sin f)$$

$$+ \frac{3}{2} \frac{J_2 r_e^2}{a^2 \beta^4} \left(1 - \frac{3}{2} \sin^2 i \right) \left[\frac{1}{e} \left(1 - \frac{1}{4} e^2 \right) \sin f + \frac{1}{2} \sin 2f + \frac{1}{12} e \sin 3f \right]$$

$$- \frac{3}{2} \frac{J_2 r_e^2}{a^2 \beta^4} \frac{1}{e} \left[\frac{1}{4} \sin^2 i + \frac{1}{2} e^2 \left(1 - \frac{15}{8} \sin^2 i \right) \right] \sin (f + 2\omega)$$

$$- \frac{3}{2} \frac{J_2 r_e^2}{a^2 \beta^4} \left[\frac{1}{16} e \sin^2 i \sin (2\omega - f) + \frac{1}{2} \left(1 - \frac{5}{2} \sin^2 i \right) \sin (2f + 2\omega) \right.$$

$$- \frac{1}{e} \left(\frac{7}{12} \sin^2 i - \frac{e^2}{6} \left(1 - \frac{19}{8} \sin^2 i \right) \right) \sin (3f + 2\omega)$$

$$\left. - \frac{3}{8} \sin^2 i \sin (4f + 2\omega) - \frac{1}{16} e \sin^2 i \sin (5f + 2\omega) + \frac{3}{8} \sin^2 i \sin 2\omega \right].$$

$$(8.67)$$

Now we need to obtain the periodics for the fast variable. For the formula in Eq. 8.36 specialized to our problem, we have $\omega_\alpha = n$, so

$$n \frac{\partial \phi}{\partial M} = \eta_j \frac{\partial n}{\partial x_j} - \frac{\partial n}{\partial x_j} \langle \eta_j \rangle + u_1 - \langle u_1 \rangle,$$

which reduces to

$$n \frac{\partial \phi}{\partial M} = \delta n_{SP} - \langle \delta n_{SP} \rangle + u_1 - \langle u_1 \rangle. \tag{8.68}$$

The periodic for δn_{SP} was constructed by choice of the constant of integration to have zero mean. Thus we have

$$\langle \delta n_{SP} \rangle = -\frac{3}{na^2} \langle \mathcal{R} \rangle + \frac{3}{na^2} \langle \langle \mathcal{R} \rangle \rangle = 0,$$

and Eq. 8.68 reduces to

$$n \frac{\partial \phi}{\partial M} = \delta n_{SP} + u_1 - \langle u_1 \rangle. \tag{8.69}$$

Using Eq. 8.57 in Eq. 8.69, we can integrate to obtain

$$\delta M_{SP} = -\frac{3}{n^2 a^2} \int [\mathcal{R} - \langle \mathcal{R} \rangle] dM + \frac{1}{n} \int u_1 dM - \frac{1}{n} \int \langle u_1 \rangle \, dM. \qquad (8.70)$$

From the last equation of Eqs. 8.39, we can substitute for the functions u_1 and $\langle u_1 \rangle$ to write

$$\delta M_{SP} = -\frac{3}{n^2 a^2} \int [\mathcal{R} - \langle \mathcal{R} \rangle] dM - \left(\frac{\beta^2}{n^2 a^2 e} \right) \int \left[\frac{\partial \mathcal{R}}{\partial e} - \left\langle \frac{\partial \mathcal{R}}{\partial e} \right\rangle \right] dM$$

$$+ \left(\frac{3}{na^2} \right) \int \left[\left(\frac{\partial \mathcal{R}}{\partial n} \right)_M - \left\langle \left(\frac{\partial \mathcal{R}}{\partial n} \right)_M \right\rangle \right] dM. \qquad (8.71)$$

In Eq. 8.71, we can bring the partial derivative outside the integral sign since all quantities under the integral sign are treated as constants except for the mean anomaly. Thus, we have

$$\delta M_{SP} = -\frac{3}{n^2 a^2} \int [\mathcal{R} - \langle \mathcal{R} \rangle] dM - \left(\frac{\beta^2}{n^2 a^2 e} \right) \frac{\partial}{\partial e} \int [\mathcal{R} - \langle \mathcal{R} \rangle] dM$$

$$+ \left(\frac{3}{na^2} \right) \frac{\partial}{\partial n} \int [\langle \mathcal{R} \rangle - \langle \mathcal{R} \rangle] dM. \qquad (8.72)$$

Using Eq. 8.58, we can compute the partial derivative

$$\frac{\partial}{\partial n} \int [\mathcal{R} - \langle \mathcal{R} \rangle] dM = \frac{1}{2} \frac{n J_2 r_e^2}{\beta^3} \left(2 - 3\sin^2 i \right) (f - M + e \sin f)$$

$$+ \frac{1}{4} \frac{n J_2 r_e^2}{\beta^3} \left(3\sin^2 i \right)$$

$$\times \left[\sin 2u + \frac{e}{3} \sin (3f + 2\omega) + e \sin (f + 2\omega) \right]. \qquad (8.73)$$

Substituting Eq. 8.47 into the first term of Eqs. 8.72, substituting Eq. 8.63 into the second term of Eq. 8.72, and substituting Eq. 8.73 into the third term of Eq. 8.72, we find

$$\delta M_{SP} = -\frac{3}{4} \frac{J_2 r_e^2}{a^2 \beta^3} \left(2 - 3\sin^2 i \right) (f - M + e \sin f)$$

$$- \frac{3}{8} \frac{J_2 r_e^2}{a^2 \beta^3} \left(3\sin^2 i \right) \left[\sin 2u + \frac{e}{3} \sin (3f + 2\omega) + e \sin (f + 2\omega) \right]$$

$$- \frac{3}{4} \frac{J_2 r_e^2}{a^2 \beta^3} \left(2 - 3\sin^2 i \right) (f - M + e \sin f)$$

$$- \frac{3}{2} \frac{J_2 r_e^2}{a^2 \beta^3 e} \left(1 - \frac{3}{2} \sin^2 i\right) \left[\left(1 - \frac{1}{4} e^2\right) \sin f + \frac{1}{2} e \sin 2f + \frac{1}{12} e^2 \sin 3f\right]$$

$$- \frac{3}{8} \frac{n^2 J_2 r_e^2}{a^2 \beta^3 e} \left(\sin^2 i\right) \left[\sin \left(f + 2\omega\right) \left(-1 + \frac{7}{4} e^2\right) + 3e \sin \left(2f + 2\omega\right)\right.$$

$$+ \left(\frac{7}{3} + \frac{11}{12} e^2\right) \sin \left(3f + 2\omega\right) + \frac{3}{2} e \sin \left(4f + 2\omega\right) + \frac{1}{4} e^2 \sin \left(5f + 2\omega\right)$$

$$\left. - \frac{1}{4} e^2 \sin \left(2\omega - f\right) - \frac{3}{2} e \sin 2\omega\right]$$

$$+ \frac{3}{2} \frac{J_2 r_e^2}{a^2 \beta^3} \left(2 - 3\sin^2 i\right) \left(f - M + e \sin f\right)$$

$$+ \frac{3}{4} \frac{J_2 r_e^2}{a^2 \beta^3} \left(3\sin^2 i\right) \left[\sin 2u + \frac{e}{3} \sin \left(3f + 2\omega\right) + e \sin \left(f + 2\omega\right)\right],$$

which simplifies to

$$\delta M_{SP} = - \frac{3}{2} \frac{J_2 r_e^2}{a^2 \beta^3} \frac{1}{e} \left(1 - \frac{3}{2} \sin^2 i\right) \left[\left(1 - \frac{1}{4} e^2\right) \sin f + \frac{1}{2} e \sin 2f + \frac{1}{12} e^2 \sin 3f\right]$$

$$- \frac{3}{2} \frac{J_2 r_e^2}{a^2 \beta^3} \frac{1}{e} \left(\sin^2 i\right) \left[-\frac{1}{4} \sin \left(f + 2\omega\right) \left(1 + \frac{5}{4} e^2\right)\right.$$

$$+ \frac{7}{12} \left(1 - \frac{1}{28} e^2\right) \sin \left(3f + 2\omega\right) + \frac{3}{8} e \sin \left(4f + 2\omega\right)$$

$$\left. + \frac{1}{16} e^2 \sin \left(5f + 2\omega\right) - \frac{1}{16} e^2 \sin \left(2\omega - f\right) - \frac{3}{8} e \sin 2\omega\right]. \qquad (8.74)$$

In summary, we have

$$\delta n_{SP} = - \frac{3}{2} \frac{n J_2 r_e^2}{a^2} \left[\left(\frac{a}{r}\right)^3 \left(1 - \frac{3}{2} \sin^2 i + \frac{3}{2} \sin^2 i \cos 2u\right) - \frac{1}{\beta^3} \left(1 - \frac{3}{2} \sin^2 i\right)\right]$$

$$\delta e_{SP} = \frac{1}{2} \frac{J_2 r_e^2}{a^2 \beta^4 e} \left(1 - \frac{3}{2} \sin^2 i\right) \left(1 + \frac{3e^2}{2} - \beta^3\right)$$

$$+ \frac{1}{2} \frac{J_2 r_e^2}{a^2 \beta^4} \left(1 - \frac{3}{2} \sin^2 i\right) \left[3 \left(1 + \frac{e^2}{4}\right) \cos f + \frac{3e}{2} \cos 2f + \frac{e^2}{4} \cos 3f\right]$$

$$+ \frac{1}{8} \frac{J_2 r_e^2}{a^2 \beta^4} \left(3\sin^2 i\right) \left[\left(1 + \frac{11e^2}{4}\right) \cos \left(f + 2\omega\right) + \frac{e^2}{4} \cos \left(-f + 2\omega\right)\right.$$

$$+ 5e \cos \left(2f + 2\omega\right) + \frac{1}{3} \left(7 + \frac{17e^2}{4}\right) \cos \left(3f + 2\omega\right) + \frac{3e}{2} \cos \left(4f + 2\omega\right)$$

$$+ \frac{e^2}{4} \cos (5f + 2\omega) + \frac{3e}{2} \cos 2\omega \Bigg]$$

$$\delta i_{SP} = \frac{3}{8} \frac{J_2 r_e^2}{a^2 \beta^4} (\sin 2i) \left[\cos 2u + \frac{e}{3} \cos (3f + 2\omega) + e \cos (f + 2\omega) \right] \quad (8.75)$$

$$\delta \Omega_{SP} = - \frac{3}{2} \frac{J_2 r_e^2}{a^2 \beta^4} (\cos i)$$

$$\times \left[(f - M + e \sin f) - \frac{1}{2} \sin 2u - \frac{e}{6} \sin (3f + 2\omega) - \frac{e}{2} \sin (f + 2\omega) \right]$$

$$\delta \omega_{SP} = \frac{3}{4} \frac{J_2 r_e^2}{a^2 \beta^4} \left(4 - 5\sin^2 i \right) (f - M + e \sin f)$$

$$+ \frac{3}{2} \frac{J_2 r_e^2}{a^2 \beta^4} \left(1 - \frac{3}{2}\sin^2 i \right) \left[\frac{1}{e} \left(1 - \frac{1}{4}e^2 \right) \sin f + \frac{1}{2} \sin 2f + \frac{1}{12} e \sin 3f \right]$$

$$- \frac{3}{2} \frac{J_2 r_e^2}{a^2 \beta^4} \frac{1}{e} \left[\frac{1}{4} \sin^2 i + \frac{1}{2} e^2 \left(1 - \frac{15}{8} \sin^2 i \right) \right] \sin (f + 2\omega)$$

$$- \frac{3}{2} \frac{J_2 r_e^2}{a^2 \beta^4} \left[\frac{1}{16} e \sin^2 i \sin (2\omega - f) + \frac{1}{2} \left(1 - \frac{5}{2} \sin^2 i \right) \sin (2f + 2\omega) \right.$$

$$- \frac{1}{e} \left(\frac{7}{12} \sin^2 i - \frac{e^2}{6} \left(1 - \frac{19}{8} \sin^2 i \right) \right) \sin (3f + 2\omega)$$

$$\left. - \frac{3}{8} \sin^2 i \sin (4f + 2\omega) - \frac{1}{16} e \sin^2 i \sin (5f + 2\omega) + \frac{3}{8} \sin^2 i \sin 2\omega \right]$$

$$\delta M_{SP} = - \frac{3}{2} \frac{J_2 r_e^2}{a^2 \beta^3} \frac{1}{e} \left(1 - \frac{3}{2} \sin^2 i \right) \left[\left(1 - \frac{1}{4} e^2 \right) \sin f + \frac{1}{2} e \sin 2f + \frac{1}{12} e^2 \sin 3f \right]$$

$$- \frac{3}{2} \frac{J_2 r_e^2}{a^2 \beta^3} \frac{1}{e} \left(\sin^2 i \right) \left[-\frac{1}{4} \sin (f + 2\omega) \left(1 + \frac{5}{4} e^2 \right) \right.$$

$$+ \frac{7}{12} \left(1 - \frac{1}{28} e^2 \right) \sin (3f + 2\omega) + \frac{3}{8} e \sin (4f + 2\omega)$$

$$\left. + \frac{1}{16} e^2 \sin (5f + 2\omega) - \frac{1}{16} e^2 \sin (2\omega - f) - \frac{3}{8} e \sin 2\omega \right].$$

This result completes the determination of the transformation of variables that transformed our original system of osculating differential equations (Eq. 8.39) into a new system of differential equations (Eq. 8.52) that are free of dependence on the satellite anomaly. Thus, these equations will vary more slowly than the original equations. The transformed orbital elements will not have any periodic variation during one revolution of the satellite. For this reason, the orbital elements are often referred to as singly averaged or mean elements. It is important to appreciate the fact that we have not lost knowledge of the osculating motion of the satellite. For a given state of the mean elements, Eq. 8.75 tells us precisely how to add the short-period periodics onto the mean state to recover the osculating state.

One way to think of this is that we have discovered a very clever frame that wiggles in a special way relative to the osculating world. Because the frame wiggles in a special way, the motion within that frame does not wiggle. Therefore, within this special frame the differential equations are better behaved and perhaps can be integrated.

If we just wanted a first-order solution, then the method of averaging has done what it promised, and we simply need to integrate the transformed differential Eq. 8.52. By inspection, we see that this is easy to do. But we have a bigger vision— we set out to solve this problem to second order in J_2. So, after this reflection, let us get on with it.

The second-order terms in the transformed differential equations for the slow variables are described by Eq. 8.37, repeated here for convenience:

$$M_i^{(2)} + M_j^{(1)} \frac{\partial \eta_i^{(1)}}{\partial \xi_j} + \Omega_\beta^{(1)} \frac{\partial \eta_i^{(1)}}{\partial \chi_\beta} + \omega_\beta \frac{\partial \eta_i^{(2)}}{\partial \chi_\beta} = \eta_j^{(1)} \frac{\partial f_i^{(1)}}{\partial \xi_j} + \varphi_\beta^{(1)} \frac{\partial f_i^{(1)}}{\partial \chi_\beta} + f_i^{(2)}(\boldsymbol{\xi}, \boldsymbol{\chi}).$$

(8.76)

Then we can perform a definite integral over one period of the satellite to obtain

$$M_i^{(2)} = \left\langle f_i^{(2)} \right\rangle - M_j^{(1)} \left\langle \frac{\partial \eta_i^{(1)}}{\partial \xi_j} \right\rangle - \Omega_1^{(1)} \left\langle \frac{\partial \eta_i^{(1)}}{\partial M} \right\rangle$$
$$- n \left\langle \frac{\partial \eta_i^{(2)}}{\partial M} \right\rangle + \left\langle \eta_j^{(1)} \frac{\partial f_i^{(1)}}{\partial x_j} \right\rangle + \left\langle \varphi_\beta^{(1)} \frac{\partial f_i^{(1)}}{\partial M} \right\rangle.$$

(8.77)

Now

$$\left\langle f_i^{(2)} \right\rangle = 0$$

because we have not included any second-order forces in our model, and

$$\left\langle \frac{\partial \eta_i^{(1)}}{\partial \xi_j} \right\rangle = 0 \qquad \left\langle \frac{\partial \eta_i^{(1)}}{\partial M} \right\rangle = 0 \qquad \left\langle \frac{\partial \eta_i^{(2)}}{\partial M} \right\rangle = 0$$

because we have required the periodics to have zero mean. Thus Eq. 8.76 reduces to

$$M_i^{(2)} = \left\langle \eta_j^{(1)} \frac{\partial f_i^{(1)}}{\partial x_j} \right\rangle + \left\langle \varphi_\beta^{(1)} \frac{\partial f_i^{(1)}}{\partial M} \right\rangle.$$

(8.78)

The second-order terms in the transformed differential equations for the fast variables are described by Eq. 8.38 repeated here for convenience.

$$\Omega_\alpha^{(2)} + M_j^{(1)} \frac{\partial \phi_\alpha^{(1)}}{\partial \xi_j} + \Omega_\beta^{(1)} \frac{\partial \phi_\alpha^{(1)}}{\partial \chi_\beta} + \omega_\beta \frac{\partial \phi_\alpha^{(2)}}{\partial \chi_\beta} = \eta_j^{(2)} \frac{\partial \omega_\alpha}{\partial \xi_j} + \frac{1}{2} \eta_j^{(1)} \eta_k^{(1)} \frac{\partial^2 \omega_\alpha}{\partial \xi_j \partial \xi_k}$$

$$+ \eta_j^{(1)} \frac{\partial u_\alpha^{(1)}}{\partial \xi_j} + \varphi_\beta^{(1)} \frac{\partial u_\alpha^{(1)}}{\partial \chi_\beta}$$

$$+ u_\alpha^{(2)} (\xi, \chi). \qquad (8.79)$$

Then we can perform a definite integral over one period of the satellite to obtain

$$\Omega_1^{(2)} = \left\langle u_1^{(2)} \right\rangle - M_j^{(1)} \left\langle \frac{\partial \phi_\alpha^{(1)}}{\partial \xi_j} \right\rangle - \Omega_1^{(1)} \left\langle \frac{\partial \phi_\alpha^{(1)}}{\partial M} \right\rangle - n \left\langle \frac{\partial \phi_\alpha^{(2)}}{\partial M} \right\rangle + \left\langle \eta_j^{(2)} \right\rangle \frac{\partial n}{\partial x_j}$$

$$+ \frac{1}{2} \left\langle \eta_j^{(1)} \eta_k^{(1)} \right\rangle \frac{\partial^2 n}{\partial x_j \partial x_k} + \left\langle \eta_j^{(1)} \frac{\partial u_1^{(1)}}{\partial x_j} \right\rangle + \left\langle \phi_1^{(1)} \frac{\partial u_1^{(1)}}{\partial M} \right\rangle. \qquad (8.80)$$

Now

$$\left\langle u_1^{(2)} \right\rangle = 0$$

because we have not included any second-order forces in our model, and

$$\left\langle \frac{\partial \phi_\alpha^{(1)}}{\partial \xi_j} \right\rangle = 0 \quad \left\langle \frac{\partial \phi_\alpha^{(1)}}{\partial M} \right\rangle = 0 \quad \left\langle \frac{\partial \phi_\alpha^{(2)}}{\partial M} \right\rangle = 0 \quad \left\langle \eta_j^{(2)} \right\rangle = 0.$$

because we have required the periodics to have zero mean. Also

$$\frac{\partial^2 n}{\partial x_j \partial x_k} = 0,$$

so that Eq. 8.80 reduces to

$$\Omega_1^{(2)} = \left\langle \eta_j^{(1)} \frac{\partial u_1^{(1)}}{\partial x_j} \right\rangle + \left\langle \phi_1^{(1)} \frac{\partial u_1^{(1)}}{\partial M} \right\rangle. \qquad (8.81)$$

Thus, the second-order effects for the slow and fast variables can be computed using Eqs. 8.78 and 8.81. The right-hand sides contain only the original first-order functions given in Eq. 8.39 and the first-order periodics given in Eq. 8.75. These computations are straightforward and are not repeated in detail here. Rather we provide these results from the paper by Liu and Alford (1980). The second-order secular terms are

$$\dot{n}^{(2)} = 0$$

$$\dot{e}^{(2)} = -\frac{3}{32}nJ_2^2\left(\frac{R}{p}\right)^4 \sin^2 i \left(14 - 15\sin^2 i\right) e\beta^2 \sin 2\omega$$

$$\frac{di}{dt}^{(2)} = \frac{3}{64}nJ_2^2\left(\frac{R}{p}\right)^4 \sin 2i \left(14 - 15\sin^2 i\right) e^2 \sin 2\omega$$

$$\dot{\Omega}^{(2)} = -\frac{3}{2}nJ_2^2\left(\frac{R}{p}\right)^4 \cos i \left[\frac{9}{4} + \frac{3}{2}\beta - \sin^2 i \left(\frac{5}{2} + \frac{9}{4}\beta\right) + \frac{e^2}{4}\left(1 + \frac{5}{4}\sin^2 i\right)\right.$$

$$\left. + \frac{e^2}{8}\left(7 - 15\sin^2 i\right)\cos 2\omega\right] \tag{8.82}$$

$$\dot{\omega}^{(2)} = \frac{3}{16}nJ_2^2\left(\frac{R}{p}\right)^4 \left\{48 - 103\sin^2 i + \frac{215}{4}\sin^4 i + \left(7 - \frac{9}{2}\sin^2 i - \frac{45}{8}\sin^4 i\right)e^2\right.$$

$$+ 6\left(1 - \frac{3}{2}\sin^2 i\right)\left(4 - 5\sin^2 i\right)\beta$$

$$\left. - \frac{1}{4}\left[2\left(14 - 15\sin^2 i\right)\sin^2 i - \left(28 - 158\sin^2 i + 135\sin^4 i\right)e^2\right]\cos 2\omega\right\}$$

$$\dot{M}^{(2)} = \frac{3}{2}nJ_2^2\left(\frac{R}{p}\right)^4 \left\{\left(1 - \frac{3}{2}\sin^2 i\right)^2\beta^2 + \left[\frac{5}{4}\left(1 - \frac{5}{2}\sin^2 i + \frac{13}{8}\sin^4 i\right)\right.\right.$$

$$+ \frac{5}{8}\left(1 - \sin^2 i - \frac{5}{8}\sin^4 i\right)e^2$$

$$\left.\left. + \frac{1}{16}\sin^2 i \left(14 - 15\sin^2 i\right)\left(1 - \frac{5}{2}e^2\right)\cos 2\omega\right]\beta\right\}$$

$$+ \frac{3}{8}nJ_2^2\left(\frac{R}{p}\right)^4 \frac{1}{\beta}\left\{3\left[3 - \frac{15}{2}\sin^2 i + \frac{47}{8}\sin^4 i + \left(\frac{3}{2} - 5\sin^2 i + \frac{117}{16}\sin^4 i\right)e^2\right.\right.$$

$$\left. - \frac{1}{8}\left(1 + 5\sin^2 i - \frac{101}{8}\sin^4 i\right)e^4\right]$$

$$+ \frac{e^2}{8}\sin^2 i\left[70 - 123\sin^2 i + \left(56 - 66\sin^2 i\right)e^2\right]\cos 2\omega$$

$$\left. + \frac{27}{128}e^4\sin^4 i\cos 4\omega\right\}.$$

We see that a dependence on the argument of periapsis, ω, now appears in the second-order portion of the differential equations. This dependence is due to coupling between the first-order transformation periodics giving rise to second-order terms. These second-order terms do not contain any dependence on the satellite orbital anomaly, but there remains a dependence on the argument of periapsis. This angle has period 2π and satisfies the requirements for developing a transformation of variables to remove the dependence on ω. We can characterize the transformed differential equations using Eqs. 8.52 and 8.82 to obtain

$$\dot{n} = 0$$

$$\dot{e} = \dot{e}^{(2)}\left(\xi, \omega\right)$$

$$\frac{di}{dt} = \left(\frac{di}{dt}\right)^{(2)}\left(\xi, \omega\right)$$

$$\dot{\Omega} = \left\langle \dot{\Omega}^{(1)}\left(\xi, -\right)\right\rangle + \dot{\Omega}^{(2)}\left(\xi, \omega\right) \tag{8.83}$$

$$\dot{\omega} = \left\langle \dot{\omega}^{(1)}\left(\xi, -\right)\right\rangle + \dot{\omega}^{(2)}\left(\xi, \omega\right)$$

$$\dot{M} = n + \left\langle \dot{M}^{(1)}\left(\xi, -\right)\right\rangle + \dot{M}^{(2)}\left(\xi, \omega\right),$$

where the functional notation with a variable replaced by the symbol (-) denotes that the function is explicitly free of dependence on that missing variable. Here, we no longer have a dependence on the χ variable.

Equation 8.83 contains three fast variables Ω, ω, M and three slow variables n, e, i. Although the perturbations are second order in an absolute sense, they are first order relative to the main rate terms.

If we relate these equations back to the general description of the method of averaging for the slow variables Eq. 8.22, we have

$$f_1 = 0$$

$$f_2 = \dot{e}^{(2)} \tag{8.84}$$

$$f_3 = \left(\frac{di}{dt}\right)^{(2)},$$

and for the fast variables Eq. 8.23, we have

$$\omega_1 = n + \left\langle \dot{M}^{(1)}\right\rangle \qquad u_1 = \dot{M}^{(2)}$$

$$\omega_2 = \left\langle \dot{\Omega}^{(1)}\right\rangle \qquad u_2 = \dot{\Omega}^{(2)} \tag{8.85}$$

$$\omega_3 = \left\langle \dot{\omega}^{(1)}\right\rangle \qquad u_3 = \dot{\omega}^{(2)}.$$

We have previously introduced the $\langle F \rangle$ symbol to indicate the average of the function F over one period of an angular variable. But as we see in Eq. 8.82, those averaged functions now have a dependence on another angular variable, ω. The argument of periapsis is also periodic with period 2π. From the symbolic Eq. 8.83, the rate at which ω varies is proportional to J_2. Hence, the period will be much longer than the period of the satellite, and we refer to these periodics as "long-period periodics." We seek a transformation of variables that removes the dependence on argument of periapsis. We now apply the method of averaging again; that is, we will perform a second averaging over one period of the argument of periapsis. The doubly averaged result is denoted by the symbol $\langle\langle F \rangle\rangle$.

Using Eq. 8.28 for the slow variables, we have

$$M_i = f_i - \omega_\beta \frac{\partial \eta_i}{\partial \chi_\beta}$$

$$= \langle f_i \rangle - \omega_\beta \left\langle \frac{\partial \eta_i}{\partial \chi_\beta} \right\rangle$$

$$= \langle f_i \rangle .$$

Then the transformed equations are just the average with respect to ω of the first three of Eq. 8.82. We obtain

$$\langle\langle \dot{n} \rangle\rangle^{(2)} = 0 \tag{8.86}$$

$$\langle\langle \dot{e} \rangle\rangle^{(2)} = \frac{1}{2\pi} \int_0^{2\pi} \left[-\frac{3}{32} n J_2^2 \left(\frac{R}{p}\right)^4 \sin^2 i \left(14 - 15\sin^2 i\right) e\beta^2 \right] \sin 2\omega \, d\omega$$

$$= -\frac{3}{32} n J_2^2 \left(\frac{R}{p}\right)^4 \sin^2 i \left(14 - 15\sin^2 i\right) e\beta^2 \frac{1}{2\pi} \int_0^{2\pi} \sin 2\omega \, d\omega$$

$$= 0 \tag{8.87}$$

$$\left\langle\left\langle \frac{di}{dt} \right\rangle\right\rangle^{(2)} = \frac{1}{2\pi} \int_0^{2\pi} \left[\frac{3}{64} n J_2^2 \left(\frac{R}{r}\right)^4 \sin 2i \left(14 - 15\sin^2 i\right) e^2 \sin 2\omega \right] d\omega$$

$$= \frac{3}{64} n J_2^2 \left(\frac{R}{r}\right)^4 \sin 2i \left(14 - 15\sin^2 i\right) e^2 \frac{1}{2\pi} \int_0^{2\pi} \sin 2\omega \, d\omega$$

$$= 0. \tag{8.88}$$

Using Eq. 8.29 for the fast variables, we have

$$\Omega_\alpha + \omega_\beta \frac{\partial \phi_\alpha}{\partial \chi_\beta} = \eta_j \frac{\partial \omega_\alpha}{\partial x_j} + u_\alpha \left(\boldsymbol{\xi}, \boldsymbol{\chi} \right)$$

$$\Omega_\alpha = -\omega_\beta \left\langle \frac{\partial \phi_\alpha}{\partial \chi_\beta} \right\rangle + \langle \eta_j \rangle \frac{\partial \omega_\alpha}{\partial x_j} + \langle u_\alpha \rangle$$

$$= \langle \eta_j \rangle \frac{\partial \omega_\alpha}{\partial x_j} + \langle u_\alpha \rangle$$

$$= \langle u_\alpha \rangle .$$

Then the transformed equations are just the average with respect to ω of the last three of Eq. 8.82. We obtain

$$\langle\langle \dot{\Omega} \rangle\rangle^{(2)} = -\frac{3}{2} n J_2^2 \left(\frac{R}{p} \right)^4 \cos i \left[\frac{9}{4} + \frac{3}{2}\beta - \sin^2 i \left(\frac{5}{2} + \frac{9}{4}\beta \right) + \frac{e^2}{4} \left(1 + \frac{5}{4}\sin^2 i \right) \right]$$

$$(8.89)$$

$$\langle\langle \dot{M} \rangle\rangle^{(2)} = \frac{15}{16} n J_2^2 \left(\frac{R}{p} \right)^4 \beta \left[\left(2 - 5\sin^2 i + \frac{13}{4}\sin^4 i \right) + \left(1 - \sin^2 i - \frac{5}{8}\sin^4 i \right) e^2 \right.$$

$$+ \frac{8}{5} \left(1 - \frac{3}{2}\sin^2 i \right)^2 \beta \right] + \frac{9}{8} n J_2^2 \left(\frac{R}{p} \right)^4 \frac{1}{\beta} \left[3 - \frac{15}{2}\sin^2 i + \frac{47}{8}\sin^4 i \right.$$

$$+ \left(\frac{3}{2} - 5\sin^2 i + \frac{117}{16}\sin^4 i \right) e^2 - \frac{1}{8} \left(1 + 5\sin^2 i - \frac{101}{8}\sin^4 i \right) e^4 \right]$$

$$(8.90)$$

$$\langle\langle \dot{\omega} \rangle\rangle^{(2)} = \frac{3}{4} n J_2^2 \left(\frac{R}{p} \right)^4 \left[12 - \frac{103}{4} \sin^2 i + \frac{215}{16}\sin^4 i + \left(\frac{7}{4} - \frac{9}{8}\sin^2 i - \frac{45}{32}\sin^4 i \right) e^2 \right.$$

$$+ \frac{3}{2} \left(1 - \frac{3}{2}\sin^2 i \right) \left(4 - 5\sin^2 i \right) \beta \right]. \qquad (8.91)$$

The transformed differential Eqs. 8.86–8.91 are free of the argument of periapsis, ω. The method of averaging requires that the transformation be purely periodic in the argument of periapsis angle. Although we have found the transformed equations (Eqs. 8.86–8.91), we do not yet know the transformation that produced them. That transformation will be determined by integration of explicit constraint equations provided in the method of averaging.

The periodics for the slow variables are described by Eq. 8.33 and are given by the solution of the equation

$$\delta x_{LP} = \frac{1}{\langle \dot{\omega}^{(1)} \rangle} \int [\dot{x} - \langle \dot{x} \rangle] d\omega,$$

where the subscript indicates a long-period periodic with frequency of argument of periapsis. Then we have

$$
\delta e_{LP} = \frac{1}{\langle \dot{\omega}^{(1)} \rangle} \int \left[-\frac{3}{32} n J_2^2 \left(\frac{R}{p} \right)^4 \sin^2 i \left(14 - 15\sin^2 i \right) e\beta^2 \right] \sin 2\omega d\omega
$$

$$
= \frac{1}{\langle \dot{\omega}^{(1)} \rangle} \frac{3}{64} n J_2^2 \left(\frac{R}{p} \right)^4 \sin^2 i \left(14 - 15\sin^2 i \right) e\beta^2 \cos 2\omega
$$

$$
= \frac{1}{16} J_2 \left(\frac{R}{p} \right)^2 \gamma \sin^2 i \left(14 - 15\sin^2 i \right) e\beta^2 \cos 2\omega, \tag{8.92}
$$

where

$$
\gamma = \frac{1}{4 - 5\sin^2 i}
$$

and

$$
\delta i_{LP} = \frac{1}{\langle \dot{\omega}^{(1)} \rangle} \int \left[\frac{3}{64} n J_2^2 \left(\frac{R}{p} \right)^4 \sin 2i \left(14 - 15\sin^2 i \right) e^2 \right] \sin 2\omega d\omega
$$

$$
= -\frac{1}{32} n J_2 \left(\frac{R}{p} \right)^2 \gamma \sin 2i \left(14 - 15\sin^2 i \right) e^2 \cos 2\omega. \tag{8.93}
$$

The periodics for the fast variables are described by Eq. 8.36 and are given by the solution of the equation

$$
\phi_\alpha = \frac{1}{\omega_\beta} \frac{\partial \omega_\alpha}{\partial x_j} \int [\eta_j - \langle \eta_j \rangle] d\omega + \frac{1}{\omega_\beta} \int [u_\alpha - \langle u_\alpha \rangle] d\omega. \tag{8.94}
$$

The periodics for the slow variables all have zero mean. So Eq. 8.94 reduces to

$$
\phi_\alpha = \frac{1}{\langle \dot{\omega}^{(1)} \rangle} \frac{\partial \omega_\alpha}{\partial x_j} \int \eta_j d\omega + \frac{1}{\langle \dot{\omega}^{(1)} \rangle} \int [u_\alpha - \langle u_\alpha \rangle] d\omega, \tag{8.95}
$$

and we have

$$
\delta \Omega_{LP} = \frac{1}{\langle \dot{\omega}^{(1)} \rangle} \frac{\partial \langle \dot{\Omega}^{(1)} \rangle}{\partial e} \int \delta e d\omega + \frac{1}{\langle \dot{\omega}^{(1)} \rangle} \frac{\partial \langle \dot{\Omega}^{(1)} \rangle}{\partial i} \int \delta i d\omega
$$

$$
+ \frac{1}{\langle \dot{\omega}^{(1)} \rangle} \int [u_\Omega - \langle u_\Omega \rangle] d\omega \tag{8.96}
$$

$$\delta\omega_{LP} = \frac{1}{\langle\dot\omega^{(1)}\rangle}\frac{\partial\langle\dot\omega^{(1)}\rangle}{\partial e}\int \delta e\,d\omega + \frac{1}{\langle\dot\omega^{(1)}\rangle}\frac{\partial\langle\dot\omega^{(1)}\rangle}{\partial i}\int \delta i\,d\omega$$

$$+ \frac{1}{\dot\omega}\int [u_\omega - \langle u_\omega\rangle]d\omega \tag{8.97}$$

$$\delta M_{LP} = \frac{1}{\langle\dot\omega^{(1)}\rangle}\frac{\partial\langle\dot M^{(1)}\rangle}{\partial e}\int \delta e\,d\omega + \frac{1}{\langle\dot\omega^{(1)}\rangle}\frac{\partial\langle\dot M^{(1)}\rangle}{\partial i}\int \delta i\,d\omega$$

$$+ \frac{1}{\langle\dot\omega^{(1)}\rangle}\int [u_M - \langle u_M\rangle]d\omega. \tag{8.98}$$

We need the following partial derivatives:

$$\frac{\partial\langle\dot\Omega^{(1)}\rangle}{\partial e} = -\frac{3}{2}\frac{n J_2 r_e^2}{a^2\beta^4}\cos i\left(4\frac{e}{\beta^2}\right) = 4\frac{e}{\beta^2}\langle\dot\Omega^{(1)}\rangle$$

$$\frac{\partial\langle\dot\Omega^{(1)}\rangle}{\partial i} = -\frac{3}{2}\frac{n J_2 r_e^2}{a^2\beta^4}(-\sin i) = -\frac{\sin i}{\cos i}\langle\dot\Omega^{(1)}\rangle$$

$$\frac{\partial\langle\dot\omega^{(1)}\rangle}{\partial e} = \frac{3}{4}\frac{n J_2 r_e^2}{a^2\beta^4}\left(4 - 5\sin^2 i\right)\left(4\frac{e}{\beta^2}\right) = 4\frac{e}{\beta^2}\langle\dot\omega^{(1)}\rangle$$

$$\frac{\partial\langle\dot\omega^{(1)}\rangle}{\partial i} = \frac{3}{4}\frac{n J_2 r_e^2}{a^2\beta^4}(-10\sin i\cos i) = (-5\gamma\sin 2i)\langle\dot\omega^{(1)}\rangle \tag{8.99}$$

$$\frac{\partial\langle\dot M^{(1)}\rangle}{\partial e} = \frac{3}{4}\frac{n J_2 r_e^2}{a^2\beta^3}\left(-1 + 3\cos^2 i\right)\left(3\frac{e}{\beta^2}\right) = 3\frac{e}{\beta^2}\langle\dot M^{(1)}\rangle$$

$$\frac{\partial\langle\dot M^{(1)}\rangle}{\partial i} = \frac{3}{4}\frac{n J_2 r_e^2}{a^2\beta^3}(-6\sin i\cos i) = \frac{3}{4}\frac{n J_2 r_e^2}{a^2\beta^3}(-3\sin 2i).$$

Using the results of Eqs. 8.92, 8.93, and 8.99 in Eq. 8.96, we obtain

$$\delta\Omega_{LP} = \frac{1}{\langle\dot\omega^{(1)}\rangle}4\frac{e}{\beta^2}\langle\dot\Omega^{(1)}\rangle\frac{1}{16}J_2\left(\frac{R}{p}\right)^2\gamma\sin^2 i\left(14 - 15\sin^2 i\right)e\beta^2\int \cos 2\omega\,d\omega$$

$$+ \frac{1}{\langle\dot\omega^{(1)}\rangle}\left(\frac{-\sin i}{\cos i}\right)\langle\dot\Omega^{(1)}\rangle\left(-\frac{1}{32}nJ_2\right)\left(\frac{R}{p}\right)^2$$

$$\times\gamma\sin 2i\left(14 - 15\sin^2 i\right)e^2\int \cos 2\omega\,d\omega$$

$$+ \frac{1}{\langle\dot\omega^{(1)}\rangle}\int\left[-\frac{3}{2}nJ_2^2\left(\frac{R}{p}\right)^4\cos i\frac{e^2}{8}\left(7 - 15\sin^2 i\right)\right]\cos 2\omega\,d\omega,$$

which simplifies to

$$\delta\Omega_{LP} = -\frac{1}{8}J_2\left(\frac{R}{p}\right)^2 \gamma e^2 \cos i \left[\left(7 - 15\sin^2 i\right) + \frac{5}{2}\gamma\sin^2 i\left(14 - 15\sin^2 i\right)\right]\sin 2\omega.$$
$$(8.100)$$

Using the results of Eqs. 8.92, 8.93, and 8.99 in Eq. 8.97, we obtain

$$\delta\omega_{LP} = \frac{1}{\langle\dot\omega^{(1)}\rangle}4\frac{e}{\beta^2}\langle\dot\omega^{(1)}\rangle\frac{1}{16}nJ_2\left(\frac{R}{p}\right)^2\gamma\sin^2 i\left(14 - 15\sin^2 i\right)e\beta^2\int\cos 2\omega d\omega$$

$$+\frac{1}{\langle\dot\omega^{(1)}\rangle}(-5\gamma\sin 2i)\langle\dot\omega^{(1)}\rangle\left(-\frac{1}{32}\right)nJ_2\left(\frac{R}{p}\right)^2$$

$$\times\,\gamma\sin 2i\left(14 - 15\sin^2 i\right)e^2\int\cos 2\omega d\omega$$

$$-\frac{1}{32}J_2\left(\frac{R}{p}\right)^2\gamma$$

$$\times\left[2\left(14 - 15\sin^2 i\right)\sin^2 i - \left(28 - 158\sin^2 i + 135\sin^4 i\right)e^2\right]\sin 2\omega,$$
$$(8.101)$$

which simplifies to

$$\delta\omega_{LP} = -\frac{1}{32}J_2\left(\frac{R}{p}\right)^2\gamma\left\{2\sin^2 i\left(14 - 15\sin^2 i\right)\left[1 - \gamma\left(13 - 15\sin^2 i\right)e^2\right]\right.$$

$$\left. -\left(28 - 158\sin^2 i + 135\sin^4 i\right)e^2\right\}\sin 2\omega.\qquad(8.102)$$

Using the results of Eqs. 8.92, 8.93, and 8.99 in Eq. 8.98, we obtain

$$\delta M_{LP} = \frac{1}{\langle\dot\omega^{(1)}\rangle}\frac{3}{4}\frac{nJ_2 r_e^2}{a^2\beta^3}\left(-1 + 3\cos^2 i\right)\left(3\frac{e}{\beta^2}\right)\frac{1}{16}J_2\left(\frac{R}{p}\right)^2$$

$$\times\,\gamma\sin^2 i\left(14 - 15\sin^2 i\right)e\beta^2\int\cos 2\omega d\omega$$

$$+\frac{1}{\langle\dot\omega^{(1)}\rangle}\frac{3}{4}\frac{nJ_2 r_e^2}{a^2\beta^3}(-3\sin 2i)$$

$$\times\int\left(-\frac{1}{32}\right)nJ_2\left(\frac{R}{p}\right)^2\gamma\sin 2i\left(14 - 15\sin^2 i\right)e^2 d\omega$$

$$+\frac{1}{\dot\omega^{(1)}}\int\left[\frac{3}{2}nJ_2^2\left(\frac{R}{p}\right)^4\frac{1}{16}\sin^2 i\left(14 - 15\sin^2 i\right)\left(1 - \frac{5}{2}e^2\right)\beta\right]\cos 2\omega d\omega$$

$$+ \frac{1}{\dot{\omega}^{(1)}} \int \left[\frac{3}{8} n J_2^2 \left(\frac{R}{p}\right)^4 \frac{1}{\beta} \frac{e^2}{8} \sin^2 i \left[70 - 123\sin^2 i + \left(56 - 66\sin^2 i\right) e^2 \right] \right]$$

$$\times \cos 2\omega d\omega$$

$$+ \frac{1}{\dot{\omega}^{(1)}} \int \left[\frac{3}{8} n J_2^2 \left(\frac{R}{p}\right)^4 \frac{1}{\beta} \frac{27}{128} e^4 \sin^4 i \right] \cos 4\omega d\omega, \tag{8.103}$$

which simplifies to

$$\delta M_{LP} = \frac{1}{16} J_2 \left(\frac{R}{p}\right)^2 \beta \gamma \sin^2 i \left(14 - 15\sin^2 i\right) \left(1 - e^2\right) \sin 2\omega$$

$$+ \frac{1}{32} J_2 \left(\frac{R}{p}\right)^2 \frac{1}{\beta} \gamma \sin^2 i \left[\left(70 - 123\sin^2 i\right) e^2 + \left(56 - 66\sin^2 i\right) e^4 \right] \sin 2\omega$$

$$+ \frac{27}{1024} J_2 \left(\frac{R}{p}\right)^2 \frac{1}{\beta} e^4 \gamma \sin^4 i \sin 4\omega. \tag{8.104}$$

In summary, we have

$$\delta n_{LP} = 0$$

$$\delta e_{LP} = \frac{1}{16} J_2 \left(\frac{R}{p}\right)^2 \gamma \sin^2 i \left(14 - 15\sin^2 i\right) e \beta^2 \cos 2\omega$$

$$\delta i_{LP} = -\frac{1}{32} n J_2 \left(\frac{R}{p}\right)^2 \gamma \sin 2i \left(14 - 15\sin^2 i\right) e^2 \cos 2\omega$$

$$\delta \Omega_{LP} = -\frac{1}{8} J_2 \left(\frac{R}{p}\right)^2 \gamma e^2 \cos i \left[\left(7 - 15\sin^2 i\right) + \frac{5}{2}\gamma \sin^2 i \left(14 - 15\sin^2 i\right)\right] \sin 2\omega \tag{8.105}$$

$$\delta \omega_{LP} = -\frac{1}{32} J_2 \left(\frac{R}{p}\right)^2 \gamma \left\{ 2\sin^2 i \left(14 - 15\sin^2 i\right) \left[1 - \gamma \left(13 - 15\sin^2 i\right) e^2\right] \right.$$

$$\left. - \left(28 - 158\sin^2 i + 135\sin^4 i\right) e^2 \right\} \sin 2\omega$$

$$\delta M_{LP} = \frac{1}{16} J_2 \left(\frac{R}{p}\right)^2 \beta \gamma \sin^2 i \left(14 - 15\sin^2 i\right) \left(1 - \frac{5}{2} e^2\right) \sin 2\omega$$

$$+ \frac{1}{32} J_2 \left(\frac{R}{p}\right)^2 \frac{1}{\beta} \gamma \sin^2 i \left[\left(70 - 123\sin^2 i\right) e^2 + \left(56 - 66\sin^2 i\right) e^4 \right] \sin 2\omega$$

$$+ \frac{27}{1024} J_2 \left(\frac{R}{p}\right)^2 \frac{1}{\beta} e^4 \gamma \sin^4 i \sin 4\omega.$$

This set of equations completes the determination of the transformation of variables that transformed our system of mean differential equations (Eq. 8.82) into a new system of differential equations (Eqs. 8.86–8.91) that are free of dependence on the argument of periapsis. These equations have no remaining dependence on angles other than the inclination. For this reason the orbital elements are often referred to as doubly averaged, doubly transformed, or mean elements. We emphasize the fact that we have not lost knowledge of the osculating motion of the satellite. For a given state of the doubly averaged elements, Eqs. 8.92, 8.93, 8.100, 8.102, and 8.104 tell us precisely how to add the long-period periodics onto the doubly averaged state to recover the singly averaged state. Then Eq. 8.75 tells us precisely how to add the short-period periodics onto the singly averaged state to recover the osculating state.

The transformed differential equations are the sum of the first-order parts from Eq. 8.52 and the second-order parts from Eqs. 8.86 through 8.91. We must also remember to include the original zeroth-order term in the mean anomaly. Thus, we have

$$\langle\langle \dot{n} \rangle\rangle = 0$$

$$\langle\langle \dot{e} \rangle\rangle = 0$$

$$\left\langle\left\langle \frac{di}{dt} \right\rangle\right\rangle = 0$$

$$\langle\langle \dot{\Omega} \rangle\rangle = -\frac{3}{2}\frac{n J_2 r_e^2}{a^2 \beta^4} \cos i$$

$$-\frac{3}{2} n J_2^2 \left(\frac{R}{p}\right)^4 \cos i \left[\frac{9}{4} + \frac{3}{2}\beta - \sin^2 i \left(\frac{5}{2} + \frac{9}{4}\beta\right) + \frac{e^2}{4}\left(1 + \frac{5}{4}\sin^2 i\right)\right]$$

(8.106)

$$\langle\langle \dot{\omega} \rangle\rangle = \frac{3}{4}\frac{n J_2 r_e^2}{a^2 \beta^4}\left(-1 + 5\cos^2 i\right) + \frac{3}{4} n J_2^2 \left(\frac{R}{p}\right)^4 \left[12 - \frac{103}{4}\sin^2 i + \frac{215}{16}\sin^4 i\right.$$

$$\left. + \left(\frac{7}{4} - \frac{9}{8}\sin^2 i - \frac{45}{32}\sin^4 i\right)e^2 + \frac{3}{2}\left(1 - \frac{3}{2}\sin^2 i\right)\left(4 - 5\sin^2 i\right)\beta\right]$$

$$\langle\langle \dot{M} \rangle\rangle = n + \frac{3}{4}\frac{n J_2 r_e^2}{a^2 \beta^3}\left(-1 + 3\cos^2 i\right) + \frac{15}{16} n J_2^2 \left(\frac{R}{p}\right)^4 \beta\left[\left(2 - 5\sin^2 i + \frac{13}{4}\sin^4 i\right)\right.$$

$$\left. + \left(1 - \sin^2 i - \frac{5}{8}\sin^4 i\right)e^2 + \frac{8}{5}\left(1 - \frac{3}{2}\sin^2 i\right)^2 \beta\right]$$

$$+ \frac{9}{8} n J_2^2 \left(\frac{R}{p}\right)^4 \frac{1}{\beta}\left[3 - \frac{15}{2}\sin^2 i + \frac{47}{8}\sin^4 i + \left(\frac{3}{2} - 5\sin^2 i + \frac{117}{16}\sin^4 i\right)e^2\right.$$

$$\left. - \frac{1}{8}\left(1 + 5\sin^2 i - \frac{101}{8}\sin^4 i\right)e^4\right].$$

Notice that the majority of the right-hand sides of Eq. 8.106 are terms that depend on the square of the J_2 zonal harmonic. *This dependency clearly demonstrates the danger in thinking that the method of averaging simply means "averaging" the right-hand side of the original differential equations.* Such an approach would entirely miss the presence and importance of the coupled terms. Any result of such an approach would be wrong.

The transformed differential equations can be solved to give

$$n = n_0$$

$$e = e_0$$

$$i = i_0$$

$$\Omega = \Omega_0 + \langle\langle\dot{\Omega}_0\rangle\rangle t$$

$$\omega = \omega_0 + \langle\langle\dot{\omega}_0\rangle\rangle t$$

$$M = M_0 + n_0 t + \langle\langle\dot{M}_0\rangle\rangle t.$$

The long-period periodics are provided by Eq. 8.105, and the short-period periodics are provided by Eq. 8.75. The final result is

$$n = n_0 + \delta n_{SP}$$

$$e = e_0 + \delta e_{LP} + \delta e_{SP}$$

$$i = i_0 + \delta i_{LP} + \delta i_{SP}$$

$$\Omega = \Omega_0 + \langle\langle\dot{\Omega}_0\rangle\rangle t + \delta\Omega_{LP} + \delta\Omega_{SP}$$

$$\omega = \omega_0 + \langle\langle\dot{\omega}_0\rangle\rangle t + \delta\omega_{LP} + \delta\omega_{SP}$$

$$M = M_0 + n_0 t + \langle\langle\dot{M}_0\rangle\rangle t + \delta M_{LP} + \delta M_{SP}.$$

To summarize, we began with the Lagrange planetary equations containing only the perturbation due to an oblate planet. We used the MOA to develop a transformation of variables that removed dependence on the orbital anomaly. The resulting transformed equations still contained a dependence on the argument of periapsis. We developed a second transformation of variables that removed dependence on the argument of periapsis. The resulting transformed equations were able to be integrated as a closed-form function of time. The solution to these equations is transformed back to the original state space by applying the long-period and short-period periodics.

8.4   Effects of Earth Oblateness on Orbits

Equations 8.106 describe the secular evolution of the perturbed orbit. These equations dramatically demonstrate the insight that can be gained from explicit formulas developed through analytical perturbation methods. First, we consider the equation for the rate of change of the right ascension of the ascending node (to order J_2)

$$\dot{\Omega} = -\frac{3}{2}\frac{n J_2 r_e^2}{a^2 \beta^4}\cos i. \tag{8.107}$$

We observe that the rate is directly proportional to the cosine of the inclination.

Let us consider an example of a circular 300-km altitude orbit. Figure 8.1 shows that the line of nodes regresses (moves westward) for inclinations less than 90° and progresses (moves eastward) for inclinations greater than 90°. A polar orbit with an inclination of 90° is stable and neither regresses nor progresses.

A very important application that this insight provides is in the field of Earth observation. It is desirable for satellite photography or mapping to have consistent lighting conditions. Figure 8.2a illustrates the lighting conditions over a year for a polar orbiting satellite (where $i = 90°$). Since the right ascension of the ascending node rate is zero, the orientation of the orbital plane is fixed in inertial space. As the Earth travels around the Sun, the lighting conditions change from direct overhead lighting in Summer to dawn and dusk lighting in the Fall. By the Winter there again is direct overhead lighting followed by dawn and dusk lighting in the Spring.

In contrast, Fig. 8.2b illustrates a more desirable remote sensing arrangement in which the lighting conditions are the same throughout the year. To achieve this outcome, we would need the line of nodes to rotate by 90° every 3 months. We examine Eq. 8.107 to determine if that desired rate can be achieved. We continue to use our example 300-km circular orbit and determine the required inclination. Then

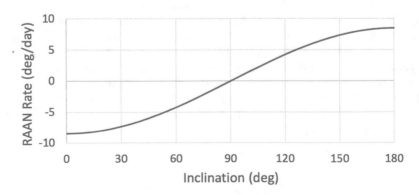

Fig. 8.1 The right ascension of the ascending node rate is primarily controlled by the inclination

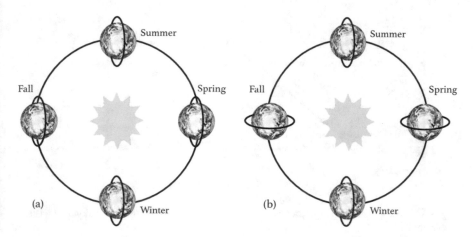

Fig. 8.2 (**a**) Orbit plane orientation for a polar orbit; (**b**) desired "Sun-synchronous" orbital plane orientation. [Earth image credit: NOAA National Environmental Satellite, Data, and Information Service (NESDIS)]

Eq. 8.107 gives

$$i = \cos^{-1}\left(-\frac{2a^2\beta^4\dot{\Omega}}{3n J_2 r_e^2}\right), \tag{8.108}$$

where

$$\dot{\Omega} = \frac{360\,\text{deg}}{365\,\text{days}} \approx 0.986\,\text{deg}/\text{day}.$$

For our example, the required inclination is 96.7°, which results in the orbital plane progressing about 1° each day. For different altitudes, the required inclination is different. Equation 8.108 allows us to compute the proper inclination as a function of altitude for this behavior. Orbits designed in this way are called "Sun-synchronous orbits." They are designed such that the orbital plane remains synchronized with the apparent motion of the Sun. The result is that a remote sensing satellite has similar lighting conditions (local time of day) throughout the year.

Next, we consider the equation for the rate of change of argument of perigee (to order J_2)

$$\dot{\omega} = \frac{3}{4}\frac{n J_2 r_e^2}{a^2\beta^4}\left(-1 + 5\cos^2 i\right). \tag{8.109}$$

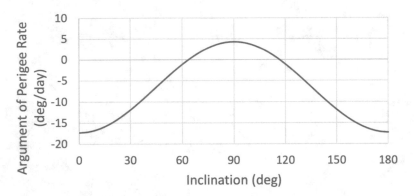

Fig. 8.3 The argument of perigee rate is also primarily controlled by the inclination and has two stable points

As with the right ascension of the ascending node above, we see that the largest effect on the argument of perigee rate is controlled by the inclination. We consider a 300-km altitude orbit with an eccentricity of 0.1.

Figure 8.3 shows that the argument of perigee either regresses or progresses depending on the inclination. The stable points are determined by solving

$$-1 + 5\cos^2 i = 0. \tag{8.110}$$

The solutions of Eq. 8.110 are

$$i = 63.43 \, \text{deg} \quad \text{and} \quad i = 116.57 \, \text{deg}.$$

These solutions are known as the critical inclinations. The prograde ($i = 63.4°$) solution, which is less expensive to reach due to the eastward rotation of the Earth, has a very important application in satellite orbit design. In our example, we used an orbit with a moderate eccentricity. That means the orbit will be closest to the Earth as it passes through perigee, and any Earth observing will be performed at a closer range and better resolution. If improved resolution was desired to occur repeatedly, then an orbit at the critical inclination would produce such benefits. On the other hand, critically inclined orbits can also maintain a long apogee dwell over a region of interest, which is commonly desired in satellite communication applications. The "Molniya orbit" is an example of a 12-h period orbit at 63.4° inclination to control the apogee to remain over a given hemisphere (e.g., the northern hemisphere).

The final secular effect due to the Earth oblateness is on the mean anomaly rate (to order J_2):

$$\dot{M} = \frac{3}{4} \frac{n J_2 r_e^2}{a^2 \beta^3} \left(-1 + 3\cos^2 i\right). \tag{8.111}$$

Depending on the inclination, this term simply increases or decreases the rate of the mean anomaly. Generally, this property has not been a factor in satellite orbit design.

References

Cefola, P. J., Folcik, Z., Di-Costanzo, R., Bernard, N., Setty, S., & San Juan, J. F. (2014). Revisiting the DSST standalone orbit propagator. *Advances in the Astronautical Sciences, 152,* 2891–2914.

Chao, C.-C., & Hoots, F. R. (2018). *Applied orbit perturbations and maintenance* (2nd ed.). El Segundo: The Aerospace Press.

Danielson, D. A., Neta, B., & Early, L. W. (1994). *Semianalytic satellite theory (SST): Mathematical algorithms.* Naval Postgraduate School. Report Number NPS-MA94-001.

Kwok, J. H. (1986). *The long-term orbit predictor (LOP).* JPL Technical Report EM 312/86-151.

Liu, J. F. .F., & Alford, R. L. (1980). Semianalytic theory for a close-earth artificial satellite. *Journal of Guidance and Control, 3*(4), 304–311.

Morrison, J. A. (1966). Generalized method of averaging and the von Zeipel Method. In R. Duncombe, & V. Szebehely (Eds.), *Progress in astronautics and aeronautics—methods in astrodynamics and celestial mechanics* (Vol. 17). London: Academic Press.

Steinberg, S. (1984). Lie series and nonlinear ordinary differential equations. *Journal of Mathematical Analysis and Applications, 101*(1), 39–63.

von Zeipel, H. (1916). Recherches sur le mouvement des petites planètes. *Arkiv för Matematik, Astronomi och Fysik, 11*(1).

Chapter 9
Periodic Solutions in Nonlinear Oscillations

9.1 Secular Terms

Modern perturbation theory is attributed to Poincaré. Early attempts by astronomers were plagued by the appearance of *terms increasing with time*. Such terms are referred to as secular terms and may affect the convergence of the solution. A large variety of perturbation methods suppressing the secular terms have been developed, including a method by Poincaré and Lindstedt. (See Meirovitch (1970), pp. 293–302.)

We often encounter nonlinear differential equations in celestial mechanics. Since we usually expect the motion to be bounded, we should be wary if we see the appearance of secular terms such as f^n or $f^n \cos mf$ in our solution, where f is the true anomaly. Secular terms will cause our solution to grow without bound. We certainly do not expect a conservative system to behave in that way.

However, note that we sometimes must keep the secular terms due to the nature of the problem. For example, in the case of Earth's oblateness, the right ascension of the ascending node and argument of perigee are explicit functions of time and grow without bound. But that secular growth is just a rotation of the line of nodes and the argument of perigee around the Earth.

Another acceptable example is the short-period oscillation in the right ascension of the ascending node due to Earth's oblateness ($\delta\Omega_{SP}$). Equation 8.66 contains the term

$$(f - M + e \sin f). \tag{9.1}$$

Both f and M will grow secularly with time, but their difference is purely periodic and bounded for all time.

The importance of the Poincaré–Lindstedt method (also known as the Lindstedt–Poincaré method) can hardly be exaggerated as it applies not only to problems in celestial mechanics but also to problems in engineering, physics, mechanical

J. M. Longuski et al., *Introduction to Orbital Perturbations*, Space Technology Library 40, https://doi.org/10.1007/978-3-030-89758-1_9

systems, and many others. In this chapter, we apply the method to the pendulum problem, the advance of Mercury's perihelion due to general relativity, and to the motion of a satellite in the equatorial plane of an oblate planet. The key equation that governs the method is given in Sect. 9.3, namely Eq. 9.26, which applies to many nonlinear dynamics problem. Analysis of this equation provides insight into cases in which time should not appear (due to known behavior of the system) or that are indeed secular from a physics point of view.

9.2 When Secular Terms Should Not Appear

If physical and mathematical deduction proves that a quantity is bounded, then the solution should be free of secular terms because their presence is due to a *defect of the method*—not because of the *nature of the problem*.

9.2.1 Example of Defective Method: Pendulum Problem

Consider the simple pendulum in Fig. 9.1. The equation of motion is

$$ml^2\ddot{\theta} + mgl \sin \theta = 0, \tag{9.2}$$

or

$$\ddot{\theta} + \frac{g}{l} \sin \theta = 0. \tag{9.3}$$

We let

$$\omega^2 = \frac{g}{l}, \tag{9.4}$$

so we have

$$\ddot{\theta} + \omega^2 \sin \theta = 0. \tag{9.5}$$

Problem: Find the solution of Eq. 9.5 for the initial conditions (Fig. 9.2)

$$t_0 = 0$$
$$\theta_0 = \eta \tag{9.6}$$
$$\left(\frac{d\theta}{dt}\right)_0 = 0.$$

Fig. 9.1 Simple pendulum in
a uniform gravity field where
the massless rod has a
constant length, l

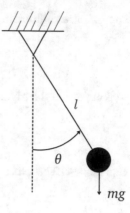

Fig. 9.2 Simple pendulum in
a uniform gravity field where
η may be large

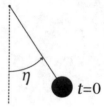

For the linear problem

$$\ddot{\theta} + \omega^2\theta = 0 \tag{9.7}$$

$$\theta = A\cos\omega t + B\sin\omega t. \tag{9.8}$$

Using the initial conditions (Eqs. 9.6) in Eq. 9.8, the solution for the linear
problem is

$$\theta = \eta\cos\omega t. \tag{9.9}$$

Now we try to improve the solution (*using a defective method* for illustrative
purposes). Let us write (by expanding $\sin\theta$ in a series)

$$\ddot{\theta} + \omega^2\left(\theta - \frac{\theta^3}{3!} + \frac{\theta^5}{5!} - \ldots\right) = 0. \tag{9.10}$$

Rearranging Eq. 9.10, we have

$$\ddot{\theta} + \omega^2\theta = \frac{\omega^2}{6}\theta^3, \tag{9.11}$$

where we have truncated the series for $\sin\theta$ to include only terms up to θ^3. Now we
integrate Eq. 9.11, considering the right-hand side as a forcing function given by the

approximate solution in Eq. 9.9:

$$\ddot{\theta} + \omega^2\theta = \frac{\omega^2}{6}\eta^3 \cos^3 \omega t. \tag{9.12}$$

Making use of the trigonometric identity

$$\cos^3 \omega t = \frac{1}{4}\left[\cos(3\omega t) + 3\cos\omega t\right], \tag{9.13}$$

Equation 9.12 becomes

$$\ddot{\theta} + \omega^2\theta = \frac{\omega^2\eta^3}{24}\left[\cos(3\omega t) + 3\cos\omega t\right]$$
$$= \theta_1 + \theta_3, \tag{9.14}$$

where

$$\theta_1 = \frac{\omega^2\eta^3}{8}\cos\omega t \tag{9.15}$$

$$\theta_3 = \frac{\omega^2\eta^3}{24}\cos(3\omega t). \tag{9.16}$$

Let the particular solution for the θ_3 forcing function be written as

$$\theta_{3p} = b\cos(3\omega t) \tag{9.17}$$

so that

$$\dot{\theta}_{3p} = -3\omega b\sin(3\omega t) \tag{9.18}$$

$$\ddot{\theta}_{3p} = -9\omega^2 b\cos(3\omega t). \tag{9.19}$$

Substituting Eqs. 9.17–9.19 into Eq. 9.14 (with θ_1 omitted) we have

$$-9\omega^2 b\cos(3\omega t) + \omega^2 b\cos(3\omega t) = \frac{\omega^2\eta^3}{24}\cos(3\omega t) \tag{9.20}$$

and

$$-8\omega^2 b = \frac{\omega^2\eta^3}{24} \tag{9.21}$$

so that

$$b = -\frac{\eta^3}{192}.$$ (9.22)

For the particular solution, θ_{3p}, we have

$$\theta_{3p} = -\frac{\eta^3}{192}\cos 3\omega t.$$ (9.23)

However, for the particular solution, θ_{1p}, since the forcing function has the same frequency as the homogenous solution, we use

$$\theta_{1p} = at\cos\omega t.$$ (9.24)

Thus our "improved solution" takes the form (from Eqs. 9.9, 9.23, and 9.24):

$$\theta = at\cos\omega t + b\cos 3\omega t + c\cos\omega t.$$ (9.25)

Because of the secular term (the first term on the right-hand side), the solution in Eq. 9.25 can only be good for short periods of time (See Meirovitch (1970).) For long periods of time, the secular term will cause the solution to diverge. We know, of course, that the amplitude of the solution for the simple pendulum cannot monotonically increase with time because the system is *conservative*.

9.3   The Lindstedt–Poincaré Method for Periodic Solutions

Many problems in dynamics are governed by nonlinear equations of the type

$$\frac{d^2\phi}{dt^2} + k^2\phi = \beta_2\frac{\phi^2}{2!} + \beta_3\frac{\phi^3}{3!} + \beta_4\frac{\phi^4}{4!} + \dots,$$ (9.26)

where $k > 0$ and β_1 are given constants. This equation is trivially satisfied by the equilibrium solution, $\phi = 0$. Let the initial conditions be

$$t_0 = 0$$

$$\phi_0 = \eta$$ (9.27)

$$\left(\frac{d\phi}{dt}\right)_0 = 0.$$

If $\beta_i = 0$ for $i = 2, 3, \dots$, then

$$\frac{d^2\phi}{dt^2} + k^2\phi = 0,$$ (9.28)

and the solution is

$$\phi = \eta \cos kt. \tag{9.29}$$

The period of oscillation for Eq. 9.28 is

$$P = \frac{2\pi}{k}. \tag{9.30}$$

For the solution of Eq. 9.26, we assume a period of oscillation of the form

$$P = \frac{2\pi}{k} \left(1 + h_1\eta + h_2\eta^2 + h_3\eta^3 + \ldots \right), \tag{9.31}$$

where η is the initial value of ϕ, assumed sufficiently small, and the h_i ($i = 1, 2, 3, \ldots$) are constant *coefficients to be determined* (which we will use to remove unwanted secular terms).

We make a change of the time scale by letting

$$t = \frac{\tau}{k} \left(1 + h_1\eta + h_2\eta^2 + h_3\eta^3 + \ldots \right). \tag{9.32}$$

The idea is that in the defective method example of the preceding section we had

$$\frac{d^2\phi}{dt^2} + k^2\phi = d \cos kt,$$

and the secular term of the form $\phi_p = ct \cos kt$ (Eq. 9.25) ruined our solution. Now we try the scheme

$$\frac{d^2\phi}{d\tau^2} + k^2\phi = (d - h_1) \cos k\tau,$$

where we set $d = h_1$ to get rid of the secular term.

From Eq. 9.32, using

$$\frac{d\tau}{dt} = \frac{k}{\left(1 + h_1\eta + h_2\eta^2 + \ldots \right)}$$

and the chain rule, we have

$$\frac{d}{dt} (\quad) = \frac{d}{d\tau} (\quad) \frac{k}{\left(1 + h_1\eta + h_2\eta^2 + h_3\eta^3 + \ldots \right)} \tag{9.33}$$

$$\frac{d^2}{dt^2} (\quad) = \frac{d^2}{d\tau^2} (\quad) \frac{k^2}{\left(1 + h_1\eta + h_2\eta^2 + h_3\eta^3 + \ldots \right)^2}. \tag{9.34}$$

Substituting Eq. 9.34 into Eq. 9.26, we obtain

$$
\frac{d^2\phi}{d\tau^2} + \left(1 + h_1\eta + h_2\eta^2 + \dots\right)^2 \phi = \left(1 + h_1\eta + h_2\eta^2 + \dots\right)^2
$$

$$
\times \left(\frac{\beta_2}{k^2}\frac{\phi^2}{2!} + \frac{\beta_3}{k^2}\frac{\phi^3}{3!} + \dots\right). \tag{9.35}
$$

We seek the solution for ϕ in the form

$$
\phi = \eta\phi_1(\tau) + \eta^2\phi_2(\tau) + \eta^3\phi_3(\tau) + \dots, \tag{9.36}
$$

where the $\phi_1(\tau)$ are periodic functions in τ of period 2π satisfying the initial conditions:

$$
\tau = 0
$$
$$
\phi_1(0) = 1
$$
$$
\phi_1'(0) = 0
$$
$$
\phi_i(0) = 0
$$
$$
\phi_i'(0) = 0 \qquad i = 2, 3, 4, \dots, \tag{9.37}
$$

where the primes indicate differentiation with respect to τ. Substituting Eq. 9.36 into Eq. 9.35, we find

$$
\eta\phi_1'' + \eta^2\phi_2'' + \eta\phi_3'' + \left(1 + h_1\eta + h_2\eta^2 + \dots\right)^2 \left(\eta\phi_1 + \eta^2\phi_2 + \dots\right)
$$

$$
= \left(1 + h_1\eta + h_2\eta^2 + \dots\right)^2
$$

$$
\times \left[\frac{\beta_2}{2!k^2}\left(\eta\phi_1 + \eta^2\phi_2 + \dots\right)^2 + \frac{\beta_3}{3!k^2}\left(\eta\phi_1 + \eta^2\phi_2 + \dots\right)^3 + \dots\right]. \tag{9.38}
$$

Expanding terms in Eq. 9.38 (keeping terms up to η^3), we have

$$
\left(1 + h_1\eta + h_2\eta^2\right)^2 = 1 + h_1\eta + h_2\eta^2 + h_1\eta + h_1^2\eta^2 + h_2\eta^2
$$

$$
= \left[1 + 2h_1\eta + \left(h_1^2 + 2h_2\right)\eta^2\right]. \tag{9.39}
$$

The expansion of the product on the left-hand side of Eq. 9.38 is

$$\left(1 + h_1\eta + h_2\eta^2\right)^2 \left(\eta\phi_1 + \eta^2\phi_2\right) = \eta\phi_1 + \eta^2\left(2h_1\phi_1\right) + \eta^3\left(h_1^2\phi_1 + 2h_2\phi_1\right)$$
$$+ \eta^2\phi_2 + \eta^3\left(2h_1\phi_2\right)$$
$$= \eta\phi_1 + \eta^2\left(\phi_2 + 2h_1\phi_1\right)$$
$$+ \eta^3\left(h_1^2\phi_1 + 2h_2\phi_1 + 2h_1\phi_2\right). \qquad (9.40)$$

The right-hand side of Eq. 9.38 is

$$(1 + h_1\eta)^2 \left[\frac{\beta_2}{2k^2}\left(\eta^2\phi_1^2 + \eta^3 2\phi_1\phi_2\right) + \frac{\beta_3}{6k^2}\eta^3\phi_1^3\right]$$
$$= \eta^2\frac{\beta_2\phi_1^2}{2k^2} + \eta^3\left(\frac{\beta_2}{k^2}\phi_1\phi_2 + \frac{\beta_2}{2k^2}h_1\phi_1^2 + \frac{\beta_3}{6k^2}\phi_1^3\right). \qquad (9.41)$$

By equating the coefficients of like powers of η, we obtain the equations for $\phi_i(\tau)$ from Eqs. 9.38–9.41:

$$\phi_1'' + \phi_1 = 0 \qquad\qquad\qquad\qquad\qquad\qquad\qquad\qquad (9.42)$$

$$\phi_2'' + \phi_2 = -2h_1\phi_1 + \frac{\beta_2}{2k^2}\phi_1^2 \qquad\qquad\qquad\qquad\qquad (9.43)$$

$$\phi_3'' + \phi_3 = -h_1^2\phi_1 - 2h_1\phi_2 - 2h_2\phi_1 + \frac{\beta_2}{k^2}\phi_1\phi_2 + \frac{h_1\beta_2}{2k^2}\phi_1^2 + \frac{\beta_3}{6k^2}\phi_1^3. \qquad (9.44)$$

Integration of Eq. 9.42 provides

$$\phi_1 = \cos\tau, \qquad\qquad\qquad\qquad\qquad\qquad\qquad\qquad (9.45)$$

where we recall the initial conditions from Eqs. 9.37:

$$\tau = 0$$
$$\phi_1(0) = 1$$
$$\phi_1'(0) = 0 \qquad\qquad\qquad\qquad\qquad\qquad\qquad\qquad (9.46)$$
$$\phi_i(0) = 0$$
$$\phi_i'(0) = 0 \qquad\qquad i = 2, 3, 4, \dots .$$

Using Eq. 9.45 in Eq. 9.43 provides

$$\phi_2'' + \phi_2 = -2h_1\cos\tau + \frac{\beta_2}{2k^2} + \frac{\beta_2}{2k^2}\cos^2\tau$$
$$= -2h_1\cos\tau + \frac{\beta_2}{2k^2} + \frac{\beta_2}{2k^2}\left[\frac{1}{2}(1 + \cos 2\tau)\right], \qquad (9.47)$$

or

$$\phi_2'' + \phi_2 = -2h_1 \cos \tau + \frac{\beta_2}{4k^2} + \frac{\beta_2}{4k^2} \cos 2\tau. \qquad (9.48)$$

For the *periodic solution*, we choose

$$h_1 = 0. \qquad (9.49)$$

From Eqs. 9.49 and 9.48 becomes

$$\phi_2'' + \phi_2 = \frac{\beta_2}{4k^2} + \frac{\beta_2}{4k^2} \cos 2\tau. \qquad (9.50)$$

For the particular solution of Eq. 9.50, we have

$$\phi_{2p} = c_1 + c_2 \cos 2\tau \qquad (9.51)$$

so that

$$\phi_{2p}' = -2c_2 \sin 2\tau \qquad (9.52)$$

$$\phi_{2p}'' = -4c_2 \cos 2\tau. \qquad (9.53)$$

Substituting Eqs. 9.51 and 9.53 into Eq. 9.50, we have

$$-4c_2 \cos 2\tau + c_2 \cos 2\tau + c_1 = \frac{\beta_2}{4k^2} \cos 2\tau + \frac{\beta_2}{4k^2} \qquad (9.54)$$

so that

$$c_1 = \frac{\beta_2}{4k^2} \qquad (9.55)$$

$$c_2 = -\frac{\beta_2}{12k^2}. \qquad (9.56)$$

Since the homogenous solution for Eq. 9.50 is $c_3 \cos \tau$, we can write the total solution as

$$\phi_{2\text{total}} = \frac{\beta_2}{4k^2} - \frac{\beta_2}{12t^2} \cos 2\tau + c_3 \cos \tau. \qquad (9.57)$$

Applying the initial conditions, Eqs. 9.46, to Eq. 9.57, we find

$$c_3 = \frac{\beta_2}{12k^2} - \frac{\beta_2}{4k^2} = -\frac{\beta_2}{6k^2}. \qquad (9.58)$$

So the solution for ϕ_2 is

$$\phi_2 = \frac{\beta_2}{4k^2} - \frac{\beta_2}{6k^2} \cos \tau - \frac{\beta_2}{12k^2} \cos 2\tau. \tag{9.59}$$

Using our solutions for h_1, ϕ_1, and ϕ_2 (Eqs. 9.49, 9.45, and 9.59) in Eq. 9.44, we have

$$\phi_3'' + \phi_3 =$$

$$- 2h_2 \cos \tau + \frac{\beta_3}{6k^2} \cos^3 \tau + \frac{\beta_2}{k^2} \left(\frac{\beta_2 \cos \tau}{4k^2} - \frac{\beta_2 \cos^2 \tau}{6k^2} - \frac{\beta_2}{12k^2} \cos \tau \cos 2\tau \right). \tag{9.60}$$

We make use of the trigonometric identities:

$$\cos^2 \tau = \frac{1}{2} + \frac{1}{2} \cos \tau \tag{9.61}$$

$$\cos^3 \tau = \frac{1}{4} \cos 3\tau + \frac{3}{4} \cos \tau \tag{9.62}$$

$$\cos \tau \cos 2\tau = \frac{1}{2} \cos 3\tau + \frac{1}{2} \cos \tau. \tag{9.63}$$

Substituting Eqs. 9.61–9.63 into Eq. 9.60 gives

$$\phi_3'' + \phi_3 = - 2h_2 \cos \tau + \frac{\beta_3}{24k^2} \cos 3\tau + \frac{3\beta_3}{24k^2} \cos \tau + \frac{\beta_2^2}{4k^2} \cos \tau - \frac{\beta_2^2}{12k^4}$$

$$- \frac{\beta_2^2}{12k^4} \cos 2\tau - \frac{\beta_2^2}{24k^4} \cos 3\tau - \frac{\beta_2^2}{24k^4} \cos \tau. \tag{9.64}$$

Collecting terms in Eq. 9.64, we obtain

$$\phi_3'' + \phi_3 = - \frac{\beta_2^2}{12k^4} + (\cos \tau) \left(\frac{\beta_3}{8k^2} + \frac{\beta_2^2}{4k^4} - \frac{\beta_2^2}{24k^4} - 2h_2 \right) + (\cos 2\tau) \left(- \frac{\beta_2^2}{12k^4} \right)$$

$$+ (\cos 3\tau) \left(\frac{\beta_3}{24k^2} - \frac{\beta_2^2}{24k^4} \right) \tag{9.65}$$

or

$$\phi_3'' + \phi_3 = - \frac{\beta_2^2}{12k^4} + \left(\frac{5\beta_2^2}{24k^4} + \frac{\beta_3}{8k^2} - 2h_2 \right) \cos \tau$$

$$- \frac{\beta_2^2}{12k^4} \cos 2\tau + \left(\frac{\beta_3}{24k^2} - \frac{\beta_2^2}{24k^4} \right) \cos 3\tau. \tag{9.66}$$

We note that if $\cos \tau$ were not zero, the system would be forced at its natural frequency, which immediately leads to terms appearing in time explicitly. In other words, secular terms would appear. To avoid the secular terms, the coefficient of $\cos \tau$ in Eq. 9.66 must be zero. Hence, we select

$$h_2 = \frac{5\beta_2^2}{48k^4} + \frac{\beta_3}{16k^2}. \tag{9.67}$$

Using Eq. 9.67 in Eq. 9.66, we have

$$\phi_3'' + \phi_3 = -\frac{\beta_2^2}{12k^4} - \frac{\beta_2^2}{12k^4} \cos 2\tau + \left(\frac{\beta_3}{24k^2} - \frac{\beta_2^2}{24k^4} \right) \cos 3\tau. \tag{9.68}$$

For the particular solution of Eq. 9.68, we write

$$\phi_{3p} = c_1 + c_2 \cos 2\tau + c_3 \cos 3\tau \tag{9.69}$$

so that

$$\phi_{3p}' = -2c_2 \sin 2\tau - 3c_3 \sin 3\tau \tag{9.70}$$

$$\phi_{3p}'' = -4c_2 \cos 2\tau - 9c_3 \cos 3\tau. \tag{9.71}$$

Substituting Eqs. 9.69 and 9.71 into Eq. 9.68, we obtain

$$-4c_2 \cos 2\tau - 9c_3 \cos 3\tau + c_2 \cos 2\tau + c_3 \cos 3\tau + c_1$$
$$= -3c_2 \cos 2\tau - 8c_3 \cos 3\tau + c_1$$
$$= -\frac{\beta_2^2}{12k^4} \cos 2\tau + \left(\frac{\beta_3}{24k^2} - \frac{\beta_2^2}{24k^4} \right) \cos 3\tau - \frac{\beta_2^2}{12k^4} \tag{9.72}$$

so that

$$c_1 = -\frac{\beta_2^2}{12k^4}$$

$$c_2 = \frac{\beta_2^2}{36k^4} \tag{9.73}$$

$$c_3 = \left(\frac{\beta_2^2}{192k^4} - \frac{\beta_3}{192k^2} \right).$$

Since the homogenous solution for Eq. 9.68 is $c_4 \cos \tau$, we can write the total solution as

$$\phi_3 = -\frac{\beta_2^2}{12k^4} + \frac{\beta_2^2}{36k^4} \cos 2\tau + \left(\frac{\beta_2^2}{192k^4} - \frac{\beta_3}{192k^2} \right) \cos 3\tau + c_4 \cos \tau. \tag{9.74}$$

Applying the initial conditions in Eqs. 9.46, to Eq. 9.74, we obtain

$$c_4 = \frac{\beta_2^2}{12k^4} - \frac{\beta_2^2}{36k^4} - \frac{\beta_2^2}{192k^4} + \frac{\beta_3}{192k^2}, \tag{9.75}$$

or

$$c_4 = \frac{29\beta_2^2}{576k^4} + \frac{\beta_3}{192k^2}. \tag{9.76}$$

So the solution for ϕ_3 is

$$\phi_3 = -\frac{\beta_2^2}{12k^4} + \left(\frac{29\beta_2^2}{576k^4} + \frac{\beta_3}{192k^2} \right) \cos \tau + \frac{\beta_2^2}{36k^4} \cos 2\tau$$

$$+ \left(\frac{\beta_2^2}{192k^4} - \frac{\beta_3}{192k^2} \right) \cos 3\tau. \tag{9.77}$$

Thus, the solution of the differential equation, Eq. 9.35,

$$\frac{d^2\phi}{d\tau^2} + \left(1 + h_1\eta + h_2\eta^2 + \dots \right)^2 \phi = \left(1 + h_1\eta + h_2\eta^2 + \dots \right)$$

$$\times \left(\frac{\beta_2}{k^2} \frac{\phi^2}{2!} + \frac{\beta_3}{k^2} \frac{\phi^3}{3!} + \dots \right) \tag{9.78}$$

to the order of η^3 included is

$$\phi = \eta\phi_1(\tau) + \eta^2\phi_2(\tau) + \eta^3\phi_3(\tau), \tag{9.79}$$

where ϕ_1, ϕ_2, and ϕ_3 are given in Eqs. 9.45, 9.59, and 9.77 and where (as mentioned previously) η is the initial value of ϕ, assumed to be sufficiently small.

The coefficient h_3 is obtained by equating to zero the coefficient of $\cos \tau$ in the differential equation for ϕ_4. Its value is

$$h_3 = -\frac{5\beta_2^3}{144k^6} - \frac{\beta_2\beta_3}{48k^4}. \tag{9.80}$$

We see from the solution of $\phi(\tau)$ that the period in τ units is 2π. The period to the order of η^3 included is, from Eqs. 9.31, 9.49, 9.67, and 9.80 for h_1, h_2, and h_3, in units of t:

$$P = \frac{2\pi}{k}\left[1 + \frac{\eta^2}{16k^2}\left(\frac{5\beta_2^2}{3k^2} + \beta_3\right) - \frac{\eta^3\beta_2}{48k^4}\left(\frac{5\beta_2^2}{3k^2} + \beta_3\right)\right], \tag{9.81}$$

where the β_i ($i = 2, 3, 4, \ldots$) are obtained from the governing nonlinear differential equation, Eq. 9.26:

$$\frac{d^2\phi}{dt^2} + k^2\phi = \beta_2\frac{\phi^2}{2!} + \beta_3\frac{\phi^3}{3!} + \beta_4\frac{\phi^4}{4!} + \ldots. \tag{9.82}$$

9.4 Application to Mercury's Orbit

We consider the following problem: Using the Lindstedt–Poincaré method, integrate the equation of motion for Mercury's orbit about the Sun including the effect of relativity, starting from an apse. Deduce the advance in the perihelion predicted by general relativity by obtaining the first-order perturbation in the true anomaly, f (Fig. 9.3).

The third of Eqs. 2.18 provides the two components of the acceleration in polar coordinates. The effect of general relativity adds a small perturbation to the radial component (See Szebehely (1989)).

$$\frac{d^2r}{dt^2} - r\dot{f}^2 = -\frac{\mu}{r^2} - \frac{\mu}{r^2}\left(\frac{3h^2}{c^2r^2}\right). \tag{9.83}$$

Using the specific angular momentum $r^2\dot{f} = h$ from Eq. 2.17 in Eq. 9.83, we obtain the scalar equation

Fig. 9.3 Mercury in orbit about the Sun where f is the true anomaly and $\mathbf{r}$ is the position vector

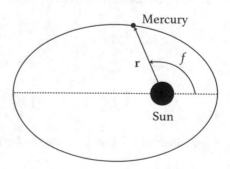

$$\frac{d^2r}{dt^2} - \frac{h^2}{r^3} = -\frac{\mu}{r^2} - \frac{\mu}{r^2}\left(\frac{3h^2}{c^2r^2}\right). \tag{9.84}$$

A common technique in dealing with equations where the variable is in the denominator is to make the substitution

$$r = \frac{1}{u}. \tag{9.85}$$

Then we have

$$\frac{dr}{dt} = \frac{dr}{df}\frac{df}{dt} = \frac{d}{df}\left(\frac{1}{u}\right)\frac{df}{dt} = -\frac{1}{u^2}\frac{du}{df}\left(hu^2\right) = -h\frac{du}{df} \tag{9.86}$$

and

$$\frac{d^2r}{dt^2} = -h\frac{d}{dt}\left(\frac{du}{df}\right) = -h\frac{d^2u}{df^2}\frac{df}{dt} = -h^2u^2\frac{d^2u}{df^2}. \tag{9.87}$$

Substituting Eqs. 9.85 and 9.87 into Eq. 9.84 gives

$$-h^2u^2\frac{d^2u}{df^2} - h^2u^3 = -\mu u^2 - \mu u^4\frac{3h^2}{c^2},$$

$$\frac{d^2u}{df^2} + u = \frac{\mu}{h^2} + \epsilon u^2, \tag{9.88}$$

where $\epsilon = 3\mu/c^2$ and c is the speed of light. Let us write Eq. 9.88 as

$$\frac{d^2u}{df^2} + u = \frac{\mu}{h^2}\left(1 + \frac{h^2}{\mu^2}\epsilon u^2\right). \tag{9.89}$$

The initial conditions are assumed to be

$$f_0 = 0$$

$$u_0 = \frac{1}{r_0} \tag{9.90}$$

$$\left(\frac{du}{df}\right)_0 = 0.$$

We assume the solution for u has the form

$$u = u_0(\tau) + \epsilon u_1(\tau), \tag{9.91}$$

where the u_i are periodic functions in τ of period 2π defined by (Eq. 9.32)

$$f = \tau (1 + h_1 \epsilon).$$ (9.92)

By the chain rule, we have

$$\frac{du}{df} = \frac{du}{d\tau}\frac{d\tau}{df} = \frac{du}{d\tau}\frac{1}{(1 + h_1 \epsilon)}$$ (9.93)

$$\frac{d^2 u}{df^2} = \frac{d^2 u}{d\tau^2}\frac{1}{(1 + h_1 \epsilon)^2}$$ (9.94)

as in Eqs. 9.33 and 9.34.

Substituting Eq. 9.94 into Eq. 9.89, we have

$$\frac{d^2 u}{d\tau^2} + (1 + h_1 \epsilon)^2 u = (1 + h_1 \epsilon)^2 \left(\frac{\mu}{h^2} + \epsilon u^2\right)$$ (9.95)

as in Eq. 9.35. Next we use Eq. 9.91 in Eq. 9.95 and retain ϵ:

$$\frac{d^2 u_0}{d\tau^2} + \epsilon \frac{d^2 u_1}{d\tau^2} + (1 + 2h_1 \epsilon)(u_0 + \epsilon u_1) = (1 + 2h_1 \epsilon)\left(\frac{\mu}{h^2} + \epsilon u_0^2\right)$$ (9.96)

or

$$\frac{d^2 u_0}{d\tau^2} + \epsilon \frac{d^2 u_1}{d\tau^2} + u_0 + \epsilon u_1 + 2\epsilon h_1 u_0 = \frac{\mu}{h^2} + \epsilon u_0^2 + 2\epsilon h_1 \frac{\mu}{h^2}.$$ (9.97)

Collecting zeroth-order and ϵ terms from Eq. 9.97, we have

$$\frac{d^2 u_0}{d\tau^2} + u_0 = \frac{\mu}{h^2}$$ (9.98)

$$\frac{d^2 u_1}{d\tau^2} + u_1 = u_0^2 + 2h_1 \frac{\mu}{h^2} - 2h_1 u_0.$$ (9.99)

Equations 9.98 and 9.99 can be compared with Eqs. 9.42–9.44.

Using the initial conditions in Eq. 9.90, we solve Eq. 9.98

$$u_0 = A \cos \tau + B \sin \tau + \frac{\mu}{h^2}$$ (9.100)

$$u_0(0)] = \frac{1}{r_0} = A + \frac{\mu}{h^2}$$

$$A = \frac{1}{r_0} - \frac{\mu}{h^2}.$$ (9.101)

Since $B = 0$, we have

$$u_0 = \left(\frac{1}{r_0} - \frac{\mu}{h^2} \right) \cos \tau + \frac{\mu}{h^2}, \qquad (9.102)$$

which leads to

$$r = \frac{1}{u_0} = \frac{1}{\dfrac{\mu}{h^2} + \left(\dfrac{1}{r_0} - \dfrac{\mu}{h^2} \right) \cos \tau} \qquad (9.103)$$

or

$$r = \frac{h^2/\mu}{1 + \left(\dfrac{h^2}{\mu r_0} - 1 \right) \cos \tau}, \qquad (9.104)$$

which is the conic equation.

For convenience, we write Eq. 9.102 as

$$u_0 = \frac{\mu}{h^2} \left(1 + \eta \cos \tau \right). \qquad (9.105)$$

Substituting Eq. 9.105 into Eq. 9.99, we obtain

$$\frac{d^2 u_1}{d\tau^2} + u_1 = \frac{\mu^2}{h^4} \left(1 + 2\eta \cos \tau + \eta^2 \cos^2 \tau \right) + 2h_1 \frac{\mu}{h^2} - 2h_1 \frac{\mu}{h^2} \left(1 + \eta \cos \tau \right). \qquad (9.106)$$

Because $\cos \tau$ may appear on both the homogeneous and forced side of the equation, secular terms may appear. To avoid secular terms in the u_1 solution, we must choose h_1 to make the coefficient of the $\cos \tau$ terms equal to zero:

$$(\cos \tau) \left(2\eta \frac{\mu^2}{h^4} - 2\eta h_1 \frac{\mu}{h^2} \right) = 0, \qquad (9.107)$$

so

$$h_1 = \frac{\mu}{h^2}. \qquad (9.108)$$

Using Eq. 9.108 in Eq. 9.92, we have

$$f = \tau + \tau \frac{\mu}{h^2} \epsilon. \qquad (9.109)$$

When the next apse occurs at

$$\tau = 2\pi, \tag{9.110}$$

the value of the true anomaly will be

$$f = 2\pi \left(1 + \epsilon \frac{\mu}{h^2}\right). \tag{9.111}$$

Thus, the advance of the periapsis due to general relativity is

$$\Delta \omega = \frac{2\pi \epsilon \mu}{h^2}, \tag{9.112}$$

where $\epsilon = 3\mu/c^2$.

9.5   Motion of a Satellite in the Equatorial Plane of an Oblate Planet

In the case of a satellite orbiting in the equatorial plane of an oblate central body, the force is central in the equatorial plane and the motion is planar, although it is no longer Keplerian. The potential function of a spheroid may be written as a series of harmonics of the form (Eq. 7.77):

$$U = \frac{\mu}{r} \left[1 - \sum_{n=2}^{\infty} J_n \left(\frac{r_e}{r}\right)^n P_n (\sin \phi)\right], \tag{9.113}$$

where $P_n (\sin \phi)$ is the Legendre polynomial:

$$P_2 (\sin \phi) = \frac{1}{2} \left(3 \sin^2 \phi - 1\right) \tag{9.114}$$

$$P_3 (\sin \phi) = \frac{1}{2} \left(5 \sin^2 \phi - 3 \sin \phi\right) \tag{9.115}$$

$$P_4 (\sin \phi) = \frac{1}{8} \left(35 \sin^4 \phi - 30 \sin^2 \phi + 3\right) \tag{9.116}$$

$$P_5 (\sin \phi) = \frac{1}{8} \left(63 \sin^5 \phi - 70 \sin^3 \phi + 15 \sin \phi\right). \tag{9.117}$$

Recall that two-body motion is confined to a plane and the force acts through the center of the massive body; this is called a central force. For an oblate planet, an object in the equatorial place is subject to a central force. However, an object out of that plane will be subject to a non-central force, e.g., one acting through O' rather than O, as shown in Fig. 9.4.

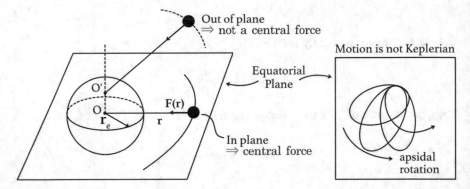

Fig. 9.4 Motion of a satellite near an oblate planet in which out-of-plane and non-Keplerian motion may occur

For motion in the equatorial plane, we set $\sin\phi = 0$ so that Eqs. 9.114–9.117 become

$$P_2 = -\frac{1}{2} \tag{9.118}$$

$$P_3 = 0 \tag{9.119}$$

$$P_4 = \frac{3}{8} \tag{9.120}$$

$$P_5 = 0. \tag{9.121}$$

Using Eqs. 9.118–9.121 in Eq. 9.113, we have

$$U = \frac{\mu}{r}\left[1 + \frac{J_2}{2}\left(\frac{r_e}{r}\right)^2 - \frac{3J_4}{8}\left(\frac{r_e}{r}\right)^4 + \dots\right]. \tag{9.122}$$

The equation of motion is

$$\ddot{\mathbf{r}} = \nabla U \tag{9.123}$$

or in scalar form:

$$\ddot{r} - r\dot{f}^2 = -\frac{\mu}{r^2}\left[1 + \frac{3J_2}{2}\left(\frac{r_e}{r}\right)^2 - \frac{15J_4}{8}\left(\frac{r_e}{r}\right)^4 + \dots\right] \tag{9.124}$$

$$\frac{1}{r}\frac{d}{dt}\left(r^2\dot{f}\right) = 0, \tag{9.125}$$

where the second equation leads to the conservation of angular momentum. We integrate the equations of motion using the initial conditions at an apse:

$$t = 0$$

$$r = r_0$$

$$V = V_0 \qquad (9.126)$$

$$\frac{dr}{dt} = 0.$$

From Eq. 9.125, we have

$$r^2 \dot{f} = h \qquad (9.127)$$

or

$$\dot{f} = \frac{h}{r^2}, \qquad (9.128)$$

and we note that the value of the specific angular momentum is

$$h = r_0 V_0. \qquad (9.129)$$

We use a change of variables, letting

$$r = \frac{1}{u}. \qquad (9.130)$$

Then we can write

$$\frac{dr}{dt} = \frac{dr}{df}\frac{df}{dt} = hu^2 \frac{d}{df} u^{-1} = hu^2 \left(-u^{-2}\frac{du}{df} \right) \qquad (9.131)$$

or

$$\frac{dr}{dt} = -h\frac{du}{df} \qquad (9.132)$$

and

$$\frac{d^2r}{dt^2} = -h^2 u^2 \frac{d^2u}{df^2}. \qquad (9.133)$$

Substituting Eqs. 9.130 and 9.133 into Eq. 9.124, we have

$$-h^2 u^2 \frac{d^2u}{df^2} - \frac{1}{u}\left(h^2 u^4 \right) = -\mu u^2 \left[1 + \frac{3J_2}{2}(r_e u)^2 - \frac{15J_4}{8}(r_e u)^4 + \dots \right],$$

$$(9.134)$$

or, after dividing through by $-h^2 u^2$,

$$\frac{d^2u}{df^2} + u = \frac{\mu}{h^2} \left[1 + \frac{3J_2}{2} (r_e u)^2 - \frac{15J_4}{8} (r_e u)^4 + \ldots \right],$$ (9.135)

where we use the initial conditions

$$f_0 = 0$$

$$u_0 = \frac{1}{r_0}$$ (9.136)

$$\left(\frac{du}{df} \right)_0 = 0.$$

9.5.1   Lindstedt–Poincaré Method

When

$$J_2 = J_4 = 0$$ (9.137)

we have the unperturbed trajectory. Since J_4 is of the order J_2^2, we apply the Lindstedt–Poincaré method as follows:

$$u = u_0 (\tau) + J_2 u_1 (\tau) + J_2^2 u_2 (\tau) + J_4 u_3 (\tau) + \ldots$$ (9.138)

where the u_i are periodic functions in τ where τ (of period 2π) is usually defined by

$$f = \tau \left(1 + h_1 J_2 + h_2 J_2^2 + h_3 J_4 + \ldots \right).$$ (9.139)

However, in this example we use a more general version by assuming a transformation of the independent variable of the form:

$$f = \tau + J_2 f_1 (\tau) + J_2^2 f_2 (\tau) + J_4 f_3 (\tau) + \ldots \, ,$$ (9.140)

where the functions $f_i (\tau)$ are to be *selected to eliminate the secular terms*. (Previously, we used the constants h_i to eliminate the secular terms. This method provides more flexibility.)

From the chain rule, we have

$$\frac{du}{df} = \frac{du}{d\tau} \frac{d\tau}{df}$$ (9.141)

$$\frac{d^2 u}{df^2} = \frac{d^2 u}{d\tau^2} \left(\frac{d\tau}{df} \right)^2 + \frac{du}{d\tau} \frac{d^2 \tau}{df^2}. \tag{9.142}$$

Let

$$\frac{d(\quad)}{d\tau} \equiv (\quad)' \tag{9.143}$$

so that

$$\frac{d\tau}{df} = \frac{1}{f'}. \tag{9.144}$$

We can write

$$\frac{d^2 \tau}{df^2} = \frac{d}{df} \left(\frac{d\tau}{df} \right) = \frac{d}{d\tau} \left(\frac{d\tau}{df} \right) \frac{d\tau}{df}$$

$$= \frac{1}{f'} \frac{d}{d\tau} \left(\frac{1}{f'} \right) = -\frac{1}{f'^3} \frac{df'}{d\tau} = -\frac{f''}{f'^3}. \tag{9.145}$$

Using Eqs. 9.143–9.145 in Eq. 9.142, we obtain

$$\frac{d^2 u}{df^2} = \frac{u''}{f'^2} - \frac{u' f''}{f'^3}. \tag{9.146}$$

Substituting Eq. 9.146 into Eq. 9.135 provides

$$\frac{u''}{f'^2} - \frac{u' f''}{f'^3} + u = \frac{\mu}{h^2} \left[1 + \frac{3J_2}{2} (r_e u)^2 - \frac{15J_4}{8} (r_e u)^4 + \dots \right]. \tag{9.147}$$

We use the initial conditions

$$\tau = 0$$

$$u_0(0) = \frac{1}{r_0}$$

$$u_1 = u_2 = \dots = 0$$

$$f_1 = f_2 = \dots = 0$$

$$\frac{du_0}{d\tau} = \frac{du_1}{d\tau} = \frac{du_2}{d\tau} = \dots = 0. \tag{9.148}$$

Multiplying Eq. 9.147 by f'^3 provides

$$u'' f' - u' f'' + u f'^3 = \frac{\mu}{h^2} f'^3 \left[1 + \frac{3 J_2}{2} (r_e u)^2 - \frac{15 J_4}{8} (r_e u)^4 + \ldots \right]. \quad (9.149)$$

Substituting the series from Eqs. 9.138 and 9.140 into Eq. 9.147, we obtain

$$\left(u_0'' + J_2 u_1'' + J_2^2 u_2'' + \ldots \right) \left(1 + J_2 f_1' + J_2^2 f_2' + \ldots \right)$$

$$- \left(u_0' + J_2 u_1' + J_2^2 u_2' + \ldots \right) \left(J_2 f_1'' + J_2^2 f_2'' + \ldots \right)$$

$$+ \left(u_0 + J_2 u_1 + J_2^2 u_2 + \ldots \right) \left(1 + J_2 f_1' + J_2^2 f_2' + \ldots \right)^3$$

$$= \frac{\mu}{h^2} \left(1 + J_2 f_1' + \ldots \right)^3$$

$$\times \left[1 + \frac{3 J_2}{2} r_e^2 (u_0 + J_2 u_1 + \ldots)^2 - \frac{15 J_4}{8} r_e^4 (u_0 + J_2 u_1 + \ldots)^4 + \ldots \right]. \quad (9.150)$$

Equating coefficients of like powers in J_2, J_2^2, and J_4 (where we do not show the terms corresponding to J_4 here):

$$u_0'' + u_0 = \frac{\mu}{h^2} \quad (9.151)$$

$$u_1'' + u_1 = \frac{\mu}{h^2} \left(3 f_1' + \frac{3}{2} r_e^2 u_0^2 \right) - u_0'' f_1' + u_0' f_1'' - 3 u_0 f_1' \quad (9.152)$$

$$u_2'' + u_2 = \frac{\mu}{h^2} \left(3 r_e^2 u_0 u_1 + \frac{9}{2} r_e^2 f_1' u_0^2 \right) - u_0'' f_2' + u_0' f_2'' - 3 u_0 f_2' - 3 u_0 f_1'^2 - u_1'' f_1'$$

$$+ u_1 f_1'' - 3 u_1 f_1' + \frac{\mu}{h^2} \left(3 f_1'^2 + 3 f_2' \right). \quad (9.153)$$

Integrating the first equation (Eq. 9.151), we obtain the same equation as in Eq. 9.105:

$$u_0 = \frac{\mu}{h^2} \left(1 + \eta \cos \tau \right), \quad (9.154)$$

which is essentially the conic equation.

Before substituting Eq. 9.154 into Eq. 9.152, we note

$$u_0' = -\frac{\mu}{h^2} \eta \sin \tau \quad (9.155)$$

$$u_0'' = -\frac{\mu}{h^2} \eta \cos \tau \quad (9.156)$$

$$(1 + \eta \cos \tau)^2 = 1 + 2\eta \cos \tau + \eta^2 \cos^2 \tau$$

$$= 1 + 2\eta \cos \tau + \frac{\eta^2}{2} + \frac{\eta^2}{2} \cos 2\tau. \tag{9.157}$$

Substituting Eqs. 9.154–9.157 into Eq. 9.152, we obtain

$$u_1'' + u_1 = \frac{\mu}{h^2} 3 f_1' + \frac{\mu}{h^2} \left(\frac{3}{2} r_e^2 \right) \left(\frac{\mu}{h^2} \right)^2 \left(1 + 2\eta \cos \tau + \frac{\eta^2}{2} + \frac{\eta^2}{2} \cos 2\tau \right)$$

$$+ \frac{\mu}{h^2} \eta \cos \tau f_1' - \frac{\mu}{h^2} \eta \sin \tau f_1'' - 3 \frac{\mu}{h^2} (1 + \eta \cos \tau) f_1', \tag{9.158}$$

or

$$u_1'' + u_1 = \frac{\mu}{h^2} \left[\left(\frac{\mu}{h^2} \right)^2 \left(\frac{3}{2} r_e^2 \right) \left(1 + \frac{\eta^2}{2} + \frac{\eta^2}{2} \cos 2\tau \right) \right]$$

$$+ \frac{\mu}{h^2} \cos \tau \left[\left(\frac{\mu}{h^2} \right)^2 \left(\frac{3}{2} r_e^2 \right) 2\eta - 2\eta f_1' \right] - \frac{\mu}{h^2} \sin \tau \left(\eta f_1'' \right). \tag{9.159}$$

If the forcing function, $\cos \tau$, were allowed to be nonzero, then the system would be forced at its natural frequency, which would again lead to a secular term. To avoid the secular term, we must have no terms in $\cos \tau$ and $\sin \tau$ in Eq. 9.159, so

$$\left[\left(\frac{\mu}{h^2} \right)^2 \left(3 r_e^2 \right) - 2 f_1' \right] \cos \tau - f_1'' \sin \tau = 0, \tag{9.160}$$

or

$$f_1'' \sin \tau + 2 f_1' \cos \tau = \left(\frac{\mu}{h^2} \right)^2 \left(3 r_e^2 \right) \cos \tau. \tag{9.161}$$

We note that if we had assumed the less general form in Eq. 9.139:

$$f = \tau (1 + J_2 h_1 + \ldots)$$

compared with Eq. 9.140

$$f = \tau + J_2 f_1 (\tau) + \ldots,$$

then we would have found

$$f_1 = h_1 \tau$$

$$h_1 = \left(\frac{\mu}{h^2} \right)^2 \left(\frac{3}{2} r_e^2 \right)$$

$$f_1' = h_1$$

$$f_1'' = 0.$$

From Eq. 9.161, we see that the solution for f_1 is

$$f_1 = \left(\frac{\mu}{h^2}\right)^2 \left(\frac{3}{2}r_e^2\right)\tau. \tag{9.162}$$

Applying Eq. 9.160 to 9.159, we have

$$u_1'' + u_1 = \frac{\mu}{h^2}\left[\left(\frac{\mu}{h^2}\right)^2 \left(\frac{3}{2}r_e^2\right)\left(1 + \frac{\eta^2}{2} + \frac{\eta^2}{2}\cos 2\tau\right)\right]. \tag{9.163}$$

To solve Eq. 9.163, we write it in a more convenient form:

$$u_1'' + u_1 = q_1 + q_2 \cos 2\tau, \tag{9.164}$$

where

$$q_1 = \frac{\mu}{h^2}\left[\left(\frac{\mu}{h^2}\right)^2 \left(\frac{3}{2}r_e^2\right)\left(1 + \frac{\eta^2}{2}\right)\right] \tag{9.165}$$

$$q_2 = \frac{\mu}{h^2}\left[\left(\frac{\mu}{h^2}\right)^2 \left(\frac{3}{2}r_e^2\right)\frac{\eta^2}{2}\right]. \tag{9.166}$$

The particular solution of Eq. 9.164 is

$$u_{1p} = c_1 q_1 + c_2 q_2 \cos 2\tau \tag{9.167}$$

and

$$u_{1p}' = -2c_2 q_2 \sin 2\tau \tag{9.168}$$

$$u_{1p}'' = -4c_2 q_2 \cos 2\tau. \tag{9.169}$$

Substituting Eqs. 9.167 and 9.169 into Eq. 9.164 provides

$$-4c_2 q_2 \cos 2\tau + c_2 q_2 \cos 2\tau + c_1 q_1 = q_1 + q_2 \cos 2\tau. \tag{9.170}$$

From Eq. 9.170, we have

$$c_1 = 1 \tag{9.171}$$

$$c_2 = -\frac{1}{3}. \tag{9.172}$$

The complete solution of Eq. 9.164 has the form

$$u_{1total} = q_1 - \frac{1}{3}q_2 \cos 2\tau + c_3 \cos \tau. \tag{9.173}$$

From the initial conditions (Eqs. 9.148), we have

$$u_{1total}(0) = q_1 - \frac{1}{3}q_2 + c_3 = 0 \tag{9.174}$$

so that

$$c_3 = \frac{1}{3}q_2 - q_1. \tag{9.175}$$

Substituting Eqs. 9.165 and 9.166 into Eq. 9.175, we find

$$c_3 = \frac{\mu}{h^2}\left[\left(\frac{\mu}{h^2}\right)^2\left(\frac{3}{2}r_e^2\right)\right]\left[\frac{1}{3}\frac{\eta^2}{2} - 1 - \frac{\eta^2}{2}\right]$$

$$= -\frac{\mu}{h^2}\left[\left(\frac{\mu}{h^2}\right)^2\left(\frac{3}{2}r_e^2\right)\right]\left(1 + \frac{\eta^2}{3}\right). \tag{9.176}$$

Using Eqs. 9.165, 9.166, 9.171, 9.172, and 9.176 in Eq. 9.173, we obtain

$$u_{1total} = \frac{\mu}{h^2}\left[\left(\frac{\mu}{h^2}\right)^2\left(\frac{3}{2}r_e^2\right)\right]\left[1 + \frac{\eta^2}{2} - \frac{\eta^2}{6}\cos 2\tau - \left(1 + \frac{\eta^2}{3}\right)\cos \tau\right], \tag{9.177}$$

where we recall from Eq. 9.154 that

$$u_0 = \frac{\mu}{h^2}(1 + \eta \cos \tau), \tag{9.178}$$

which is the conic equation and therefore

$$\eta = e. \tag{9.179}$$

The solution to order J_2 is, from Eq. 9.138

$$u = u_0(\tau) + J_2 u_1(\tau), \tag{9.180}$$

where $u = 1/r$ and $u_0(\tau)$ and $u_1(\tau)$ are given in Eqs. 9.177–9.179.
    The solution for f to order J_2 is, from Eqs. 9.140 and 9.162:

$$f = \tau + J_2 f_1 = \tau + J_2\left[\left(\frac{\mu}{h}\right)^2\left(\frac{3}{2}r_e^2\right)\right]\tau$$

$$f = \tau + \left(\frac{\mu}{h^2}\right)^2 \left(\frac{3}{2} J_2 r_e^2\right) \tau. \tag{9.181}$$

Since we have done this development for the special case of $i = 0$, Eq. 9.181 predicts the perturbation on the sum of $f + \omega + \Omega$. The individual angles are not uniquely defined for zero inclination. However, we can separate the perturbation effects into the portion that affects the period and the portion that affects $\omega + \Omega$. The next apse occurs at

$$\tau = 2\pi, \tag{9.182}$$

so we find from Eqs. 9.181 and 9.182 that

$$f = 2\pi \left[1 + \left(\frac{\mu}{h^2}\right)^2 \frac{3}{2} J_2 r_e^2\right]. \tag{9.183}$$

Thus the advance of the angle $(\omega + \Omega)$ in one revolution is

$$(\Delta\omega + \Delta\Omega)_{J_2} = 2\pi \left(\frac{\mu}{h^2}\right)^2 \frac{3}{2} J_2 r_e^2, \tag{9.184}$$

and the rate of advance is

$$\frac{(\Delta\omega + \Delta\Omega)_{J_2}}{P} = n \left(\frac{\mu}{h^2}\right)^2 \frac{3}{2} J_2 r_e^2 = \frac{3}{2} \frac{n J_2 r_e^2}{a^2 \beta^4}, \tag{9.185}$$

where we have used the relations $P = 2\pi/n$ and $h^2 = \beta^2 \mu a$.

We can compare this equation to the result using the method of averaging. Taking Eqs. 8.107 and 8.109 with $i = 0$, we obtain

$$\dot{\omega}_{J_2} + \dot{\Omega}_{J_2} = \frac{3}{4} \frac{n J_2 r_e^2}{a^2 \beta^4} (4) - \frac{3}{2} \frac{n J_2 r_e^2}{a^2 \beta^4} (1) = \frac{3}{2} \frac{n J_2 r_e^2}{a^2 \beta^4}. \tag{9.186}$$

Thus, we have an agreement between the results using the Lindstedt–Poincaré method and the method of averaging.

We recall from Eq. 8.106 that there was also a perturbation in the mean anomaly rate due to J_2. The effect simply changes the period of the satellite and was accounted for when we took τ to be one period in Eq. 9.183.

References

Meirovitch, L. (1970). *Methods of analytical dynamics*. New York: McGraw-Hill Book Company.
Szebehely, V. G. (1989). *Adventures in celestial mechanics, a first course in the theory of orbits*. Austin: University of Texas Press.

Chapter 10
The Earth–Moon System

Thus far, we have methodically developed and presented a set of tools that allows the reader to formulate a perturbation problem in a straightforward manner using the Lagrange planetary equations. We have presented a rigorous approach for the method of averaging, which allows construction of a transformation of variables that removes dependence on the fast variables, making the resulting equations simpler and perhaps more amenable to analytical solution.

As one illustration of the theory of perturbations, we consider the motion of the Moon around the Earth under the disturbing influence of the Sun. The complete theory is very complex (see Brown (1896), also published by Dover in 1960).

An equivalent problem was examined by Chao and Hoots (2018) (pp. 33–40). They used the method of averaging to consider the effect of a third body on the motion of a satellite. Their results apply if the "satellite" is the Moon and the "third body" is the Sun. They provide the transformed rate equations averaged over the period of the fast variable as well as all the periodics in the transformation. However, many details of the dependencies on the geometry of the perturbing body are embedded in notation that obscures the true geometric interactions.

Sometimes we are more interested in understanding how the complex dynamical behavior depends on the relative geometry of the orbiting body and the perturbing body than we are in actually computing a numerical result. Such insight is one of the virtues of approaching perturbations with an analytical treatment rather than a numerical integration treatment.

In this chapter, we take an approach aimed at explicitly exposing these dependencies in a way that terms and geometry dependencies can be examined and understood in a physically interpretable way. We may sacrifice the compact nature of the solution described above, and we make some simplifying assumptions of the behavior of the elements of the problem. Yet these approximations result in revealing how the geometry affects the motion and providing valuable insight into the long-term evolution of the dynamics.

© The Author(s), under exclusive license to Springer Nature Switzerland AG 2022   243
J. M. Longuski et al., *Introduction to Orbital Perturbations*, Space Technology
Library 40, https://doi.org/10.1007/978-3-030-89758-1_10

Fig. 10.1 The Earth–Moon System in which the Sun and Moon "orbit" the Earth

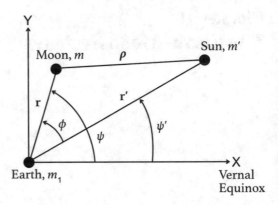

We let

$$m_1 = \text{mass of Earth (primary)}$$

$$m = \text{mass of Moon (perturbed)}$$

$$m' = \text{mass of Sun (perturbing).}$$

The Moon's orbit is inclined about 5° to the plane of the ecliptic; for simplicity, we assume this inclination is negligible. That is, we assume m_1, m, and m' are coplanar. Further, we assume that the Earth's orbit about the Sun is circular (i.e., we neglect the eccentricity, which is about 0.06). We note that in Fig. 10.1, we take the point of view that the Sun orbits the Earth.

Let the angles ψ and ψ' be the celestial longitudes of the Moon and of the Sun (respectively), as measured from the Vernal Equinox. (See Fig. 10.1.) Recalling Eq. 6.21, the disturbing function is

$$\mathcal{R} = \frac{Gm'}{r'} \left[1 + \left(\frac{r}{r'}\right)^2 P_2 + \left(\frac{r}{r'}\right)^3 P_3 + \left(\frac{r}{r'}\right)^4 P_4 + \ldots \right]. \tag{10.1}$$

For our planar problem we write

$$\phi = \psi - \psi'. \tag{10.2}$$

Also, since Earth's orbit is assumed to be circular,

$$r' = a'. \tag{10.3}$$

Neglecting the Legendre polynomials P_3 and above, Eq. 10.1 becomes

$$\mathcal{R} = \frac{Gm'}{r'} \left\{ 1 + \left(\frac{r}{a'}\right)^2 P_2 \left[\cos\left(\psi - \psi'\right)\right] \right\}. \tag{10.4}$$

From the list of Legendre coefficients in Eqs. 6.18, we have

$$P_2\left[\cos\left(\psi - \psi'\right)\right] = \frac{1}{4} + \frac{3}{4}\cos 2\left(\psi - \psi'\right). \tag{10.5}$$

Since

$$\left(\frac{r}{a'}\right)^2 = \left(\frac{a}{a'}\right)^2\left(\frac{r}{a}\right)^2, \tag{10.6}$$

$\mathcal{R}$ can be written as

$$\mathcal{R} = \frac{Gm'}{a'}\left[1 + \frac{1}{4}\left(\frac{a}{a'}\right)^2\left(\frac{r}{a}\right)^2 + \frac{3}{4}\left(\frac{a}{a'}\right)^2\left(\frac{r}{a}\right)^2\cos 2\left(\psi - \psi'\right)\right]. \tag{10.7}$$

Neglecting terms in Eq. 6.40 containing e' (since we have assumed $e' = 0$), we have

$$\left(\frac{r}{a'}\right)^2 = \left(\frac{a}{a'}\right)^2\left(1 + \frac{3e^2}{2} - 2e\cos M - \frac{e^2}{2}\cos 2M\right) \tag{10.8}$$

to second order in e. Because we have assumed that the three bodies, m_1, m, and m' are coplanar, we have

$$\psi = \Omega + \omega + f, \tag{10.9}$$

where f is the true anomaly of the Moon. Thus, the last term in Eq. 10.7 is

$$\frac{r^2}{a^2}\cos 2\left(\psi - \psi'\right) = \left(\frac{r}{a}\right)^2\cos\left[2f + 2\left(\Omega + \omega - \psi'\right)\right], \tag{10.10}$$

which can be expanded to

$$\left(\frac{r}{a}\right)^2\cos\left[2f + 2\left(\Omega + \omega - \psi'\right)\right] =$$

$$= \left(\frac{r}{a}\right)^2\left[\cos 2f\cos 2\left(\Omega + \omega - \psi'\right) - \sin 2f\sin 2\left(\Omega + \omega - \psi'\right)\right]$$

$$= \left(\frac{r}{a}\right)^2\left[\left(2\cos^2 f - 1\right)\cos 2\left(\Omega + \omega - \psi'\right) - 2\sin f\cos f\sin 2\left(\Omega + \omega - \psi'\right)\right]$$

$$= 2\left(\frac{r}{a}\right)^2\cos^2 f\cos 2\left(\Omega + \omega - \psi'\right) - \left(\frac{r}{a}\right)^2\cos 2\left(\Omega + \omega - \psi'\right)$$

$$- 2\left[\left(\frac{r}{a}\right)\sin f\right]\left[\left(\frac{r}{a}\right)\cos f\right]\sin 2\left(\Omega + \omega - \psi'\right). \tag{10.11}$$

Recalling Eqs. 2.52 and 2.56:

$$\xi = r \cos f = a \cos E - ae \tag{10.12}$$

$$\eta = r \sin f = a\sqrt{1 - e^2} \sin E, \tag{10.13}$$

and using Eqs. 6.29 and 6.35 for small e:

$$\cos E = -\frac{1}{2}e + \left(1 - \frac{3}{8}e^2\right) \cos M + \frac{1}{2}e \cos 2M + \frac{3}{8}e^2 \cos 3M \tag{10.14}$$

$$\sin E = \left(1 - \frac{e^2}{8}\right) \sin M + \frac{e}{2} \sin 2M + \frac{3e^2}{8} \sin 3M. \tag{10.15}$$

We can write expansions in the mean anomaly, M, as follows:

$$\frac{r}{a} \cos f = -\frac{3e}{2} + \left(1 - \frac{3e^2}{8}\right) \cos M + \frac{e}{2} \cos 2M + \frac{3e^2}{8} \cos 3M \tag{10.16}$$

$$\frac{r}{a} \sin f = \sqrt{1 - e^2} \sin E$$

$$= \left(1 - \frac{e^2}{2} + \dots\right)\left[\left(1 - \frac{e^2}{8}\right) \sin M + \frac{e}{2} \sin 2M + \frac{3e^2}{8} \sin 3M\right]$$

$$= \left(1 - \frac{5e^2}{8}\right) \sin M + \frac{e}{2} \sin 2M + \frac{3e^2}{8} \sin 3M, \tag{10.17}$$

where we retain terms up to order e^2. We see from Eq. 10.11 that we need to expand $(r/a)^2 \cos^2 f$ and $[(r/a) \sin f][(r/a) \cos f]$ in terms of e.

Squaring Eq. 10.16, we obtain

$$\left(\frac{r}{a}\right)^2 \cos^2 f = \left[-\frac{3e}{2} + \left(1 - \frac{3e^2}{8}\right) \cos M + \frac{e}{2} \cos 2M + \frac{3e^2}{8} \cos 3M\right]^2$$

$$= \frac{9e^2}{4} - \frac{3e}{2} \cos M - \frac{3e^2}{4} \cos 2M - \frac{3e}{2} \cos M + \left(1 - \frac{3e^2}{4}\right) \cos^2 M$$

$$+ \frac{e}{2} \cos M \cos 2M + \frac{3e^2}{8} \cos 3M \cos M - \frac{3e^2}{4} \cos 2M$$

$$+ \frac{e}{2} \cos M \cos 2M + \frac{e^2}{4} \cos^2 2M + \frac{3e^2}{8} \cos M \cos 3M$$

$$= \frac{9e^2}{4} - 3e \cos M - \frac{3e^2}{2} \cos 2M + \left(1 - \frac{3e^2}{4}\right) \cos^2 M$$

$$+ e \cos M \cos 2M + \frac{3e^2}{4} \cos M \cos 3M + \frac{e^2}{4} \cos^2 2M.$$

$$\tag{10.18}$$

We use the identity $\cos^2 M = \frac{1}{2}(1 + \cos 2M)$ to write

$$\left(1 - \frac{3e^2}{4}\right)\cos^2 M = \left(1 - \frac{3e^2}{4}\right)\frac{1}{2}(1 + \cos 2M)$$

$$= \frac{1}{2} - \frac{3e^2}{8} + \left(\frac{1}{2} - \frac{3e^2}{8}\right)\cos 2M, \qquad (10.19)$$

and

$$\frac{1}{4}e^2 \cos^2 2M = \frac{e^2}{8}(1 + \cos 4M)$$

$$= \frac{e^2}{8} + \frac{e^2}{8}\cos 4M. \qquad (10.20)$$

Substituting Eqs. 10.19 and 10.20 into Eq. 10.18, we obtain

$$\left(\frac{r}{a}\right)^2 \cos^2 f = \frac{1}{2} - \frac{3e^2}{8} + \frac{e^2}{8} + \frac{9e^2}{4} - 3e\cos M - \frac{3e^2}{2}\cos 2M$$

$$+ \left(\frac{1}{2} - \frac{3e^2}{8}\right)\cos 2M$$

$$+ e\cos M \cos 2M + \frac{3e^2}{4}\cos M \cos 3M + \frac{e^2}{8}\cos 4M$$

$$= \frac{1}{2} + 2e^2 - 3e\cos M + \left(\frac{1}{2} - \frac{15e^2}{8}\right)\cos 2M$$

$$+ e\cos M \cos 2M + \frac{3e^2}{4}\cos M \cos 3M + \frac{e^2}{8}\cos 4M. \qquad (10.21)$$

Next, we write the product of Eqs. 10.16 and 10.17

$$\left[\frac{r}{a}\sin f\right]\left[\frac{r}{a}\cos f\right] = \left[\left(1 - \frac{5e^2}{8}\right)\sin M + \frac{e}{2}\sin 2M + \frac{3e^2}{8}\sin 3M\right]$$

$$\times \left[-\frac{3e}{2} + \left(1 - \frac{3e^2}{8}\right)\cos M + \frac{e}{2}\cos 2M + \frac{3e^2}{8}\cos 3M\right]$$

$$= -\frac{3e}{2}\sin M - \frac{3e^2}{4}\sin 2M + \left(1 - e^2\right)\sin M \cos M$$

$$+ \frac{e}{2}\cos M \sin 2M + \frac{3e^2}{8}\cos M \sin 3M + \frac{e}{2}\cos 2M \sin M$$

$$+ \frac{e^2}{4}\cos 2M \sin 2M + \frac{3e^2}{8}\sin M \cos 3M. \qquad (10.22)$$

To further simplify Eqs. 10.21 and 10.22, we use the trigonometric identities:

$$\cos A \cos B = \frac{1}{2} \cos (A + B) + \frac{1}{2} \cos (A - B) \tag{10.23}$$

$$\sin A \cos B = \frac{1}{2} \sin (A + B) + \frac{1}{2} \sin (A - B), \tag{10.24}$$

so that we may rewrite the following terms:

$$\cos M \cos 2M = \frac{1}{2} \cos 3M + \frac{1}{2} \cos M$$

$$\cos M \cos 3M = \frac{1}{2} \cos 4M + \frac{1}{2} \cos 2M$$

$$\sin M \cos M = \frac{1}{2} \sin 2M - \frac{1}{2} \sin 0 = \frac{1}{2} \sin 2M$$

$$\sin 2M \cos M = \frac{1}{2} \sin 3M + \frac{1}{2} \sin M$$

$$\sin 3M \cos M = \frac{1}{2} \sin 4M + \frac{1}{2} \sin 2M \tag{10.25}$$

$$\sin M \cos 2M = \frac{1}{2} \sin 3M - \frac{1}{2} \sin M$$

$$\sin 2M \cos 2M = \frac{1}{2} \sin 4M - \frac{1}{2} \sin 0 = \frac{1}{2} \sin 4M$$

$$\sin M \cos 3M = \frac{1}{2} \sin 4M - \frac{1}{2} \sin 2M,$$

where we have used the properties $\cos(-M) = \cos M$ and $\sin(-M) = -\sin M$. Using Eqs. 10.25 in Eq. 10.21, we find

$$\left(\frac{r}{a}\right)^2 \cos^2 f = \frac{1}{2} + 2e^2 - 3e \cos M + \left(\frac{1}{2} - \frac{15e^2}{8}\right) \cos 2M$$

$$+ \frac{e}{2} \cos 3M + \frac{e}{2} \cos M + \frac{3e^2}{8} \cos 4M + \frac{3e^2}{8} \cos 2M + \frac{e^2}{8} \cos 4M$$

$$= \frac{1}{2} + 2e^2 - \frac{5e}{2} \cos M + \left(\frac{1}{2} - \frac{3e^2}{2}\right) \cos 2M + \frac{e}{2} \cos 3M + \frac{e^2}{2} \cos 4M. \tag{10.26}$$

Using Eqs. 10.25 in Eq. 10.22, we obtain

$$[(r/a)\sin f][(r/a)\cos f] = -\frac{3e}{2}\sin M - \frac{3e^2}{4}\sin 2M + \left(\frac{1}{2} - \frac{e^2}{2}\right)\sin 2M$$

$$+\frac{e}{4}\sin 3M + \frac{e}{4}\sin M + \frac{3e^2}{16}\sin 4M + \frac{3e^2}{16}\sin 2M$$

$$+\frac{e}{4}\sin 3M - \frac{e}{4}\sin M + \frac{e^2}{8}\sin 4M + \frac{3e^2}{16}\sin 4M$$

$$-\frac{3e^2}{16}\sin 2M$$

$$= -\frac{3e}{2}\sin M + \left(\frac{1}{2} - \frac{5e^2}{4}\right)\sin 2M + \frac{e}{2}\sin 3M$$

$$+\frac{e^2}{2}\sin 4M. \tag{10.27}$$

We also have, from Eq. 10.8,

$$\left(\frac{r}{a}\right)^2 = \left(\frac{r}{a'}\right)^2\left(\frac{a'}{a}\right)^2 = 1 + \frac{3e^2}{2} - 2e\cos M - \frac{e^2}{2}\cos 2M. \tag{10.28}$$

Substituting Eqs. 10.26, 10.27, and 10.28 into Eqs. 10.10 and 10.11, we obtain

$$\left(\frac{r}{a}\right)^2\cos 2\left(\psi - \psi'\right) = \left[\cos 2\left(\Omega + \omega - \psi'\right)\right]$$

$$\times\left[1 + 4e^2 - 5e\cos M + \left(1 - 3e^2\right)\cos 2M + e\cos 3M\right.$$

$$+ e^2\cos 4M - 1 - \frac{3e^2}{2} + 2e\cos M + \frac{e^2}{2}\cos 2M\bigg]$$

$$+\left[\sin 2\left(\Omega + \omega - \psi'\right)\right]$$

$$\times\left[3e\sin M + \left(-1 + \frac{5e^2}{2}\right)\sin 2M - e\sin 3M - e^2\sin 4M\right]$$

$$= \left[\cos 2\left(\Omega + \omega - \psi'\right)\right]$$

$$\times\left[\frac{3e^2}{2} - 3e\cos M + \left(1 - \frac{5e^2}{2}\right)\cos 2M + e\cos 3M\right.$$

$$+ e^2\cos 4M\bigg]$$

$$+\left[\sin 2\left(\Omega + \omega - \psi'\right)\right]$$

$$\times\left[3e\sin M + \left(-1 + \frac{5e^2}{2}\right)\sin 2M - e\sin 3M - e^2\sin 4M\right]. $$

$$\tag{10.29}$$

We can further simplify Eq. 10.29, so we define

$$\Omega^* \equiv 2\left(\Omega + \omega - \psi'\right). \tag{10.30}$$

We make use of the identities

$$\cos \Omega^* \cos B = \frac{1}{2}\cos\left(\Omega^* + B\right) + \frac{1}{2}\cos\left(\Omega^* - B\right) \tag{10.31}$$

$$\sin \Omega^* \sin B = \frac{1}{2}\cos\left(\Omega^* - B\right) - \frac{1}{2}\cos\left(\Omega^* + B\right), \tag{10.32}$$

so that

$$\cos \Omega^* \cos M = \frac{1}{2}\cos\left(\Omega^* + M\right) + \frac{1}{2}\cos\left(\Omega^* - M\right)$$

$$\cos \Omega^* \cos 2M = \frac{1}{2}\cos\left(\Omega^* + 2M\right) + \frac{1}{2}\cos\left(\Omega^* - 2M\right)$$

$$\cos \Omega^* \cos 3M = \frac{1}{2}\cos\left(\Omega^* + 3M\right) + \frac{1}{2}\cos\left(\Omega^* - 3M\right)$$

$$\cos \Omega^* \cos 4M = \frac{1}{2}\cos\left(\Omega^* + 4M\right) + \frac{1}{2}\cos\left(\Omega^* - 4M\right)$$

$$\sin \Omega^* \cos M = \frac{1}{2}\cos\left(\Omega^* - M\right) - \frac{1}{2}\cos\left(\Omega^* + M\right) \tag{10.33}$$

$$\sin \Omega^* \cos 2M = \frac{1}{2}\cos\left(\Omega^* - 2M\right) - \frac{1}{2}\cos\left(\Omega^* + 2M\right)$$

$$\sin \Omega^* \cos 3M = \frac{1}{2}\cos\left(\Omega^* - 3M\right) - \frac{1}{2}\cos\left(\Omega^* + 3M\right)$$

$$\sin \Omega^* \cos 4M = \frac{1}{2}\cos\left(\Omega^* - 4M\right) - \frac{1}{2}\cos\left(\Omega^* + 4M\right).$$

Using Eqs. 10.33 in Eq. 10.29, we find

$$\left(\frac{r}{a}\right)^2 \cos 2\left(\psi - \psi'\right)$$

$$= \frac{5e^2}{2}\cos \Omega^* - \frac{3e}{2}\cos\left(\Omega^* + M\right) - \frac{3e}{2}\cos\left(\Omega^* - M\right)$$

$$+ \left(\frac{1}{2} - \frac{5e^2}{4}\right)\cos\left(\Omega^* + 2M\right) + \left(\frac{1}{2} - \frac{5e^2}{4}\right)\cos\left(\Omega^* - 2M\right)$$

$$+ \frac{e}{2}\cos\left(\Omega^* + 3M\right) + \frac{e}{2}\cos\left(\Omega^* - 3M\right)$$

$$+ \frac{e^2}{2}\cos\left(\Omega^* + 4M\right) + \frac{e^2}{2}\cos\left(\Omega^* - 4M\right)$$

$$+ \frac{3e}{2} \cos \left(\Omega^* - M \right) - \frac{3e}{2} \cos \left(\Omega^* + M \right)$$

$$+ \left(-\frac{1}{2} + \frac{5e^2}{4} \right) \cos \left(\Omega^* - 2M \right) + \left(\frac{1}{2} - \frac{5e^2}{4} \right) \cos \left(\Omega^* + 2M \right)$$

$$+ \left(-\frac{e}{2} \right) \cos \left(\Omega^* - 3M \right) - \left(-\frac{e}{2} \right) \cos \left(\Omega^* + 3M \right)$$

$$+ \left(-\frac{e^2}{2} \right) \cos \left(\Omega^* - 4M \right) - \left(-\frac{e^2}{2} \right) \cos \left(\Omega^* + 4M \right)$$

$$= \frac{5e^2}{2} \cos \Omega^* - 3e \cos \left(\Omega^* + M \right) + \left(1 - \frac{5e^2}{2} \right) \cos \left(\Omega^* + 2M \right)$$

$$+ e \cos \left(\Omega^* + 3M \right) + e^2 \cos \left(\Omega^* + 4M \right). \tag{10.34}$$

Thus we have a remarkable simplification where all the $\cos \left(\Omega^* - nM \right)$ terms vanish.

From Eq. 10.7, we can write

$$\mathcal{R} = \frac{Gm'a^2}{a'^3} \left[\frac{a'^2}{a^2} + \frac{1}{4} \left(\frac{r}{a} \right)^2 + \frac{3}{4} \left(\frac{r}{a} \right)^2 \cos 2 \left(\psi - \psi' \right) \right]. \tag{10.35}$$

Recalling Eq. 10.28, we have

$$\left(\frac{r}{a} \right)^2 = 1 + \frac{3e^2}{2} - 2e \cos M - \frac{e^2}{2} \cos 2M. \tag{10.36}$$

Substituting Eqs. 10.34 and 10.36 into Eq. 10.35, we finally obtain the disturbing function for the effect of the Sun on the Moon's orbit (in our simplified model):

$$\mathcal{R} = \frac{Gm'a^2}{a'^3} \left\{ \left(\frac{a'}{a} \right)^2 + \frac{1}{4} + \frac{3e^2}{8} - \frac{e}{2} \cos M - \frac{e^2}{8} \cos 2M \right.$$

$$+ \frac{15e^2}{8} \cos \left[2 \left(\Omega + \omega - \psi' \right) \right]$$

$$- \frac{9e}{4} \cos \left[2 \left(\Omega + \omega - \psi' \right) + M \right]$$

$$+ \frac{3}{4} \cos \left[2 \left(\Omega + \omega - \psi' \right) + 2M \right]$$

$$- \frac{15e^2}{8} \cos \left[2 \left(\Omega + \omega - \psi' \right) + 2M \right]$$

$$+ \frac{3e}{4} \cos \left[2 \left(\Omega + \omega - \psi' \right) + 3M \right]$$

$$\left. + \frac{3e^2}{4} \cos \left[2 \left(\Omega + \omega - \psi' \right) + 4M \right] \right\}. \tag{10.37}$$

Brown points out that the first term in the series, Gm'/a', may be omitted since it does not contain the coordinates of the Moon. In Eq. 10.37, we have ignored terms of higher order than e^2, the mass of the Moon, the inclination of the Moon's orbit, and the eccentricity of Earth's orbit. Even this simplified model implies that the motion of the Moon is complex. We note that

1. Terms in $\mathcal{R}$ such as $\left(Gm'a^2/a'^3\right)\left(1/4 + 3e^2/8\right)$ give rise to secular variations in the orbital elements.
2. Terms involving only $\cos M$ and $\cos 2M$ are elliptic terms arising from a series representation of the Keplerian motion.
3. The remaining terms depend on the relative positions of the Moon and Sun; these are strictly perturbative terms—for example, the "evection" and "variation." The evection term

$$\frac{15e^2}{8}\cos\left[2\left(\Omega + \omega - \psi'\right)\right]$$

is the largest periodic perturbation in the longitude and was noticed by Ptolemy. Tycho Brahe discovered the variation term

$$\frac{3}{4}\cos\left[2\left(\Omega + \omega - \psi'\right) + 2M\right],$$

which leads to the Moon speeding up at the new- and full-Moon phases and slowing down during the first and last quarters.

10.1   Analysis of Evection

The evection term is the largest periodic perturbation in the Moon's longitude. Let the evection term for the Earth–Moon system in Eq. 10.37 be denoted by

$$A = \frac{15}{8}n'^2 a^2 e^2 \cos 2\left(\Omega + \omega - \psi'\right), \tag{10.38}$$

where

$$n'^2 = Gm'/a'^3. \tag{10.39}$$

We now write Lagrange's planetary equations for the evection term of the Moon's disturbing function. From Lagrange's planetary equations, Eqs. 5.37,

$$\dot{n} = -\frac{3}{a^2}\frac{\partial A}{\partial M} = 0. \tag{10.40}$$

From the $\dot{e}$ equation in Lagrange's planetary equations in Eqs. 5.37, we have

$$\dot{e} = \left(\frac{\beta^2}{na^2e}\right)\frac{\partial A}{\partial M} - \left(\frac{\beta}{na^2e}\right)\frac{\partial A}{\partial \omega}$$

$$= -\left(\frac{\beta}{na^2e}\right)\frac{15}{8}n'^2a^2e^2\,(-2)\sin 2\left(\Omega + \omega - \psi'\right), \tag{10.41}$$

so that

$$\dot{e} = \frac{15}{4}\frac{n'^2}{n}e\beta \sin 2\left(\Omega + \omega - \psi'\right)$$

$$\approx \frac{15}{4}\frac{n'^2}{n}e\left(1 - \frac{1}{2}e^2\right)\sin 2\left(\Omega + \omega - \psi'\right), \tag{10.42}$$

where after dropping terms of order e^3 and higher, we obtain

$$\dot{e} = \frac{15}{4}\frac{n'^2e}{n}\sin 2\left(\Omega + \omega - \psi'\right). \tag{10.43}$$

From Lagrange's planetary equations, Eqs. 5.37, we have

$$\dot{M} = n - \left(\frac{\beta^2}{na^2e}\right)\frac{\partial A}{\partial e} + \frac{3}{a^2}\left(\frac{\partial A}{\partial n}\right)_M, \tag{10.44}$$

where

$$\frac{\partial A}{\partial e} = \frac{15}{4}n'^2a^2e\cos 2\left(\Omega + \omega - \psi'\right) \tag{10.45}$$

$$\frac{\partial A}{\partial n} = \frac{15}{4}n'^2ae^2\cos 2\left(\Omega + \omega - \psi'\right)\left(\frac{\partial a}{\partial n}\right)$$

and using

$$\frac{\partial a}{\partial n} = -\frac{2a}{3n}$$

gives

$$\frac{\partial A}{\partial n} = -\frac{5}{2}\frac{n'^2}{n}a^2e^2\cos 2\left(\Omega + \omega - \psi'\right), \tag{10.46}$$

so

$$\dot{M} = n - \left[\left(\frac{\beta^2}{na^2e}\right)\frac{15}{4}n'^2a^2e + \frac{3}{a^2}\left(\frac{5}{2}\right)\frac{n'^2}{n}a^2e^2\right]\cos 2\left(\Omega + \omega - \psi'\right)$$

$$= n - \frac{15}{4} \frac{n'^2}{n} \left[1 - e^2 + 2e^2\right] \cos 2 \left(\Omega + \omega - \psi'\right)$$

$$= n - \frac{15}{4} \frac{n'^2}{n} \left(1 + e^2\right) \cos 2 \left(\Omega + \omega - \psi'\right). \tag{10.47}$$

From the $\dot\omega$ equation in Lagrange's planetary equations (Eqs. 5.37), we have

$$\dot\omega = \frac{\beta}{na^2 e} \frac{\partial A}{\partial e} - \left(\frac{\cos i}{na^2 \beta \sin i}\right) \frac{\partial A}{\partial i}. \tag{10.48}$$

Substituting Eq. 10.45 into Eq. 10.48 gives

$$\dot\omega = \frac{\beta}{na^2 e} \frac{15}{4} n'^2 a^2 e \cos 2 \left(\Omega + \omega - \psi'\right). \tag{10.49}$$

Using

$$\beta = \sqrt{1 - e^2} \approx 1 - \frac{1}{2} e^2 \tag{10.50}$$

in Eq. 10.49, we have

$$\dot\omega = \frac{15}{4} \left(1 - \frac{e^2}{2}\right) \frac{n'^2}{n} \cos 2 \left(\Omega + \omega - \psi'\right). \tag{10.51}$$

From the $\dot\Omega$ equation in Lagrange's planetary equations (Eqs. 5.37), we obtain

$$\dot\Omega = \left(\frac{1}{na^2 \beta \sin i}\right) \frac{\partial A}{\partial i}, \tag{10.52}$$

but the inclination does not appear in Eq. 10.38, so

$$\dot\Omega = 0. \tag{10.53}$$

From the di/dt equation in Lagrange's planetary equations (Eqs. 5.37), we have

$$\frac{di}{dt} = \left(\frac{\cos i}{na^2 \beta \sin i}\right) \frac{\partial A}{\partial \omega} - \left(\frac{1}{na^2 \beta \sin i}\right) \frac{\partial A}{\partial \Omega}. \tag{10.54}$$

We note that

$$\frac{\partial A}{\partial \omega} = -\frac{15}{4} n'^2 a^2 e^2 \sin 2 \left(\Omega + \omega - \psi'\right) \tag{10.55}$$

$$\frac{\partial A}{\partial \Omega} = -\frac{15}{4} n'^2 a^2 e^2 \sin 2 \left(\Omega + \omega - \psi'\right), \tag{10.56}$$

so that

$$\frac{\partial A}{\partial \omega} - \frac{\partial A}{\partial \Omega} = 0 \qquad (10.57)$$

and thus

$$\frac{di}{dt} = 0. \qquad (10.58)$$

Gathering results from Eqs. 10.40, 10.43, 10.47, 10.51, 10.53, and 10.58, we have Lagrange's planetary equations for Lunar Evection:

$$\dot{n} = 0$$

$$\dot{e} = \frac{15}{4} \frac{n'^2 e}{n} \sin 2 \left(\Omega + \omega - \psi' \right)$$

$$\dot{M} = n - \frac{15}{4} \frac{n'^2}{n} \left(1 + e^2 \right) \cos 2 \left(\Omega + \omega - \psi' \right)$$

$$\dot{\omega} = \frac{15}{4} \frac{n'^2}{n} \left(1 - \frac{1}{2} e^2 \right) \cos 2 \left(\Omega + \omega - \psi' \right) \qquad (10.59)$$

$$\dot{\Omega} = 0$$

$$\frac{di}{dt} = 0.$$

To integrate the equations for $\dot{e}$, $\dot{M}$, $\dot{\omega}$ in Eqs. 10.59, we keep in mind that the longitude of the Sun is given by

$$\psi' = \psi'(t) = \Omega' + \omega' + n'(t - t_0), \qquad (10.60)$$

where n', Ω', and ω' refer to the Sun's *apparent motion* around the Earth. Thus

$$e = e_{\text{osc}} - \frac{15}{4} \frac{n'^2 e}{n} \left[\cos 2 \left(\Omega + \omega - \psi' \right) \right] \left(\frac{-1}{2n'} \right) \qquad (10.61)$$

so that

$$e = e_{\text{osc}} + \frac{15}{8} \frac{n'}{n} e_{\text{osc}} \cos 2 \left(\Omega_{\text{osc}} + \omega_{\text{osc}} - \psi' \right). \qquad (10.62)$$

Similarly

$$M = M_{\text{osc}} + \frac{15}{8} \frac{n'}{n} \left(1 + e_{\text{osc}}^2 \right) \sin 2 \left(\Omega_{\text{osc}} + \omega_{\text{osc}} - \psi' \right), \qquad (10.63)$$

where M refers to the real (perturbed) orbit of the Moon and M_{osc} denotes the osculating orbit: $M_{\text{osc}} = n_{\text{osc}}(t - T_0)$. For $i = 0$ we can treat $\Omega + \omega$ as a variable. From Eqs. 10.48 and 10.53, we can write

$$\dot{\omega} = \frac{d}{dt}(\Omega + \omega), \qquad (10.64)$$

so integrating the $\dot{\omega}$ equation provides

$$\Omega + \omega = (\Omega + \omega)_{\text{osc}} - \frac{15}{8}\left(1 - \frac{e_{\text{osc}}^2}{2}\right)\left(\frac{n'}{n}\right)\sin 2\left(\Omega_{\text{osc}} + \omega_{\text{osc}} - \psi'\right). \quad (10.65)$$

The "osc" subscripts in Eqs. 10.62, 10.63, and 10.65 refer to the osculating orbit of reference. Summarizing, the nonzero perturbations are

$$\delta(e) = \frac{15}{8}\left(\frac{n'}{n}\right) e_{\text{osc}} \cos 2\left(\Omega_{\text{osc}} + \omega_{\text{osc}} - \psi'\right)$$

$$\delta(M) = \frac{15}{8}\left(\frac{n'}{n}\right)\left(1 + e_{\text{osc}}^2\right)\sin 2\left(\Omega_{\text{osc}} + \omega_{\text{osc}} - \psi'\right) \qquad (10.66)$$

$$\delta(\Omega + \omega) = -\frac{15}{8}\left(\frac{n'}{n}\right)\left(1 - \frac{e_{\text{osc}}^2}{2}\right)\sin 2\left(\Omega_{\text{osc}} + \omega_{\text{osc}} - \psi'\right).$$

The perturbation to first order in longitude, $\delta(\psi)$, can be found from

$$\psi = \Omega + \omega + f \qquad (10.67)$$

and from the series expansion for the true anomaly

$$f = M + 2e\sin M + \frac{5}{4}e^2\sin 2M + \ldots, \qquad (10.68)$$

which is called the equation of the center. From Eqs. 10.67 and 10.68, we can write

$$\delta(\psi) = \delta(\Omega + \omega) + \delta(M + 2e\sin M)$$
$$= \delta(\Omega + \omega) + (1 + 2e\cos M)\,\delta M + 2(\sin M)\,\delta e, \qquad (10.69)$$

where we have dropped terms of higher order than e. Substituting Eqs. 10.66 into Eq. 10.69, we obtain

$$\delta(\psi) = -\frac{15}{8}\left(\frac{n'}{n}\right)\left(1 - \frac{e_{\text{osc}}^2}{2}\right)\sin 2\left(\Omega_{\text{osc}} + \omega_{\text{osc}} - \psi'\right)$$

$$+ \frac{15}{8}\left(\frac{n'}{n}\right)\left(1 + e_{\text{osc}}^2\right)\sin 2\left(\Omega_{\text{osc}} + \omega_{\text{osc}} - \psi'\right)$$

$$+ \frac{15}{4} \left(\frac{n'}{n} \right) e_{\text{osc}} \left(1 + e_{\text{osc}}^2 \right) \cos M \sin 2 \left(\Omega_{\text{osc}} + \omega_{\text{osc}} - \psi' \right)$$

$$+ \frac{15}{4} \left(\frac{n'}{n} \right) e_{\text{osc}} \sin M \cos 2 \left(\Omega_{\text{osc}} + \omega_{\text{osc}} - \psi' \right). \tag{10.70}$$

Retaining only first-order terms in e_{osc} (i.e., dropping the e_{osc}^2 terms) in Eq. 10.70, the first two terms cancel leaving us with

$$\delta \left(\psi \right) = \frac{15}{4} \left(\frac{n'}{n} \right) e_{\text{osc}} \Big[\cos M \sin 2 \left(\Omega_{\text{osc}} + \omega_{\text{osc}} - \psi' \right)$$

$$+ \sin M \cos 2 \left(\Omega_{\text{osc}} + \omega_{\text{osc}} - \psi' \right) \Big]. \tag{10.71}$$

Noting the trigonometric identity

$$\sin \left(A + B \right) = \sin A \cos B + \cos A \sin B \tag{10.72}$$

allows us to rewrite Eq. 10.71 as

$$\delta \left(\psi \right) = \frac{15}{4} \left(\frac{n'}{n} \right) e_{\text{osc}} \sin \left[2 \left(\Omega_{\text{osc}} + \omega_{\text{osc}} - \psi' \right) + M \right]. \tag{10.73}$$

The ratio of the mean motion of the Sun to that of the Moon may be expressed as

$$\frac{n'}{n} = \frac{P}{P'} = \frac{P_{\text{Moon}}}{P_{\text{Earth}}} = \frac{27.322}{365.25} = 0.07480. \tag{10.74}$$

Using

$$e_{\text{osc}} = 0.0549 \tag{10.75}$$

for the eccentricity of the Moon's orbit, we obtain for the coefficient in Eq. 10.73:

$$\frac{15}{4} \left(\frac{n'}{n} \right) e_{\text{osc}} = \frac{15}{4} (0.0748) (0.0549) = 15.4 \times 10^{-3} = 0.882°. \tag{10.76}$$

To find the period, we note that the only terms in Eq. 10.73 that contain the time, t, are from $2\psi'$ (Eq. 10.60) and from M. The time terms reduce to $\left(n - 2n' \right) t$, so the period of the perturbation in the longitude is

$$P_{\text{longitude}} = \frac{2\pi}{n - 2n'}, \tag{10.77}$$

where

$$n = \frac{2\pi}{27.322 \text{ d}}$$

$$2n' = \frac{4\pi}{365.25 \text{ d}},$$ (10.78)

and we have

$$P_{\text{longitude}} = 32.13 \text{ days.}$$ (10.79)

Thus, the period of the perturbation in the Moon's longitude due to the evection term in $\mathcal{R}$ is about one month and has an amplitude of about one degree. (Note: the apparent width of the Moon is about 1/2 degree, so we see that this effect is readily visible from Earth.)

10.2   Other Secular Effects

According to Chobotov [2002], the equation of nodal regression is

$$\dot{\Omega} = -\frac{3}{8} \frac{n_3^2}{n} \frac{\left(1 + \frac{3}{2} e^2\right)}{\sqrt{1 - e^2}} \cos i \left(3 \cos^2 i_3 - 1\right),$$ (10.80)

where n_3 and i_3 are the mean motion and inclination of the disturbing body with respect to Earth's equatorial plane (which correspond to n' and i'). (Note: Chobotov references Brouwer and Clemence (1961).) Assuming that

$$i_3 = 0$$ (10.81)

and $i \approx 0$, and ignoring e^2, Eq. 10.80 becomes

$$\dot{\Omega} = -\frac{3}{8} \frac{n'^2}{n} (3 - 1)$$

$$= -\frac{3}{4} \frac{n'^2}{n}.$$ (10.82)

Integrating Eq. 10.82, we obtain

$$\delta(\Omega) = -\frac{3}{4} \frac{n'^2}{n} (t - t_0) = -\frac{3}{4} \frac{n'}{n} n' \delta t,$$ (10.83)

where $n' = 2\pi/P_{\text{Earth}}$ and P_{Earth} is the period of Earth's orbital motion around the Sun. Since $n = 2\pi/P_{\text{Moon}}$, for the Moon, Eq. 10.83 can be written as

$$\frac{\delta\left(\Omega\right)}{2\pi} = -\frac{3}{4}\left(\frac{P_{Moon}}{P_{Earth}}\right)\frac{n'}{2\pi}\delta t = -\frac{3}{4}\left(\frac{P_{Moon}}{P_{Earth}}\right)\frac{\delta t}{P_{Earth}}$$

$$= -\frac{3}{4}\left(\frac{27.322\text{ d}}{365.25\text{ d}}\right)\frac{\delta t}{P_{Earth}} = -0.05610\frac{\delta t}{P_{Earth}}$$

(10.84)

so

$$\delta\Omega = 2\pi \quad \text{when} \quad 0.05610\frac{\delta t}{P_{Earth}} = 1,$$

(10.85)

or

$$\delta t = \frac{365.25\text{ d}}{0.05610} = 6511\text{ d} = 17.83\text{ years},$$

(10.86)

i.e., about 18 years. The minus sign in Eq. 10.84 indicates that the node of the Moon's orbit regresses, i.e., moves westward along the ecliptic, completing a revolution in about 18 years. This motion is considered to be a secular perturbation in the sense that Ω is a linear function of time.

Now referring again to Chobotov (2002), we have the equation for the rate of change of the argument of perigee:

$$\dot{\omega} = \frac{3}{4}\frac{n_3^2}{n}\frac{\left(1 - \frac{3}{2}\sin^2 i_3\right)}{\sqrt{1 - e^2}}\left(2 - \frac{5}{2}\sin^2 i + \frac{e^2}{2}\right).$$

(10.87)

Assuming $i_3 = 0$, $i \approx 0$, ignoring e^2, and using $n' = n_3$, we obtain

$$\dot{\omega} = \frac{3}{4}\frac{n'^2}{n}2 = \frac{3}{2}\frac{n'^2}{n} = -2\dot{\Omega}$$

(10.88)

that is, the rate of change of the argument of perigee is twice the rate of node regression.

Brown (1896) presents a more detailed analysis, where the principal secular effects are:

(a) The line of nodes regresses at an average rate of one revolution in 18.6 years.
(b) The line of apsides (the major axis of the orbit) advances (i.e., $\dot{\omega} > 0$) at an average rate of one revolution in 8.9 years.

If, at a given new Moon, the Sun is sufficiently close to the line of nodes for a solar eclipse to occur, then one Saros later (6,585 days, or 18 years, and 11 days) another solar eclipse will occur. Lunar eclipses repeat in a similar manner.

References

Brouwer, D., & Clemence, G. M. (1961). *Methods of celestial mechanics*. New York: Academic Press.

Brown, E. W. (1896). *An introductory treatise on lunar theory*. New York: Cambridge University Press.

Chao, C.-C., & Hoots, F. R. (2018). *Applied orbit perturbations and maintenance* (2nd ed.). El Segundo: The Aerospace Press.

Chobotov, V. A. (Ed.) (2002). *Orbital mechanics* (3rd ed.). Reston: American Institute of Aeronautics and Astronautics, Inc.

Chapter 11
Effects of General Relativity

From his general theory of relativity, Einstein predicted three effects that could be tested by observation: (a) the motion of the perihelion of Mercury, (b) the deflection of light by the Sun, and (c) the gravitational red shift of spectral lines (See Weinberg (1972)). We examine the first two of these effects in this chapter.

11.1 Motion of Mercury's Perihelion

Observations of Mercury over many years revealed that its perihelion was advancing more rapidly than could be accounted for by the Newtonian theory. Einstein's general theory of relativity (GR) predicted the excess advance of 43" of arc per century within the errors of measurement.

According to GR, the acceleration can be expressed as

$$F(r) = -\frac{\mu}{r^2}\left(1 + \frac{3h^2}{c^2 r^2}\right),\tag{11.1}$$

where

$$\mu = G(m_1 + m_2),\tag{11.2}$$

m_1 is the mass of the Sun, m_2 is the mass of the planet, h is the specific angular momentum given in Eq. 2.17

$$h = r^2 \dot{f},\tag{11.3}$$

and c is the speed of light.

© The Author(s), under exclusive license to Springer Nature Switzerland AG 2022
J. M. Longuski et al., *Introduction to Orbital Perturbations*, Space Technology
Library 40, https://doi.org/10.1007/978-3-030-89758-1_11

The ratio h/r is the transverse velocity of the planet in its orbit. Thus, Eq. 11.1 states that, for a nearly circular orbit, the relativistic acceleration, $3h^2/(c^2r^2)$, is proportional to the ratio of the squares of the speed of the planet to the speed of light:

$$\frac{3h^2}{c^2r^2} = 3\frac{\left(r\dot{f}\right)^2}{c^2}.$$

The first term in Eq. 11.1 is the Newtonian acceleration. For the Earth, the ratio of the GR correction to the Newtonian acceleration is about 3×10^{-8}. The perturbative acceleration

$$R = -\frac{3\mu h^2}{c^2r^4} \tag{11.4}$$

is purely in the radial direction. Using the Lagrange planetary equations, (Eqs. 5.103), we find

$$\dot{n} = \left(\frac{3e}{a\beta}\sin f\right)\frac{3\mu h^2}{c^2r^4}$$

$$\dot{e} = -\left(\frac{\beta}{na}\sin f\right)\frac{3\mu h^2}{c^2r^4}$$

$$\frac{di}{dt} = 0$$

$$\dot{\Omega} = 0 \tag{11.5}$$

$$\dot{\omega} = \left(\frac{\beta\cos f}{nae}\right)\frac{3\mu h^2}{c^2r^4}$$

$$\dot{M} = n - \left(\frac{\beta^2}{nae}\cos f - \frac{2r}{na^2}\right)\frac{3\mu h^2}{c^2r^4}.$$

Equations 11.5 can be treated by the method of averaging with the mean anomaly as the only fast variable. In anticipation of integration, we rearrange the terms in Eqs. 11.5 to obtain

$$\dot{n} = \frac{9en^4 a^6 \beta}{c^2} \sin f \frac{1}{r^4}$$

$$\dot{e} = -\frac{3n^3 a^6 \beta^3}{c^2} \sin f \frac{1}{r^4}$$

$$\frac{di}{dt} = 0$$

$$\dot{\Omega} = 0 \qquad\qquad (11.6)$$

$$\dot{\omega} = \frac{3n^3 a^6 \beta^3}{ec^2} \cos f \frac{1}{r^4}$$

$$\dot{M} = n - \frac{3n^3 a^6 \beta^4}{ec^2} \cos f \frac{1}{r^4} + \frac{6n^3 a^5 \beta^2}{c^2} \frac{1}{r^3}.$$

To transform the differential equations, we need the following integrals:

$$\int \sin f \frac{1}{r^4} dM \qquad \int \cos f \frac{1}{r^4} dM \qquad \int \frac{1}{r^3} dM. \qquad (11.7)$$

The integration variable can be changed to true anomaly using Eq. 4.47

$$dM = \frac{r^2}{a^2 \beta} df.$$

Then the integrals can be performed to give

$$\int \sin f \frac{1}{r^4} dM = \frac{1}{a^4 \beta^5} \int \sin f \, (1 + e \cos f)^2 df$$

$$= -\frac{1}{3a^4 e \beta^5} (1 + e \cos f)^3$$

$$\int \cos f \frac{1}{r^4} dM = \frac{1}{a^4 \beta^5} \int \cos f \, (1 + e \cos f)^2 df$$

$$= \frac{1}{a^4 \beta^5} \left[ef + \left(1 + \frac{3}{4} e^2 \right) \sin f + \frac{1}{2} e \sin 2f + \frac{1}{12} e^2 \sin 3f \right]$$

$$\int \frac{1}{r^3} dM = \frac{1}{a^3 \beta^3} \int (1 + e \cos f) df = \frac{1}{a^3 \beta^3} (f + e \sin f).$$

$$(11.8)$$

The corresponding definite integrals are

$$\frac{1}{2\pi} \int_0^{2\pi} \sin f \frac{1}{r^4} dM = 0$$

$$\frac{1}{2\pi} \int_0^{2\pi} \cos f \frac{1}{r^4} dM = \frac{e}{a^4 \beta^5} \tag{11.9}$$

$$\frac{1}{2\pi} \int_0^{2\pi} \frac{1}{r^3} dM = \frac{1}{a^3 \beta^3}.$$

Using Eqs. 11.9 in Eqs. 11.6, the transformed rate of the slow variables can be calculated. Furthermore, we select a constant of integration for the mean motion periodic such that the mean motion periodic has zero average value. Given that approach, the transformed rate of the fast variable is just the average of the right-hand side of Eq. 11.6. Thus, the transformed rates for the slow and fast variables are

$$\langle \dot{n} \rangle = 0$$

$$\langle \dot{e} \rangle = 0$$

$$\left\langle \frac{di}{dt} \right\rangle = 0 \tag{11.10}$$

$$\langle \dot{\Omega} \rangle = 0$$

$$\langle \dot{\omega} \rangle = \frac{3n^3 a^6 \beta^3}{ec^2} \frac{e}{a^4 \beta^5} = \frac{3n^3 a^2}{c^2 \beta^2}$$

$$\langle \dot{M} \rangle = n - \frac{3n^3 a^6 \beta^4}{ec^2} \frac{e}{a^4 \beta^5} + \frac{6n^3 a^5 \beta^2}{c^2} \frac{1}{a^3 \beta^3} = n - \frac{3n^3 a^2}{c^2 \beta} + \frac{6n^3 a^2}{c^2 \beta}$$

$$= n + \frac{3n^3 a^2}{c^2 \beta}.$$

The short-period periodics (orbital period of the planet) are given by Eq. 8.33. Using Eqs. 11.8 and 11.10, we obtain

$$\delta n_{SP} = \frac{1}{n} \int \left[\frac{9en^4 a^6 \beta}{c^2} \sin f \frac{1}{r^4} - \langle \dot{n} \rangle \right] dM + c_n$$

$$= \frac{1}{n} \frac{9en^4 a^6 \beta}{c^2} \int \left(\sin f \frac{1}{r^4} \right) dM + c_n$$

$$= -\frac{3n^3a^2}{c^2\beta^4}(1 + e\cos f)^3 + c_n.$$

Since we desire the periodic to have zero average value, we need

$$c_n = \frac{3n^3a^2}{c^2\beta^4}\left\langle(1 + e\cos f)^3\right\rangle = \frac{3n^3a^2}{c^2\beta^4}\frac{1}{2\pi}\int_0^{2\pi}(1 + e\cos f)^3 dM$$

$$= \frac{3n^3a^2}{c^2\beta^4}\frac{1}{2\pi}\int_0^{2\pi}(1 + e\cos f)^3\frac{r^2}{a^2\beta}df = \frac{3n^3a^2}{c^2\beta^4}\frac{1}{2\pi}\int_0^{2\pi}(1 + e\cos f)\beta^3 df$$

$$= \frac{3n^3a^2}{c^2\beta}.$$

Thus, the periodic for the mean motion is

$$\delta n_{SP} = -\frac{3n^3a^2}{c^2\beta^4}(1 + e\cos f)^3 + \frac{3n^3a^2}{c^2\beta}$$

$$= \frac{3n^3a^2}{c^2\beta^4}\left[\beta^3 - (1 + e\cos f)^3\right]. \tag{11.11}$$

Similarly, we have for the eccentricity

$$\delta e_{SP} = -\frac{1}{n}\int\left[\frac{3n^3a^6\beta^3}{c^2}\sin f\frac{1}{r^4} - \langle\dot{e}\rangle\right]dM + c_e$$

$$= -\frac{1}{n}\frac{3n^3a^6\beta^3}{c^2}\int\left[\sin f\frac{1}{r^4}\right]dM + c_e$$

$$= \frac{n^2a^2}{ec^2\beta^2}\left[(1 + e\cos f)^3\right] + c_e.$$

Since we desire the periodic to have zero average value, we need

$$c_e = -\frac{n^2a^2}{ec^2\beta^2}\left\langle(1 + e\cos f)^3\right\rangle = -\frac{n^2a^2}{ec^2\beta^2}\frac{1}{2\pi}\int_0^{2\pi}(1 + e\cos f)^3 dM$$

$$= -\frac{n^2a^2}{ec^2\beta^2}\frac{1}{2\pi}\int_0^{2\pi}(1 + e\cos f)^3\frac{r^2}{a^2\beta}df = -\frac{n^2a^2}{ec^2\beta^2}\frac{1}{2\pi}\int_0^{2\pi}(1 + e\cos f)\beta^3 df$$

$$= -\frac{n^2 a^2 \beta}{ec^2}.$$

Thus, the periodic for eccentricity is

$$\delta e_{SP} = \frac{n^2 a^2}{ec^2 \beta^2} \left[(1 + e \cos f)^3 \right] - \frac{n^2 a^2 \beta}{ec^2}$$

$$= \frac{n^2 a^2}{ec^2 \beta^2} \left[(1 + e \cos f)^3 - \beta^3 \right]. \tag{11.12}$$

Since there is no perturbation effect in inclination and longitude of the ascending node, the transformed rates are identical to the original rates and the periodics are

$$\delta i_{SP} = 0 \tag{11.13}$$

$$\delta \Omega_{SP} = 0. \tag{11.14}$$

The argument of perihelion periodic is

$$\delta \omega_{SP} = \frac{1}{n} \int \left[\frac{3n^3 a^6 \beta^3}{ec^2} \cos f \frac{1}{r^4} - \langle \dot{\omega} \rangle \right] dM + c_\omega,$$

and using Eqs. 11.8 and 11.10, we find

$$\delta \omega_{SP} = \frac{3n^2 a^6 \beta^3}{ec^2} \int \left[\cos f \frac{1}{r^4} \right] dM - \frac{1}{n} \int \left[\frac{3n^3 a^2}{c^2 \beta^2} \right] dM + c_\omega$$

$$= \frac{3n^2 a^6 \beta^3}{ec^2} \frac{1}{a^4 \beta^5} \left[ef + \left(1 + \frac{3}{4} e^2 \right) \sin f + \frac{1}{2} e \sin 2f + \frac{1}{12} e^2 \sin 3f \right]$$

$$- \frac{3n^2 a^2}{c^2 \beta^2} \int dM + c_\omega$$

$$= \frac{3n^2 a^2}{ec^2 \beta^2} \left[e (f - M) + \left(1 + \frac{3}{4} e^2 \right) \sin f + \frac{1}{2} e \sin 2f + \frac{1}{12} e^2 \sin 3f \right] + c_\omega.$$

We can see by inspection that this function has zero mean, so we have

$$c_\omega = 0$$

and

$$\delta \omega_{SP} = \frac{3n^2 a^2}{ec^2 \beta^2} \left[e (f - M) + \left(1 + \frac{3}{4} e^2 \right) \sin f + \frac{1}{2} e \sin 2f + \frac{1}{12} e^2 \sin 3f \right]. \tag{11.15}$$

Finally, we examine the periodic for the fast variable. Using Eq. 8.36, we obtain

$$n\frac{\partial \phi}{\partial M} = \delta n_{SP} - \langle \delta n_{SP} \rangle + u_1 - \langle u_1 \rangle$$

and substituting the specific functions from Eqs. 11.6 and 11.11 gives

$$\delta M_{SP} = \frac{1}{n} \int [\delta n_{SP} - \langle \delta n_{SP} \rangle] dM + \frac{1}{n} \int [u_1 - \langle u_1 \rangle] dM$$

$$= \frac{1}{n}\frac{3n^3 a^2}{c^2 \beta^4} \int \left[\beta^3 - (1 + e \cos f)^3 \right] dM$$

$$+ \frac{1}{n} \int \left[-\frac{3n^3 a^6 \beta^4}{ec^2} \cos f \frac{1}{r^4} + \frac{6n^3 a^5 \beta^2}{c^2} \frac{1}{r^3} \right] dM - \frac{1}{n} \int \left[\frac{3n^3 a^2}{c^2 \beta} \right] dM$$

$$= \frac{3n^2 a^2}{c^2 \beta^4} \int \beta^3 dM - \frac{3n^2 a^2}{c^2 \beta^4} \int (1 + e \cos f)^3 dM$$

$$- \frac{3n^2 a^6 \beta^4}{ec^2} \int \cos f \frac{1}{r^4} dM + \frac{6n^2 a^5 \beta^2}{c^2} \int \frac{1}{r^3} dM - \frac{3n^2 a^2}{c^2 \beta} \int dM.$$

Using the integrals in Eqs. 11.8, we find

$$\delta M_{SP} = \frac{3n^2 a^2}{c^2 \beta} \int dM - \frac{3n^2 a^2}{c^2 \beta} \int dM - \frac{3n^2 a^2}{c^2 \beta^4} (f + e \sin f) \beta^3$$

$$- \frac{3n^2 a^6 \beta^4}{ec^2} \frac{1}{a^4 \beta^5} \left[ef + \left(1 + \frac{3}{4}e^2 \right) \sin f + \frac{1}{2}e \sin 2f + \frac{1}{12}e^2 \sin 3f \right]$$

$$+ \frac{6n^2 a^5 \beta^2}{c^2} \frac{1}{a^3 \beta^3} (f + e \sin f)$$

$$= \frac{3n^2 a^2}{c^2 \beta} (e \sin f) - \frac{3n^2 a^2}{ec^2 \beta} \left[\left(1 + \frac{3}{4}e^2 \right) \sin f + \frac{1}{2}e \sin 2f + \frac{1}{12}e^2 \sin 3f \right].$$

$$(11.16)$$

In summary, the average rates are

$$\langle \dot{n} \rangle = 0$$

$$\langle \dot{e} \rangle = 0$$

$$\left\langle \frac{di}{dt} \right\rangle = 0$$

$$\langle \dot{\Omega} \rangle = 0 \qquad\qquad\qquad (11.17)$$

$$\langle \dot{\omega} \rangle = \frac{3n^3 a^2}{c^2 \beta^2}$$

$$\langle \dot{M} \rangle = n + \frac{3n^3 a^2}{c^2 \beta}.$$

Thus, the mean motion, eccentricity, inclination, and longitude of the ascending node exhibit no secular variation due to general relativity. It is interesting to note that there is a secular effect on mean anomaly that will slightly change the period of Mercury each revolution. This period change is nearly identical to the effect on argument of perihelion.

The periodics are

$$\delta n_{SP} = \frac{3n^3 a^2}{c^2 \beta^4} \left[\beta^3 - (1 + e \cos f)^3 \right]$$

$$\delta e_{SP} = \frac{n^2 a^2}{e c^2 \beta^2} \left[(1 + e \cos f)^3 - \beta^3 \right]$$

$$\delta i_{SP} = 0$$

$$\delta \Omega_{SP} = 0$$

$$\delta \omega_{SP} = \frac{3n^2 a^2}{e c^2 \beta^2} \left[e \, (f - M) + \left(1 + \frac{3}{4} e^2 \right) \sin f + \frac{1}{2} e \sin 2f + \frac{1}{12} e^2 \sin 3f \right]$$

$$\delta M_{SP} = \frac{3n^2 a^2}{c^2 \beta} \, (e \sin f) - \frac{3n^2 a^2}{e c^2 \beta} \left[\left(1 + \frac{3}{4} e^2 \right) \sin f + \frac{1}{2} e \sin 2f + \frac{1}{12} e^2 \sin 3f \right].$$

$$(11.18)$$

Together Eqs. 11.17 and 11.18 predict the secular and periodic behavior of the orbital elements of Mercury due to the effect of general relativity. Because the effects are inversely proportional to the square of the speed of light, they are small, but the secular effects will accumulate over time.

Following the suggestion of Einstein, we examine the effect on the motion of the perihelion using Eq. 11.17

$$\langle \dot{\omega} \rangle = \frac{3n^3 a^2}{c^2 \beta^2}.$$

Using the definition

$$n = \frac{2\pi}{P},$$

we can replace the mean motion by the orbital period to get

$$\langle \dot{\omega} \rangle = \frac{3n^3 a^2}{c^2 \beta^2} = \frac{24\pi^3 a^2}{c^2 \beta^2 P^3}. \tag{11.19}$$

The units are radians per time. The result in Eq. 11.19 matches the result for precession of the perihelion using the Lindstedt–Poincaré method in Eq. 9.112, where we note $\langle \dot{\omega} \rangle \times P = \Delta \omega$. Figure 11.1 illustrates the effect on the perihelion of a planet.

If we choose the time unit to be seconds, then the conversion factor of radians per second to arc seconds per century is

$$\text{factor} = \frac{100\text{yr}}{\text{century}} \times \frac{31.6 \times 10^6 \text{ sec}}{\text{yr}} \times \frac{57.3 \text{ deg}}{\text{rad}} \times \frac{3600 \text{ arcsec}}{\text{deg}}$$

$$= 6.51 \times 10^{14} \text{arcsec} \times \text{sec/century}.$$

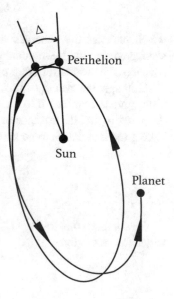

Fig. 11.1 Advance of the perihelion due to general relativity. The effect on Mercury's Orbit of 43 seconds of arc per century established the correctness of Einstein's theory (Figure adapted from Rindler (1969))

Table 11.1 Comparison of theoretical and observed centennial precessions of planetary orbits (adapted from Weinberg (1972))

Planet	a $\times 10^6$ (km)	e	Period (days)	$\Delta\phi$ general relativity	Observed (sec per century)
Mercury	57.9	0.206	88	43.0	43.1 ± 0.45
Venus	108	0.0068	225	8.6	8.4 ± 4.8
Earth	150	0.0167	365	3.8	5.0 ± 1.2
Icarus	161	0.827	408	10.3	9.8 ± 0.8

In the case of Mercury, we use the following values:

$$a = 5.79 \times 10^{10} \text{ m}$$

$$e = 0.206$$

$$P = 88.0 \text{ days} = 7.60 \times 10^6 \text{ sec} \tag{11.20}$$

$$c = 3.00 \times 10^8 \text{ m/sec}$$

$$1 \text{ yr} = 31.6 \times 10^6 \text{ sec}.$$

From Eqs. 11.19–11.20, we obtain

$$\Delta\omega_{\text{sec/cent}} = 42.9''/\text{century}, \tag{11.21}$$

which matches the observed precession of the perihelion of Mercury that is not accounted for by Newtonian gravity. Table 11.1 provides a comparison of the advance of perihelia for other planetary bodies.

Although we usually think of general relativity effects as being small, it is instructive to compare it to a non-relativistic effect such as oblateness. We can use Eq. 8.109 to find the advance in perihelion of Mercury due to the Sun oblateness J_2. Taking the inclination to be zero, we find

$$\dot{\omega}_{J_2} = 3\frac{n J_2 r_{\text{Sun}}^2}{a^2 \beta^4}. \tag{11.22}$$

We compare this to Eq. 11.19 for the advance of the perihelion of Mercury due to general relativity:

$$\dot{\omega}_{\text{GR}} = \frac{3n^3 a^2}{c^2 \beta^2}.$$

We examine the ratio

$$\frac{\dot{\omega}_{J_2}}{\dot{\omega}_{\text{GR}}} = \frac{3n J_2 r_{\text{Sun}}^2}{a^2 \beta^4} \frac{c^2 \beta^2}{3n^3 a^2}$$

$$= \frac{J_2 r_{Sun}^2}{a^4 \beta^2} \frac{c^2}{n^2}. \tag{11.23}$$

In addition to the data in Eqs. 11.20, we also need the data for the Sun:

$$J_2 = 1.7 \times 10^{-7}$$

$$r_{Sun} = 6.9 \times 10^8 \text{ m}$$

to compute the ratio shown in Eq. 11.23. The computation gives us

$$\frac{\dot{\omega}_{J_2}}{\dot{\omega}_{GR}} \approx 1 \times 10^{-3}.$$

Thus, the perturbation due to the Sun's J_2 is far smaller than the effect of Einstein's general theory of relativity!

11.2   The Newtonian Gravity-Assist Deflection Equation

Einstein predicted a second effect of GR—the deflection of light by the gravitational field of the Sun. Before examining that effect, we first look at the Newtonian effect on the motion of a spacecraft passing near a gravitating body. This gravity-assist effect is well known by space mission designers. Gravity-assist maneuvers have made possible missions to Jupiter, Saturn, Uranus, Neptune, and Pluto and are planned for future missions in the Solar System.

We assume a spacecraft is on a hyperbolic trajectory as it approaches a planet. The point of closest approach to the planet is r_p and the velocity relative to the planet is denoted by v_∞. The symbol v_∞, or "v-infinity" is the hyperbolic excess speed relative to the gravity-assist body. The geometry of the encounter is illustrated in Fig. 11.2. The angle ψ is called the turn angle or the deflection angle.

Figure 11.2 illustrates the way the $\mathbf{v}_\infty$ vector turns with respect to the gravity-assist planet so that the vector addition of $\mathbf{v}_\infty$ and the planet velocity may result in an increase or decrease of the velocity with respect to the primary body (e.g., the Sun).

We note that the heliocentric energy of the gravity-assist body changes due to the flyby but by an infinitesimal amount due to it having much larger mass than the spacecraft that is gaining energy (i.e., conservation of energy is not violated, and there is no free lunch).

In the two-body problem, the general expression for the eccentricity of a conic section is, from Eq. 2.48,

Fig. 11.2 Turn angle, ψ, due to a gravity-assist flyby

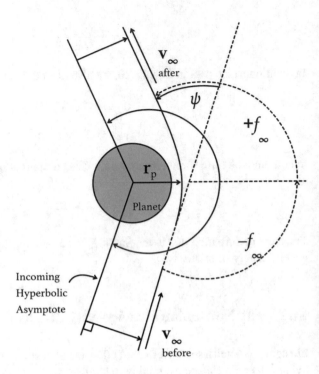

$$e = \sqrt{1 + \frac{2\mathcal{E}h^2}{\mu^2}}, \tag{11.24}$$

where h is the angular momentum, μ is the gravitational parameter ($\mu = Gm$) for the gravity-assist body, and $\mathcal{E}$ is the specific energy.

From Eq. 2.38, the specific energy of a conic orbit, expressed in terms of the position and velocity at periapsis, is

$$\mathcal{E} = \frac{v_p^2}{2} - \frac{\mu}{r_p}, \tag{11.25}$$

and the specific angular momentum (from Eq. 2.39) also expressed at periapsis is

$$h = r_p v_p. \tag{11.26}$$

So we have

$$e = \sqrt{1 + \frac{2}{\mu^2} r_p^2 v_p^2 \left(\frac{1}{2} v_p^2 - \frac{\mu}{r_p}\right)}$$

$$= \sqrt{1 + \frac{r_p^2 v_p^4}{\mu^2} - \frac{2 r_p v_p^2}{\mu}} \qquad , \qquad (11.27)$$

$$= \sqrt{\left(\frac{r_p v_p^2}{\mu} - 1\right)^2}$$

and the eccentricity is given by

$$e = \left| \frac{r_p v_p^2}{\mu} - 1 \right| . \qquad (11.28)$$

The total specific energy for any orbit is

$$\mathcal{E} = \frac{v^2}{2} - \frac{\mu}{r}$$

and, taking the limit as r increases without bound, we see that

$$\lim_{r \to \infty} \mathcal{E} = \frac{1}{2} v_\infty^2 - \frac{\mu}{\infty} = \frac{1}{2} v_\infty^2 . \qquad (11.29)$$

Since the specific energy is constant, we can equate Eqs. 11.25 and 11.29 to write

$$\mathcal{E} = \frac{1}{2} v_p^2 - \frac{\mu}{r_p} = \frac{1}{2} v_\infty^2 . \qquad (11.30)$$

So the velocity at periapsis is

$$v_p^2 = v_\infty^2 + 2 \frac{\mu}{r_p} ,$$

and substituting this into Eq. 11.28, we obtain

$$e = \left| \frac{r_p}{\mu} \left(v_\infty^2 + 2 \frac{\mu}{r_p}\right) - 1 \right| = \left| \frac{r_p}{\mu} v_\infty^2 + 2 - 1 \right|$$

$$= \frac{r_p}{\mu} v_\infty^2 + 1 , \qquad (11.31)$$

where we have dropped the absolute value since we can tell, by inspection, that the expression is always positive. From Fig. 11.2, we see that

$$\psi = 2f_\infty - 180°, \tag{11.32}$$

where f_∞ gives the asymptotic true anomaly obtained by setting the denominator of the conic equation to zero. That is,

$$1 + e \cos f_\infty = 0,$$

so that

$$\cos f_\infty = -\frac{1}{e} \tag{11.33}$$

and from Eq. 11.32

$$\sin \frac{\psi}{2} = \sin\left(f_\infty - 90^o\right) = -\cos f_\infty. \tag{11.34}$$

Using Eq. 11.33 in Eq. 11.34, we obtain

$$\sin \frac{\psi}{2} = \frac{1}{e}. \tag{11.35}$$

We see from Fig. 11.3 that the change in velocity is

$$\Delta v = 2v_\infty \sin\left(\frac{\psi}{2}\right). \tag{11.36}$$

Using Eqs. 11.31 and 11.35, we obtain

$$\Delta v = \frac{2v_\infty}{\left(\dfrac{r_p v_\infty^2}{\mu} + 1\right)}, \tag{11.37}$$

and the deflection angle is

Fig. 11.3 Gravity-assist Δv
due to turn angle, ψ

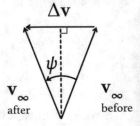

$$\psi = 2 \sin^{-1}\left[\left(1 + \frac{r_p v_\infty^2}{\mu}\right)^{-1}\right]. \tag{11.38}$$

11.3 The General Deflection Equation in General Relativity

Longuski et al. (2001) present "the general deflection equation" as follows:

$$\Delta\phi_{\text{def}} = 2\gamma\epsilon\left(\frac{x}{2+x}\right)^{1/2} + \epsilon\pi\frac{(2+2\gamma-\beta)}{2+x}$$

$$+ 2\left[1 + \epsilon\frac{(2+2\gamma-\beta)}{(2+x)}\right]\sin^{-1}\left(\frac{1}{1+x}\right), \tag{11.39}$$

where

$$\epsilon \equiv \frac{GM}{r_p c^2} \equiv \frac{\mu}{r_p c^2}$$

$$x \equiv \frac{v_\infty^2}{\epsilon}\frac{1}{c^2}$$

are both dimensionless, and the β and γ are post-Newtonian parameters (which are equal to unity in Einstein's theory).

For the non-relativistic (NR) or "Newtonian" deflection of a spacecraft, we drop the ϵ terms from Eq. 11.39:

$$\Delta\phi_{\text{NR}} = 2\sin^{-1}\left(\frac{1}{1+x}\right), \tag{11.40}$$

where when we substitute the expression for x and ϵ, we obtain the familiar expression for the deflection of a spacecraft during a gravity-assist flyby as in Eq. 11.38:

$$\Delta\phi_{\text{NR}} = 2\sin^{-1}\left(\frac{1}{1 + \frac{r_p v_\infty^2}{\mu}}\right). \tag{11.41}$$

We can use the general deflection equation to compute the deflection of light. We set the asymptotic speed, v_∞, equal to the speed of light, c, so we have

$$x = \frac{1}{\epsilon}. \tag{11.42}$$

Using this expression in the first term of the general deflection equation (Eq. 11.39), we have (neglecting terms of order ϵ^2):

$$2\gamma\epsilon\left(\frac{x}{2+x}\right)^{1/2} = 2\gamma\epsilon\left(\frac{1/\epsilon}{2+1/\epsilon}\right)^{1/2} = 2\gamma\epsilon\left(\frac{1}{1+2\epsilon}\right)^{1/2} \approx 2\gamma\epsilon. \qquad (11.43)$$

Similarly, substituting into the common expression in the second and third terms yields

$$\epsilon\frac{(2+2\gamma-\beta)}{(2+x)} = \epsilon\frac{(2+2\gamma-\beta)}{2+1/\epsilon} = \frac{\epsilon^2(2+2\gamma-\beta)}{1+2\epsilon} \approx 0, \qquad (11.44)$$

where we have neglected terms of size ϵ^2. Thus, the second term of Eq. 11.39 is zero and the third term reduces to

$$2\sin^{-1}\left(\frac{1}{1+x}\right) = 2\sin^{-1}\left(\frac{1}{1+1/\epsilon}\right) = 2\sin^{-1}\left(\frac{\epsilon}{1+\epsilon}\right) \approx 2\epsilon. \qquad (11.45)$$

Adding Eqs. 11.43–11.45, we obtain

$$\Delta\phi_{\text{def}} = 2\epsilon(1+\gamma) = 4\epsilon = \frac{4GM}{r_p c^2}, \qquad (11.46)$$

where we have set $\gamma = 1$ for Einstein's theory of GR. Thus, we have

$$\Delta\phi_{\text{def}} = \frac{4GM}{r_p c^2}, \qquad (11.47)$$

which is Einstein's formula for the deflection of light by a gravitating body. (See Weinberg (1972) for more details.)

References

Longuski, J. M., Fischbach, E., & Scheeres, D. J. (2001). Deflection of spacecraft trajectories as a new test of general relativity. *Physical Review Letters, 86*(14), 2942–2945.

Rindler, W. (1969). *Essential relativity: special, general, and cosmological*. New York: Springer-Verlag.

Weinberg, S. (1972). *Gravitation and cosmology: principles and applications of the general theory of relativity*. New York: Wiley.

Chapter 12
Perturbations due to Atmospheric Drag

12.1 The Effect of Atmospheric Drag

Another significant perturbation on the motion of a satellite about an atmosphere-bearing planet is that of atmospheric drag. To analyze this problem we closely follow the work of McCuskey (1963) and Meirovitch (1970). In our analysis we assume the satellite moves in an elliptic orbit and that the perturbative effects due to the planet's oblateness can be neglected (Fig. 12.1). The acceleration (force per unit mass) due to drag is approximated by

$$F = \frac{C_D A \rho V^2}{2m},\tag{12.1}$$

where C_D is the drag coefficient, A is the cross-sectional area, m is the mass of the spacecraft, V is the orbital speed, and $\rho = \rho(r)$ is the atmospheric density at a distance r from the planet's center. The drag coefficient C_D has a value of about

$$C_D \approx 2.2\tag{12.2}$$

for a sphere that is small compared with the mean free path of the gas molecules and for which the molecules either adhere to its surface or are totally reflected.

The drag force is assumed to be tangent to the orbit in the direction opposite the instantaneous velocity (as we ignore the effect of the rotating atmosphere), so the TNW components of the force are

© The Author(s), under exclusive license to Springer Nature Switzerland AG 2022
J. M. Longuski et al., *Introduction to Orbital Perturbations*, Space Technology
Library 40, https://doi.org/10.1007/978-3-030-89758-1_12

Fig. 12.1 Orbit decay due to
drag in which significant
perturbative force occurs at
periapsis

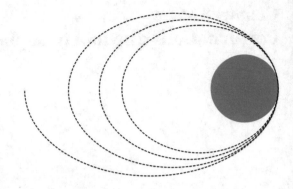

$$T = -\frac{1}{2m}C_D A\rho V^2$$

$$N = 0$$

$$W = 0. \tag{12.3}$$

Using Eqs. 12.3 in the Lagrange planetary equations Eqs. 5.114, we obtain

$$\dot{a} = -\frac{2b\rho V^2}{n\beta}\psi \tag{12.4}$$

$$\dot{e} = -\frac{2b\rho V^2\beta}{na}(\cos f + e)\frac{1}{\psi} \tag{12.5}$$

$$\frac{di}{dt} = 0 \tag{12.6}$$

$$\dot{\Omega} = 0 \tag{12.7}$$

$$\dot{\omega} = -\frac{2b\rho V^2\beta}{nae}(\sin f)\frac{1}{\psi} \tag{12.8}$$

$$\dot{M} = n + \frac{2b\rho V^2}{na^2}\sin f\left(\frac{a\beta^2}{e} + re\right)\frac{1}{\psi}, \tag{12.9}$$

where

$$b = \frac{C_D A}{2m} \tag{12.10}$$

$$\psi = \sqrt{1 + e^2 + 2e\cos f}. \tag{12.11}$$

Equation 12.9 can be expanded to

$$\dot{M} = n + \frac{2b\rho V^2 \beta^2}{nae} \sin f \left(\frac{1 + e \cos f + e^2}{1 + e \cos f} \right) \frac{1}{\psi}. \tag{12.12}$$

From Eq. 2.43, we have

$$V^2 = \mu \left(\frac{2}{r} - \frac{1}{a} \right) = \frac{n^2 a^3}{a\beta^2} \left[2(1 + e \cos f) - \beta^2 \right]$$

$$V^2 = \frac{n^2 a^2}{\beta^2} \left(1 + 2e \cos f + e^2 \right). \tag{12.13}$$

Using Eq. 12.13 in Eqs. 12.4, 12.5, 12.8, and 12.12, we obtain

$$\dot{a} = -\frac{2b\rho V^2}{n\beta} \psi = -\frac{2b\rho}{n\beta} \frac{n^2 a^2}{\beta^2} \left(1 + 2e \cos f + e^2 \right) \left(1 + 2e \cos f + e^2 \right)^{1/2}$$

$$= -2b\rho \frac{na^2}{\beta^3} \left(1 + 2e \cos f + e^2 \right)^{3/2} \tag{12.14}$$

$$\dot{e} = -\frac{2b\rho\beta}{na} \frac{n^2 a^2}{\beta^2} (\cos f + e) \left(1 + 2e \cos f + e^2 \right) \left(1 + 2e \cos f + e^2 \right)^{-1/2}$$

$$= -2b\rho \frac{na}{\beta} (\cos f + e) \left(1 + 2e \cos f + e^2 \right)^{1/2} \tag{12.15}$$

$$\dot{\omega} = -\frac{2b\rho\beta}{nae} \frac{n^2 a^2}{\beta^2} \left(1 + 2e \cos f + e^2 \right) (\sin f) \left(1 + 2e \cos f + e^2 \right)^{-1/2}$$

$$= -2b\rho \frac{na}{e\beta} \left(1 + 2e \cos f + e^2 \right)^{1/2} (\sin f) \tag{12.16}$$

$$\dot{M} = n + \frac{2b\rho\beta^2}{nae} \frac{n^2 a^2}{\beta^2} \left(1 + 2e \cos f + e^2 \right) \sin f \left(\frac{1 + e \cos f + e^2}{1 + e \cos f} \right)$$

$$\times \left(1 + 2e \cos f + e^2 \right)^{-1/2}$$

$$= n + 2b\rho \frac{na}{e} \left(1 + 2e \cos f + e^2 \right)^{1/2} \sin f \left(\frac{1 + e \cos f + e^2}{1 + e \cos f} \right). \tag{12.17}$$

To facilitate our analysis, it is necessary to convert Eqs. 12.14–12.17 from dependence on true anomaly to eccentric anomaly. Using Eq. 2.76, we see that

$$1 + 2e \cos f + e^2 = 1 + 2e \frac{\cos E - e}{1 - e \cos E} + e^2$$

$$= \frac{1 - e \cos E + 2e \cos E - 2e^2 + e^2 - e^3 \cos E}{1 - e \cos E}$$

$$= \frac{1 + e\cos E - e^2 - e^3 \cos E}{1 - e\cos E} = \frac{1 - e^2 + e\cos E \left(1 - e^2\right)}{1 - e\cos E}$$

$$= \beta^2 \frac{1 + e\cos E}{1 - e\cos E}. \tag{12.18}$$

Also using Eq. 2.76, we find

$$\cos f + e = \frac{\cos E - e}{1 - e\cos E} + e = \frac{\cos E - e + e - e^2 \cos E}{1 - e\cos E}$$

$$\cos f + e = \beta^2 \frac{\cos E}{1 - e\cos E} \tag{12.19}$$

and

$$1 + e\cos f = 1 + e\frac{\cos E - e}{1 - e\cos E} = \frac{1 - e\cos E + e\cos E - e^2}{1 - e\cos E} = \frac{\beta^2}{1 - e\cos E} \tag{12.20}$$

so that

$$\frac{1 + e\cos f + e^2}{1 + e\cos f} = \frac{\beta^2 + e^2 - e^3 \cos E}{\beta^2} = \frac{1 - e^3 \cos E}{\beta^2}. \tag{12.21}$$

We can now use the results of Eqs. 12.18, 12.19, and 12.22 to complete the transformation to eccentric anomaly. Using Eq. 12.18 in Eqs. 12.14 and 12.16 completes the transformation of the semimajor axis and argument of periapsis rate equations. Using Eqs. 12.18 and 12.19 in Eq. 12.15 completes the transformation of the eccentricity rate equation. And finally using Eqs. 12.19 and 12.21 in Eq. 12.17 completes the transformation of the mean anomaly rate equation. Thus, Eqs. 12.14–12.17 reduce to

$$\dot{a} = -2b\rho na^2 \left(\frac{1 + e\cos E}{1 - e\cos E}\right)^{3/2} \tag{12.22}$$

$$\dot{e} = -2b\rho na\beta^2 \frac{(1 + e\cos E)^{1/2}}{(1 - e\cos E)^{3/2}} \cos E \tag{12.23}$$

$$\dot{\omega} = -2b\rho \frac{na}{e} \left(\frac{1 + e\cos E}{1 - e\cos E}\right)^{1/2} \frac{\beta \sin E}{1 - e\cos E}$$

$$= -2b\rho \frac{na\beta}{e} \frac{(1 + e\cos E)^{1/2}}{(1 - e\cos E)^{3/2}} \sin E \tag{12.24}$$

$$\dot{M} = n + 2b\rho \frac{na}{e} \frac{(1 + e\cos E)^{1/2}}{(1 - e\cos E)^{3/2}} \left(1 - e^3 \cos E\right) \sin E, \tag{12.25}$$

where we have used Eq. 2.77 to substitute for $\sin f$.

12.2 Orbit Contraction Due to Atmospheric Drag

To develop the theory of orbit contraction, we follow chapter 15 of Vinh, N. X., Busemann, A., and Culp, R. D. (see Vinh et al. (1980)). We use a strictly exponential law for the density of the atmosphere

$$\rho(r) = \rho_{p0} \exp\left[\beta^* (r_{p0} - r)\right], \tag{12.26}$$

where the inverse scale height, β^*, is considered to be constant

$$\beta^* = \frac{1}{H}. \tag{12.27}$$

The quantity H, which has the dimension of length, is the scale height. The density function models an exponential decrease in atmospheric density with increasing r. We let p_0 denote a reference altitude. Then ρ_{p0} denotes the reference density and r_{p0} denotes the periapsis distance. Using the notation of Vinh et al. (1980), we introduce the dimensionless constant

$$Z_0 = \frac{\rho_{p0} C_D A r_{p0}}{2m}. \tag{12.28}$$

Using Eqs. 12.10, 12.26, and 12.28, we see that

$$\rho b = \rho_{p0} \left[\exp \beta^* (r_{p0} - r)\right] \frac{C_D A}{2m} = \frac{\rho_{p0} C_D A r_{p0}}{2m} \frac{1}{r_{p0}} \left[\exp \beta^* (r_{p0} - r)\right]$$

$$= Z_0 \frac{1}{r_{p0}} \left[\exp \beta^* (r_{p0} - r)\right]. \tag{12.29}$$

Substituting Eq. 12.29 in Eqs. 12.22–12.25, we obtain

$$\frac{da}{dt} = -2Z_0 \frac{na^2}{r_{p0}} \left(\frac{1 + e\cos E}{1 - e\cos E}\right)^{3/2} \exp\left[\beta^* (r_{p0} - r)\right] \tag{12.30}$$

$$\frac{de}{dt} = -2Z_0 \frac{na\beta^2}{r_{p0}} \cos E \frac{(1 + e\cos E)^{1/2}}{(1 - e\cos E)^{3/2}} \exp\left[\beta^* (r_{p0} - r)\right] \tag{12.31}$$

$$\frac{d\omega}{dt} = -2Z_0 \frac{an\beta}{r_{p0}e} \frac{(1 + e\cos E)^{1/2}}{(1 - e\cos E)^{3/2}} \sin E \exp\left[\beta^* (r_{p0} - r)\right] \tag{12.32}$$

$$\frac{dM}{dt} = n + 2Z_0 \frac{na\beta^2}{r_{p0}e} \frac{(1 + e\cos E)^{1/2}}{(1 - e\cos E)^{3/2}} \left(1 - e^3 \cos E\right) \sin E \exp\left[\beta^* (r_{p0} - r)\right]. \tag{12.33}$$

12.2.1   The Method of Averaging for Orbit Decay

The complete system of differential equations describing the effect of atmospheric drag is

$$\frac{da}{dt} = -2Z_0 \frac{na^2}{r_{p0}} \left(\frac{1 + e \cos E}{1 - e \cos E} \right)^{3/2} \exp\left[\beta^* \left(r_{p0} - r \right) \right]$$

$$\frac{de}{dt} = -2Z_0 \frac{na\beta^2}{r_{p0}} \cos E \frac{(1 + e \cos E)^{1/2}}{(1 - e \cos E)^{3/2}} \exp\left[\beta^* \left(r_{p0} - r \right) \right]$$

$$\frac{di}{dt} = 0$$

$$\frac{d\Omega}{dt} = 0 \qquad\qquad\qquad\qquad\qquad\qquad\qquad\qquad (12.34)$$

$$\frac{d\omega}{dt} = -2Z_0 \frac{an\beta}{r_{p0}e} \frac{(1 + e \cos E)^{1/2}}{(1 - e \cos E)^{3/2}} \sin E \exp\left[\beta^* \left(r_{p0} - r \right) \right]$$

$$\frac{dM}{dt} = n + 2Z_0 \frac{na\beta^2}{r_{p0}e} \frac{(1 + e \cos E)^{1/2}}{(1 - e \cos E)^{3/2}} \left(1 - e^3 \cos E \right) \sin E \exp\left[\beta^* \left(r_{p0} - r \right) \right].$$

This system satisfies the conditions for application of the method of averaging (MOA). The fast variable is the mean anomaly. The unitless small parameter used to define the order of the treatment can be taken as Z_0. This approach is analogous to the identification of J_2 as the small parameter in the oblateness treatment in Chap. 8. Thus, we see that all the other equations have time rate of change proportional to Z_0 and can be treated as the slow variables.

The first step in the MOA is to develop a transformation of variables that removes the dependence on the mean anomaly fast variable. For the treatment of drag, we are going to develop only the first-order secular terms. Thus, all we need to do is to determine the average time rates of change of the slow variables. We note that the argument of periapsis and mean anomaly perturbations are odd functions and have zero mean. Thus, we focus on only the semimajor axis and eccentricity equations.

We are about to embark on some significant algebraic development where we introduce a series expansion in powers of eccentricity as well as some special functions arising from integration. These functions involve the integrands of the rates of change of semimajor axis and eccentricity. We begin by examining the integrals required by the MOA and showing how a change of variables should be properly carried within the integrands.

Let Λ denote a slow variable. Using the chain rule, we can write

$$\frac{d\Lambda}{dt} = \frac{d\Lambda}{dE} \frac{dE}{dt}. \qquad\qquad\qquad\qquad\qquad\qquad (12.35)$$

Application of the MOA to this slow variable gives

$$\left\langle \frac{d\Lambda}{dt} \right\rangle = \left\langle \frac{d\Lambda}{dE} \frac{dE}{dt} \right\rangle = \frac{1}{n} \frac{1}{2\pi} \int_0^{2\pi} \left(\frac{d\Lambda}{dE} \frac{dE}{dt} \right) dM = \frac{1}{n} \frac{1}{2\pi} \int_0^{2\pi} \left(\frac{d\Lambda}{dE} \frac{dE}{dt} \right) n dt$$

$$= \frac{1}{2\pi} \int_0^{2\pi} \frac{d\Lambda}{dE} dE. \tag{12.36}$$

Therefore, the subsequent development of the average focuses on development of the integrands

$$\frac{da}{dE} \quad \text{and} \quad \frac{de}{dE}. \tag{12.37}$$

To express these integrands, we need an expression for dt/dE. We begin by differentiating Kepler's equation to find

$$M = E - e \sin E$$

$$dM = n dt = (1 - e \cos E) \, dE.$$

Then we have

$$\frac{dt}{dE} = \frac{1 - e \cos E}{n} \tag{12.38}$$

and

$$\frac{da}{dE} = \frac{da}{dt} \frac{dt}{dE} = -2Z_0 \frac{a^2}{r_{p0}} \frac{(1 + e \cos E)^{3/2}}{(1 - e \cos E)^{1/2}} \exp\left[\beta^* \left(r_{p0} - r \right) \right] \tag{12.39}$$

$$\frac{de}{dE} = \frac{de}{dt} \frac{dt}{dE} = -2Z_0 \frac{a\beta^2}{r_{p0}} \cos E \frac{(1 + e \cos E)^{1/2}}{(1 - e \cos E)^{1/2}} \exp\left[\beta^* \left(r_{p0} - r \right) \right]. \tag{12.40}$$

As can be observed from Eqs. 12.39 and 12.40, under the dissipative effect of drag, the semimajor axis decreases continuously while the eccentricity, although having an oscillatory behavior, also decreases secularly with time. The orbit undergoes a contraction and, as e decreases, tends to circularize naturally. We use the method of averaging for the integration of the equations.

For the radial distance, r, we use Eq. 2.57

$$r = a (1 - e \cos E) \tag{12.41}$$

$$r_{p0} = a_0 (1 - e_0). \tag{12.42}$$

Substituting Eqs. 12.41 and 12.42 into Eq. 12.26, we obtain

$$\rho = \rho_{p0} \exp \left[\beta^* \left(r_{p0} - r \right) \right]$$
$$= \rho_{p0} \exp \left[\beta^* \left(a_0 - a - a_0 e_0 \right) + \beta^* a e \cos E \right]. \tag{12.43}$$

Over each revolution, a is nearly constant whereas the varying quantity $\beta^* a e \cos E$ provides the fluctuation in the atmospheric density. This observation leads to a natural choice of the variable

$$x \equiv \beta^* a e \tag{12.44}$$

to replace the eccentricity. (See King-Hele (1964).)

Taking the derivative of Eq. 12.44 using Eqs. 12.39 and 12.40, we have

$$\frac{dx}{dE} = \beta^* \frac{da}{dE} e + \beta^* a \frac{de}{dE}$$

$$= \beta^* e \left\{ -2Z_0 \frac{a^2}{r_{p0}} \frac{(1 + e \cos E)^{3/2}}{(1 - e \cos E)^{1/2}} \exp \left[\beta^* \left(r_{p0} - r \right) \right] \right\}$$

$$+ \beta^* a \left\{ -2Z_0 \left(1 - e^2 \right) \frac{a}{r_{p0}} \cos E \left(\frac{1 + e \cos E}{1 - e \cos E} \right)^{1/2} \exp \left[\beta^* \left(r_{p0} - r \right) \right] \right\}$$

$$= \frac{-2\beta Z_0 e^{\beta^* (r_{p0} - r)}}{(1 - e \cos E)^{1/2}} \frac{a^2}{r_{p0}} \left\{ (1 + e \cos E)^{1/2} \left[e \left(1 + e \cos E \right) + \left(1 - e^2 \right) \cos E \right] \right\}$$

$$= \frac{-2a^2 \beta^* Z_0}{r_{p0}} \left(\frac{1 + e \cos E}{1 - e \cos E} \right)^{1/2} (e + \cos E) \exp \left[\beta^* \left(r_{p0} - r \right) \right]. \tag{12.45}$$

Thus the equation for x is

$$\frac{dx}{dE} = -\frac{2a^2 \beta^* Z_0}{r_{p0}} (e + \cos E) \left(\frac{1 + e \cos E}{1 - e \cos E} \right)^{1/2} \exp \left[\beta^* \left(r_{p0} - r \right) \right]. \tag{12.46}$$

The new dimensionless variable, x, behaves like the eccentricity; that is, during each revolution x passes through stationary values when

$$\cos E = -e \tag{12.47}$$

but, on the average, x decreases with time.

12.2.2   The Averaged Equations for Orbit Decay

Since the decaying process is slow, we can use the averaging technique in Eq. 12.36 applied to the integrands given by the right-hand side of Eqs. 12.39 and 12.46 for a and x. For the semimajor axis, we have the averaged equation

$$\left\langle \frac{da}{dt} \right\rangle = -2Z_0 \frac{a^2}{r_{p0}} \exp\left[\beta^* (a_0 - a - a_0 e_0)\right]$$

$$\times \frac{1}{2\pi} \int_0^{2\pi} \frac{(1 + e \cos E)^{3/2}}{(1 - e \cos E)^{1/2}} \exp(x \cos E) \, dE, \qquad (12.48)$$

where we have replaced $\beta^* ae$ by x inside the integral. Similarly, from Eq. 12.46, we obtain the averaged equation for x (the new dimensionless variable that replaces the eccentricity, e)

$$\left\langle \frac{dx}{dt} \right\rangle = -\frac{2a^2 \beta^* Z_0}{r_{p0}} \exp\left[\beta^* (a_0 - a - a_0 e_0)\right]$$

$$\times \frac{1}{2\pi} \int_0^{2\pi} (e + \cos E) \left(\frac{1 + e \cos E}{1 - e \cos E}\right)^{1/2} \exp(x \cos E) \, dE. \qquad (12.49)$$

The purpose of introducing the new variable (Eq. 12.44)

$$x = \beta^* ae \qquad (12.50)$$

becomes clear when we consider the definition of the nth-order modified Bessel function, $I_n(x)$, of the first kind:

$$I_n(x) \equiv \frac{1}{2\pi} \int_0^{2\pi} \cos nE \, \exp(x \cos E) \, dE. \qquad (12.51)$$

The modified Bessel function is a solution to the modified Bessel differential equation. Given the order n and a value for x, one can obtain the numeric value by looking it up in a table or writing a subroutine to compute it. See Abramowitz and Stegun (1972) for details on these functions.

For small eccentricity, the integrands of Eqs. 12.48 and 12.49 can be expanded in power series in $e \cos E$. To simplify the notation, we let $Q = e \cos E$. From the binomial theorem, we have

$$(1 + Q)^n = 1 + nQ + \frac{n(n-1) Q^2}{2!} + \frac{n(n-1)(n-2) Q^3}{3!} + \cdots$$

$$(1 - Q)^{-n} = 1 + nQ + \frac{n(n+1) Q^2}{2!} + \frac{n(n+1)(n+2) Q^3}{3!} + \cdots \qquad (12.52)$$

For $n = 3/2$, we have for the first of Eqs. 12.52

$$(1 + Q)^{3/2} = 1 + \frac{3}{2}Q + \frac{3}{8}Q^2 - \frac{3}{48}Q^3 + \cdots \tag{12.53}$$

To find $(1 - Q))^{-1/2}$, we use the second of Eqs. 12.52

$$(1 - Q)^{-1/2} = 1 + \frac{1}{2}Q + \frac{3}{8}Q^2 + \frac{5}{16}Q^3 + \cdots \tag{12.54}$$

Multiplying Eq. 12.53 by 12.54, we find

$$(1 + Q)^{3/2}(1 - Q)^{-1/2} = 1 + \frac{3}{2}Q + \frac{3}{8}Q^2 - \frac{1}{16}Q^3 + \frac{1}{2}Q + \frac{3}{4}Q^2 + \frac{3}{16}Q^3$$

$$+ \frac{3}{8}Q^2 + \frac{9}{16}Q^3 + \frac{5}{16}Q^3$$

$$= 1 + 2Q + \frac{3}{2}Q^2 + Q^3, \tag{12.55}$$

where we have retained terms to the third order in Q.

For the term $(1 + Q)^{1/2}$, we use the first of Eqs. 12.52

$$(1 + Q)^{1/2} = 1 + \frac{1}{2}Q - \frac{1}{8}Q^2 + \frac{1}{16}Q^3 + \cdots \tag{12.56}$$

Multiplying Eq. 12.54 by Eq. 12.56, we obtain

$$(1 - Q)^{-1/2}(1 + Q)^{1/2} = 1 + \frac{1}{2}Q + \frac{3}{8}Q^2 + \frac{5}{16}Q^3 + \frac{1}{2}Q + \frac{1}{4}Q^2 + \frac{3}{16}Q^3$$

$$- \frac{1}{8}Q^2 - \frac{1}{16}Q^3 + \frac{1}{16}Q^3$$

$$= 1 + Q + \frac{1}{2}Q^2 + \frac{1}{2}Q^3, \tag{12.57}$$

where we have again retained terms only to Q^3. Replacing Q in Eq. 12.55 with $e \cos E$, we obtain the expansion of the integrand in Eq. 12.48

$$\frac{(1 + e \cos E)^{3/2}}{(1 - e \cos E)^{1/2}} = 1 + 2e \cos E + \frac{3}{2}e^2 \cos^2 E + e^3 \cos^3 E. \tag{12.58}$$

We use the trigonometric identities for $\cos^2 E$ and $\cos^3 E$:

$$\cos^2 E = \frac{1}{2}(1 + \cos 2E) \tag{12.59}$$

$$\cos^3 E = \frac{1}{4}\left(3\cos E + \cos 3E\right). \tag{12.60}$$

Substituting Eqs. 12.59 and 12.60 into Eq. 12.58 provides

$$\frac{(1+e\cos E)^{3/2}}{(1-e\cos E)^{1/2}} = 1 + 2e\cos E + \frac{3}{4}e^2\left(1+\cos 2E\right) + \frac{1}{4}e^3\left(3\cos E + \cos 3E\right). \tag{12.61}$$

Substituting Eqs. 12.61 into Eq. 12.48 and using the definition of the modified Bessel function, $I_n(x)$, in Eq. 12.51, we have the averaged equation for a

$$\left\langle\frac{da}{dt}\right\rangle = -2Z_0\frac{a^2}{r_{p0}}\exp\left[\beta^*\left(a_0 - a - a_0 e_0\right)\right]$$

$$\times\left[I_0 + 2eI_1 + \frac{3}{4}e^2\left(I_0 + I_2\right) + \frac{1}{4}e^3\left(3I_1 + I_3\right)\right]. \tag{12.62}$$

Similarly, we replace Q in Eq. 12.57 with $e\cos E$ and, according to the integrand in Eq. 12.49, multiply the result by $e + \cos E$ to obtain

$$(e+\cos E)\left(\frac{1+e\cos E}{1-e\cos E}\right)^{1/2} = (e+\cos E)$$

$$\times\left(1 + e\cos E + \frac{1}{2}e^2\cos^2 E + \frac{1}{2}e^3\cos^3 E\right)$$

$$= \cos E + e\cos^2 E + \frac{1}{2}e^2\cos^3 E + \frac{1}{2}e^3\cos^4 E$$

$$+ e + e^2\cos E + \frac{1}{2}e^3\cos^2 E$$

$$= \cos E + e\left[1 + \frac{1}{2} + \frac{1}{2}\cos 2E\right]$$

$$+ e^2\left[\cos E + \frac{1}{8}\left(3\cos E + \cos 3E\right)\right]$$

$$+ e^3\left[\frac{1}{16}\left(3 + 4\cos 2E + \cos 4E\right)\right.$$

$$+ \frac{1}{4}\left(1 + \cos 2E\right)\bigg], \tag{12.63}$$

where we have made use of Eqs. 12.59 and 12.60 and the trigonometric identity

$$\cos^4 E = \frac{1}{8}\left(3 + 4\cos 2E + \cos 4E\right). \tag{12.64}$$

From Eqs. 12.49, 12.51, and 12.63, we obtain the averaged equation for x

$$\left\langle \frac{dx}{dt} \right\rangle = -2Z_0 \frac{a^2 \beta^*}{r_{p0}} \exp\left[\beta^* (a_0 - a - a_0 e_0)\right]$$

$$\times \left[I_1 + \frac{1}{2} e \left(3 I_0 + I_2\right) + \frac{1}{8} e^2 \left(11 I_1 + I_3\right) + \frac{1}{16} e^3 \left(7 I_0 + 8 I_2 + I_4\right) \right].$$

$$(12.65)$$

Vinh et al. point out that Eqs. 12.62 and 12.65 were given by King-Hele (1964) and by others where they truncated the equations to order e^4, formed the equation da/dx, and then integrated it separately for the cases where x is very large and where x is very small. Vinh et al. (1980) integrate the da/dx equation without that asymptotic simplification—*obtaining a solution uniformly valid for any x*. To find the da/dx equation, we divide Eq. 12.62 by Eq. 12.65. A useful source for series operations is Abramowitz and Stegun (1972). Let

$$s_1 = 1 + a_1 e + a_2 e^2 + a_3 e^3 + a_4 e^4 + \ldots \qquad (12.66)$$

$$s_2 = 1 + b_1 e + b_2 e^2 + b_3 e^3 + b_4 e^4 + \ldots \qquad (12.67)$$

$$s_3 = 1 + c_1 e + c_2 e^2 + c_3 e^3 + c_4 e^4 + \ldots \qquad (12.68)$$

Then for

$$s_3 = s_1 / s_2, \qquad (12.69)$$

we have

$$c_1 = a_1 - b_1$$

$$c_2 = a_2 - (b_1 c_1 + b_2)$$

$$c_3 = a_3 - (b_1 c_2 + b_2 c_1 + b_3)$$

$$c_4 = a_4 - (b_1 c_3 + b_2 c_2 + b_3 c_1 + b_4). \qquad (12.70)$$

Let us write the square bracket term in Eq. 12.62 after dividing out the lead term, I_0, and associate the remaining series with Eq. 12.66 so that

$$a_1 = \frac{2 I_1}{I_0}$$

$$a_2 = \frac{3}{4} \frac{(I_0 + I_2)}{I_0}$$

$$a_3 = \frac{1}{4} \frac{(3 I_1 + I_3)}{I_0}. \qquad (12.71)$$

Similarly, dividing out I_1 in Eq. 12.65 in the square brackets, and associating the resulting series with Eq. 12.67, we have

$$b_1 = \frac{1}{2}\frac{(3I_0 + I_2)}{I_1}$$

$$b_2 = \frac{1}{8}\frac{(11I_1 + I_3)}{I_1}$$

$$b_3 = \frac{1}{16}\frac{(7I_0 + 8I_2 + I_4)}{I_1}. \tag{12.72}$$

Thus from Eqs. 12.70–12.72, we have

$$c_1 = \frac{2I_1}{I_0} - \frac{(3I_0 + I_2)}{2I_1}$$

$$c_2 = \frac{3(I_0 + I_2)}{4I_0} - \left\{ \frac{(3I_0 + I_2)}{2I_1} \left[\frac{2I_1}{I_0} - \frac{(3I_0 + I_2)}{2I_1} \right] + \frac{(11I_1 + I_3)}{8I_1} \right\}$$

$$c_3 = \frac{(3I_1 + I_3)}{4I_0} - \frac{(11I_1 + I_3)}{8I_1} \left[\frac{2I_1}{I_0} - \frac{(3I_0 + I_2)}{2I_1} \right] - \frac{(7I_0 + 8I_2 + I_4)}{16I_1}$$

$$- \frac{(3I_0 + I_2)}{2I_1} \left(\frac{3(I_0 + I_2)}{4I_0} - \left\{ \frac{(3I_0 + I_2)}{2I_1} \left[\frac{2I_1}{I_0} - \frac{(3I_0 + I_2)}{2I_1} \right] + \frac{(11I_1 + I_3)}{8I_1} \right\} \right). \tag{12.73}$$

Let us define the ratios of the Bessel functions (as in Vinh et al. (1980)):

$$y_n \equiv \frac{I_n}{I_1}, \qquad n \neq 1. \tag{12.74}$$

Using this definition, we can write the c_1 term in Eq. 12.73 as

$$c_1 = \frac{2}{y_0} - \frac{3}{2}y_0 - \frac{1}{2}y_2. \tag{12.75}$$

For the c_2 term in Eq. 12.73, we have

$$c_2 = \frac{3}{4} + \frac{3}{4}\frac{y_2}{y_0} - \left(\frac{3}{2}y_0 + \frac{1}{2}y_2 \right) \left(\frac{2}{y_0} - \frac{3}{2}y_0 - \frac{1}{2}y_2 \right) - \frac{11}{8} - \frac{1}{8}y_3$$

$$= \frac{3}{4} + \frac{3}{4}\frac{y_2}{y_0} - \frac{11}{8} - \frac{1}{8}y_3 - 3 + \frac{9}{4}y_0^2 + \frac{3}{4}y_0 y_2 - \frac{y_2}{y_0} + \frac{3}{4}y_2 y_0 + \frac{1}{4}y_2^2$$

$$= -\frac{29}{8} - \frac{1}{4}\frac{y_2}{y_0} - \frac{1}{8}y_3 + \frac{9}{4}y_0^4 + \frac{6}{4}y_0 y_2 + \frac{1}{4}y_2^2$$

$$= \frac{1}{8}\left(-29 - 2\frac{y_2}{y_0} - y_3 + 18y_0^2 + 12y_0y_2 + 2y_2^2\right). \tag{12.76}$$

The result of dividing Eq. 12.62 by Eq. 12.65 is given by Vinh et al. (1980) as

$$\beta^*\frac{da}{dx} = y_0 + \frac{1}{2}e\left(4 - 3y_0^2 - y_0y_2\right)$$

$$+ \frac{1}{8}e^2\left[2y_0\left(3y_0 + y_2\right)^2 - 29y_0 - 2y_2 - y_0y_3\right]$$

$$+ \frac{1}{16}e^3\left[-32 + 113y_0^2 + 38y_0y_2 - y_0y_4 + 2y_2^2 + 6y_0^2y_3\right.$$

$$\left. + 2y_0y_2y_3 - 2y_0\left(3y_0 + y_2\right)^3\right], \tag{12.77}$$

where we note that the leading term on the right-hand side, y_0, is due to the I_0 of Eq. 12.62 divided by the I_1 of Eq. 12.65; also the inverse scale height, β^*, has been moved to the left-hand side in Eq. 12.77.

The Bessel functions satisfy the recurrence formula (from Abramowitz and Stegun (1972)):

$$I_{n-1}\left(x\right) - I_{n+1}\left(x\right) = \frac{2n}{x}I_n\left(x\right). \tag{12.78}$$

Hence, any function $y_n\left(x\right)$ can be expressed in terms of $y_0\left(x\right)$ and x. For example, setting $n = 1$ in Eq. 12.78, we have

$$I_0 - I_2 = \frac{2}{x}I_1 \tag{12.79}$$

so that (dividing by I_1)

$$y_0 - y_2 = \frac{2}{x}$$

$$y_2 = y_0 - \frac{2}{x}. \tag{12.80}$$

Setting $n = 2$ in Eq. 12.78 provides

$$I_1 - I_3 = \frac{4}{x}I_2 \tag{12.81}$$

so that

$$I_3 = I_1 - \frac{4}{x}I_2 \tag{12.82}$$

and

$$y_3 = 1 - \frac{4}{x}y_2$$

$$= 1 - \frac{4}{x}\left(y_0 - \frac{2}{x}\right)$$

$$= 1 - \frac{4y_0}{x} + \frac{8}{x^2}. \tag{12.83}$$

After completing a similar process to find y_4, we have the results:

$$y_2(x) = y_0 - \frac{2}{x}$$

$$y_3(x) = 1 + \frac{8}{x^2} - \frac{4}{x}y_0 \tag{12.84}$$

$$y_4(x) = -\frac{8}{x} - \frac{48}{x^3} + y_0 + \frac{24}{x^2}y_0.$$

Let

$$z \equiv \frac{a}{a_0} \tag{12.85}$$

be the dimensionless semimajor axis. Then, from the definition of x in Eq. 12.44, we have for the eccentricity

$$e = \frac{x}{a\beta^*} = \frac{x/a_0}{\beta^* a/a_0} = \frac{1}{\beta^* a_0}\frac{x}{z}, \tag{12.86}$$

or

$$e = \epsilon\frac{x}{z}, \tag{12.87}$$

where

$$\epsilon \equiv \frac{1}{\beta^* a_0} = \frac{H}{a_0}. \tag{12.88}$$

For constant scale height, ϵ is a constant; it is typically a small quantity on the order of 10^{-3} (as $a_0 \gg H$ usually). Using the first equation of Eqs. 12.84 in the first-order term in e in Eq. 12.77, we have

$$\frac{1}{2}e\left(4 - 3y_0^2 - y_0 y_2\right) = \frac{1}{2}e\left(4 - 3y_0^2 - y_0^2 + \frac{2y_0}{x}\right)$$

$$= e \left(2 - 2y_0^2 + \frac{y_0}{x} \right). \tag{12.89}$$

Using the first two equations of Eqs. 12.84 in the second-order term (e^2) in Eq. 12.77, we have

$$\frac{1}{8}e^2 \left[2y_0 \, (3y_0 + y_2)^2 - 29y_0 - 2y_2 - y_0 y_3 \right]$$

$$= \frac{1}{8}e^2 \left[2y_0 \left(3y_0 + y_0 - \frac{2}{x} \right)^2 - 29y_0 - 2y_0 + \frac{4}{x} - y_0 - \frac{8y_0}{x^2} + \frac{4}{x}y_0^2 \right]$$

$$= \frac{1}{8}e^2 \left[2y_0 \left(16y_0^2 - 16\frac{y_0}{x} + \frac{4}{x^2} \right) - 32y_0 + \frac{4}{x} - \frac{8y_0}{x^2} + \frac{4}{x}y_0^2 \right]$$

$$= \frac{1}{8}e^2 \left[32y_0^3 - 32\frac{y_0^2}{x} + \frac{8y_0}{x^2} - 32y_0 + \frac{4}{x} - \frac{8y_0}{x^2} + \frac{4y_0^2}{x} \right]$$

$$= \frac{1}{8}e^2 \left[32y_0^3 - 28\frac{y_0^2}{x} - 32y_0 + \frac{4}{x} \right], \tag{12.90}$$

or

$$\frac{1}{8}e^2 \left[2y_0 \, (3y_0 + y_2)^2 - 29y_0 - 2y_2 - y_0 y_3 \right] = \frac{1}{2}e^2 \left(8y_0^3 - 7\frac{y_0^2}{x} - 8y_0 + \frac{1}{x} \right). \tag{12.91}$$

Using Eqs. 12.85, 12.87, 12.88, 12.89, and 12.91 to rewrite Eq. 12.77, we obtain one of the fundamental equations for the problem of orbit decay

$$\frac{dz}{dx} = \epsilon y_0 + \epsilon^2 \frac{x}{z} \left(2 + \frac{y_0}{x} - 2y_0^2 \right) + \epsilon^3 \frac{x^2}{2z^2} \left(\frac{1}{x} - 8y_0 - \frac{7y_0^2}{x} + 8y_0^3 \right)$$

$$+ \epsilon^4 \frac{x^3}{2z^2} \left(-4 + \frac{1}{x^2} + 20y_0^2 - 10\frac{y_0}{x} + \frac{4y_0}{x^3} - 5\frac{y_0^2}{x^2} + 20\frac{y_0^3}{x} - 16y_0^4 \right), \tag{12.92}$$

where the ϵ^4 term is given by Vinh et al. (1980).

We see that the true nature of Eq. 12.92 is a nonlinear differential equation. Since ϵ is a very small quantity, we need not go further with the expansion. Indeed, the solution of Eq. 12.92 can be considered the exact solution of Eq. 12.77 truncated to the order e^4.

12.2.3 Integration by Poincaré's Method of Small Parameters

Previously, we derived the nonlinear differential equation for orbit contraction due to atmospheric drag (Eq. 12.92):

$$\frac{dz}{dx} = \epsilon y_0 + \epsilon^2 \frac{x}{z} \left(2 + \frac{y_0}{x} - 2y_0^2 \right) + \epsilon^3 \frac{x^2}{2z^2} \left(\frac{1}{x} - 8y_0 - \frac{7y_0^2}{x} + 8y_0^3 \right),$$

where we have dropped the ϵ^4 term. We recall from Eqs. 12.74, 12.51, 12.85, 12.44, and 12.88:

$$y_n = I_n/I_1, \qquad n \neq 1$$

$$I_n = \frac{1}{2\pi} \int_0^{2\pi} \cos(nE) \exp(x \cos E) \, dE \qquad (12.93)$$

$$z = a/a_0$$

$$x = \beta^* a e \qquad (12.94)$$

$$\epsilon = \frac{1}{\beta^* a_0}. \qquad (12.95)$$

Poincaré's method for integration of a nonlinear differential equation containing a small parameter is a rigorous mathematical technique, proven to be convergent for small values of the parameter ϵ. We assume a solution for $z(x)$ of the form

$$z = z_0 + \epsilon z_1 + \epsilon^2 z_2 + \epsilon^3 z_3 + \epsilon^4 z_4 + \dots$$

$$= \sum_{k=0}^{\infty} \epsilon^k z_k(x). \qquad (12.96)$$

Substituting Eq. 12.96 into Eq. 12.92, we find

$$\frac{dz}{dx} = \frac{dz_0}{dx} + \epsilon \frac{dz_1}{dx} + \epsilon^2 \frac{dz_2}{dx} + \epsilon^3 \frac{dz_3}{dx}$$

$$= \epsilon y_0 + \epsilon^2 (z_0 + \epsilon z_1)^{-1} x \left(2 + \frac{y_0}{x} - 2y_0^2 \right) + \epsilon^3 \frac{x^2}{2z_0^2} \left(\frac{1}{x} - 8y_0 - \frac{7y_0^2}{x} + 8y_0^3 \right),$$

$$(12.97)$$

where we retain ϵ^3 terms.

We note that in Eq. 12.97, we have

$$(z_0 + \epsilon z_1)^{-1} = \left[z_0 \left(1 + \epsilon \frac{z_1}{z_0} \right) \right]^{-1}$$

$$= \frac{1}{z_0} \left(1 - \epsilon \frac{z_1}{z_0} \right)$$

$$= \frac{1}{z_0} - \epsilon \frac{z_1}{z_0^2}, \tag{12.98}$$

where we have dropped the ϵ^2 terms. Thus, using Eq. 12.98 in Eq. 12.97, the ϵ^2 term becomes

$$\epsilon^2 (z_0 + \epsilon z_1)^{-1} x \left(2 + \frac{y_0}{x} - 2y_0^2 \right) = \epsilon^2 \frac{x}{z_0} \left(2 + \frac{y_0}{x} - 2y_0^2 \right)$$

$$- \epsilon^3 \frac{z_1 x}{z_0^2} \left(2 + \frac{y_0}{x} - 2y_0^2 \right), \tag{12.99}$$

where we see that an ϵ^3 term appears.

Equating coefficients of like powers in ϵ (using Eqs. 12.97 and 12.99), we obtain the differential equations for $z_k(x)$:

$$\frac{dz_0}{dx} = 0 \tag{12.100}$$

$$\frac{dz_1}{dx} = y_0 \tag{12.101}$$

$$\frac{dz_2}{dx} = \frac{x}{z_0} \left(2 + \frac{y_0}{x} - 2y_0^2 \right) \tag{12.102}$$

$$\frac{dz_3}{dx} = -\frac{x z_1}{z_0^2} \left(2 + \frac{y_0}{x} - 2y_0^2 \right) + \frac{x^2}{2z_0^2} \left(\frac{1}{x} - 8y_0 - \frac{7y_0^2}{x} + 8y_0^3 \right). \tag{12.103}$$

We also have the initial conditions

$$z_0(x_0) = 1$$

$$z_1(x_0) = z_2(x_0) = \ldots = 0. \tag{12.104}$$

The integration of Eqs. 12.100–12.103 is accomplished by successive quadratures. Its success depends on whether or not the integrals can be expressed in terms of known functions. It has been found that the following recurrence formula is useful:

$$\int p(x) y_0^{n+1} dx = -\frac{p(x)}{n} y_0^n + \int p(x) y_0^{n-1} dx + \int \left[\frac{p(x)}{x} + \frac{p'(x)}{n} \right] y_0^n dx, \tag{12.105}$$

where $n \neq 0$ and $p(x)$ is any arbitrary function.

To derive Eq. 12.105 for Bessel functions, we use the relation given by Abramowitz and Stegun (1972)

$$x I_n' (x) + n I_n (x) = n I_{n-1} (x).$$ (12.106)

For $n = 1$, Eq. 12.106 provides

$$x I_1' + I_1 = x I_0.$$ (12.107)

Dividing Eq. 12.107 by $x I_1$, we have

$$y_0 = \frac{I_1'}{I_1} + \frac{1}{x},$$ (12.108)

where we recall that $y_0 \equiv I_0/I_1$. When $n = 0$, Eq. 12.106 gives

$$x I_0' = x I_{-1} = x I_1,$$ (12.109)

where we note from the second of Eqs. 12.93 that $\cos(-E) = \cos(E)$, so $I_{-1} = I_1$. Differentiating $y_0 = I_0/I_1$ with respect to x, we obtain

$$\frac{d}{dx} \left(I_0 I_1^{-1} \right) = \frac{1}{I_1} I_0' - \frac{I_0}{I_1^2} I_1'.$$ (12.110)

Using $I_0' = I_1$ from Eq. 12.109 and $I_1'/I_1 = y_0 - 1/x$ from Eq. 12.108 in Eq. 12.110, we obtain

$$y_0' = 1 - y_0 \left(y_0 - \frac{1}{x} \right),$$ (12.111)

or

$$y_0' = 1 - y_0^2 + \frac{y_0}{x}.$$ (12.112)

Next we consider

$$\int p(x)\, d \left(\frac{y_0^n}{n} \right) = \frac{p(x)}{n} y_0^n - \int \frac{p'(x)}{n} y_0^n dx,$$ (12.113)

or

$$\int p(x) y_0^{n-1} y_0' dx = \int p(x) y_0^{n-1} \left(1 + \frac{y_0}{x} - y_0^2 \right) dx$$

$$= \frac{p(x)}{n} y_0^n - \int \frac{p'(x)}{n} y_0^n dx.$$ (12.114)

Multiplying out the terms in the second integrand in Eq. 12.114, we have

$$\int \left[p(x) y_0^{n-1} + p(x) \frac{y_0^n}{x} - p(x) y_0^{n+1} \right] dx = \frac{p(x)}{n} y_0^n - \int \frac{p'(x)}{n} y_0^2 dx.$$
(12.115)

Isolating the term, $\int p(x) y_0^{n+1} dx$, in Eq. 12.115, we have

$$\int p(x) y_0^{n+1} dx = -\frac{p(x)}{n} y_0^n + \int p(x) y_0^{n-1} dx + \int \left[\frac{p(x)}{x} y_0^n + \frac{p'(x)}{n} y_0^n \right] dx,$$
(12.116)

which we see confirms the recurrence formula in Eq. 12.105.

Now we can proceed with the integration of Eqs. 12.100–12.103. The value of z_0 is constant (by Eq. 12.100) and from the initial condition set by Eqs. 12.104, we have simply

$$z_0 = 1.$$
(12.117)

We solve for z_1 in Eq. 12.101 by integrating Eq. 12.108:

$$\int dz_1 = \int y_0 dx$$

$$= \int \frac{I_1'}{I_1} dx + \frac{1}{x} dx$$

$$= \ln I_1(x) - \ln I_1(x_0) + \ln x - \ln x_0.$$
(12.118)

Since (from Eqs. 12.104) the initial condition for z_1 is zero, we have from Eq. 12.118

$$z_1(x) = \ln \left[\frac{x I_1(x)}{x_0 I_1(x_0)} \right].$$
(12.119)

With $z_0 = 1$, the equation for z_2 (Eq. 12.102) is

$$\frac{dz_2}{dx} = 2x + y_0 - 2x y_0^2.$$
(12.120)

Integrating Eq. 12.120, we have

$$z_2(x) = x^2 + \ln [x I_1(x)] - 2 \int x y_0^2 dx.$$
(12.121)

By the recurrence formula, Eq. 12.105, with $p(x) = x$ and $n = 1$, we have

$$\int x y_0^2 dx = -x y_0 + \int x dx + \int 2 y_0 dx$$

$$= -xy_0 + \frac{1}{2}x^2 + 2\ln[xI_1(x)], \tag{12.122}$$

where we have used the results of Eqs. 12.118 and 12.119 to integrate $2y_0$. Substituting Eq. 12.122 into Eq. 12.121, we obtain

$$z_2(x) = x^2 + \ln[xI_1(x)] + 2xy_0 - x^2 - 4\ln[xI_1(x)]. \tag{12.123}$$

Applying the initial conditions from Eqs. 12.104, we have from Eq. 12.123

$$z_2(x) = 2xy_0(x) - 2x_0y_0(x_0) - 3\ln\left[\frac{xI_1(x)}{x_0I_1(x_0)}\right]. \tag{12.124}$$

The integrations for obtaining $z_3(x)$ and $z_4(x)$ are performed similarly, but they are much more laborious. It is found that the $z_k(x)$ can be expressed in terms of two functions:

$$A(x) \equiv x\frac{I_0(x)}{I_1(x)} = xy_0$$

$$B(x) \equiv \ln[xI_1(x)]. \tag{12.125}$$

The final solution is

$$z_0(x) = 1 \tag{12.126}$$

$$z_1(x) = B - B_0 \tag{12.127}$$

$$z_2(x) = 2(A - A_0) - 3(B - B_0) \tag{12.128}$$

$$z_3(x) = \frac{7}{2}\left(x^2 - x_0^2\right) - \frac{13}{2}(A - A_0) - 2\left(A^2 - A_0^2\right)$$

$$+ 13(B - B_0) - 2A(B - B_0) + \frac{3}{2}(B - B_0)^2, \tag{12.129}$$

where A_0 and B_0 are the values of A and B evaluated at $x = x_0$. The solution for $z_4(x)$ is given by Vinh et al. (1980) on p. 289. The solution for $z_5(x)$ is given by Longuski (1979).

The semimajor axis of the orbit under contraction is

$$\frac{a}{a_0} = 1 + \epsilon z_1 + \epsilon^2 z_2 + \epsilon^3 z_3 + \epsilon^4 z_4. \tag{12.130}$$

Using x as a parameter ($x = \beta^* ae$), we can express other quantities of interest. The eccentricity is given by Eq. 12.87 as

$$e = \epsilon\frac{x}{z}, \tag{12.131}$$

where

$$\epsilon \equiv \frac{1}{\beta^* a_0}. \tag{12.132}$$

The drop in periapsis (in units of scale heights) is obtained from

$$\frac{r_{p0} - r_p}{H} = \beta^* \left[r_{p0} - a\,(1 - e) \right]$$

$$= \beta^* a_0 - \beta^* a_0 e_0 + \beta^* a e - \beta^* a_0 \left(1 + \epsilon z_1 + \epsilon^2 z_2 + \ldots \right)$$

$$= x - x_0 - \frac{1}{\epsilon} \left(\epsilon z_1 + \epsilon^2 z_2 + \ldots \right), \tag{12.133}$$

or

$$\frac{r_{p0} - r_p}{H} = (x - x_0) - \left(z_1 + \epsilon z_2 + \epsilon^2 z_3 + \epsilon^3 z_4 \right). \tag{12.134}$$

Since the eccentricity is (from Eq. 12.131) $e = \epsilon x / z$, we can write the ratio of the eccentricity as

$$\frac{e}{e_0} = \frac{\epsilon x / z}{\epsilon x_0 / z_0} = \frac{x}{x_0} \frac{z_0}{z} \tag{12.135}$$

and since $z_0 = 1$,

$$\frac{e}{e_0} = \frac{x}{x_0 z}. \tag{12.136}$$

During the process of orbit contraction, the drag force is most significant near periapsis. The strong braking force at periapsis has the effect of drastically reducing the apoapsis distance while the periapsis distance remains nearly constant. To show this effect we can calculate the ratios of the apsidal distances as functions of the variable x. For the periapsis ratio we have

$$\frac{r_p}{r_{p0}} = \frac{a\,(1 - e)}{a_0\,(1 - e_0)} = \frac{z - \epsilon x}{(1 - e_0)}. \tag{12.137}$$

Similarly, for the ratio of the apoapsis distances, we have

$$\frac{r_a}{r_{a0}} = \frac{a\,(1 + e)}{a_0\,(1 + e_0)} = \frac{z + \epsilon x}{(1 + e_0)}. \tag{12.138}$$

To calculate the drop in the apoapsis (in scale heights), we can use the formula (similar to Eq. 12.133):

$$\frac{r_{a_0} - r_a}{H} = \beta^* \left[a_0 (1 + e_0) - a (1 + e) \right]$$

$$= \beta^* a_0 + \beta^* a_0 e_0 - \beta^* a e - \beta^* a_0 \left(1 + \epsilon z_1 + \epsilon^2 z_2 + \ldots \right)$$

$$= x - x_0 - \beta^* a_0 \left(\epsilon z_1 + \epsilon^2 z_2 + \ldots \right), \tag{12.139}$$

or

$$\frac{r_{a_0} - r_a}{H} = (x_0 - x) - \left(z_1 + \epsilon z_2 + \epsilon^2 z_3 + \epsilon^3 z_4 \right). \tag{12.140}$$

Finally, the orbital period is simply (from Kepler's third law):

$$\frac{P}{P_0} = \left(\frac{a}{a_0} \right)^{3/2} = z^{3/2} (x). \tag{12.141}$$

For each initial value $\epsilon = H/a_0$ and initial eccentricity e_0, we can calculate the initial value $x_0 \equiv \beta^* a_0 e_0 = e_0/\epsilon$. Then we can compute the expressions a/a_0, e/e_0, P/P_0, etc. as functions of x. Subsequently, they can be cross-plotted in any combination.

For example, Figs. 12.2, 12.3, and 12.4 show the normalized semimajor axis, orbital period, periapsis radius, and apoapsis radius plotted against the normalized eccentricity. These figures also compare the analytical averaged solutions against a numerical solution generated by integrating Newton's second law (Eq. 1.34). The numerical integration is performed in Cartesian coordinates. The conversion to and from orbital elements is carried out by a well-known algorithm (see Vallado (2013)). The analytical solutions are created using x as a parameter and evaluating the expressions for the other quantities of interest starting from Eq. 12.125. The comparisons show very good agreement. The solutions are nearly linear until small values of eccentricity are reached. The sample orbit shown in these plots used the values $e_0 = 0.165$, $a_0 = 7860$ km, $P_0 = 6934$ s, $r_{a_0} = 9160$ km, $r_{p_0} = 6560$ km, $H = 29.74$ km. (Note: if a_0 is specified, specifying the P_0 is not required.) Values corresponding to the Earth are assumed.

A higher-order analysis of the general orbit decay theory based on the normalized elements, $z(x)$, and the dimensionless constant, Z_0, is presented by Longuski (1979).

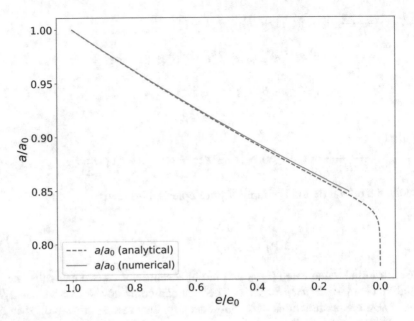

Fig. 12.2 The relationship between normalized semimajor axis and normalized eccentricity. See Eqs. 12.130 and 12.136. The relationship is nearly linear until the orbit nears final decay

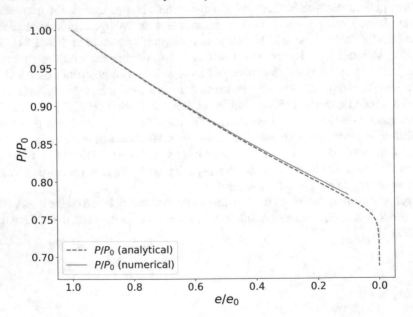

Fig. 12.3 The relationship between normalized orbital period and normalized eccentricity. See Eqs. 12.141 and 12.136. This relationship mirrors that of semimajor axis and eccentricity

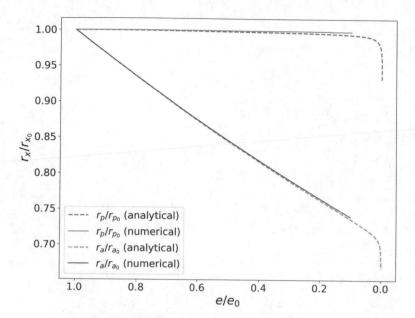

Fig. 12.4 The relationships between both the normalized apoapsis and periapsis and normalized eccentricity. See Eqs. 12.138, 12.137, and 12.136. Apoapsis decreases significantly due to the large drag force at periapsis, while periapsis is nearly constant as eccentricity approaches zero

References

Abramowitz, M., & Stegun, I. A. (1972). *Handbook of mathematical functions*. New York: Dover Publications.

King-Hele, D. (1964). *Theory of satellite orbits in an atmosphere*. London: Butterworths.

Longuski, J. M. (1979). Analytic theory of orbit contraction and ballistic entry into planetary atmospheres. Ph.D. Thesis, University of Michigan, Ann Arbor.

McCuskey, S. W. (1963). *Introduction to celestial mechanics*. Reading: Addison-Wesley Publishing Company, Inc.

Meirovitch, L. (1970). *Methods of analytical dynamics*. New York: McGraw-Hill Book Company.

Vallado, D. A. (2013). *Fundamentals of astrodynamics and applications* (4th ed.). El Segundo: Microcosm Press.

Vinh, N. X., Busemann, A., & Culp, R. D. (1980). *Hypersonic and planetary flight mechanics*. Ann Arbor: The University of Michigan Press.

Chapter 13
Introduction to Orbit Determination

13.1 Connecting Models to the Real World

We now consider the problem of using an orbit model to predict the motion of a satellite in the real world. Suppose we have assembled a credible analytical orbit model by combining the effects of the second zonal harmonic developed in Chap. 8 with the effects of atmospheric drag developed in Chap. 11. Further, let us assume we have obtained some tracking data on a satellite of interest, and we have an initial estimate of that satellite's orbit.

We now must determine the initial conditions for use in our orbit model. Even if our tracking instrument can provide a position and velocity at some instant of time, that will not fulfill our need for initial conditions. Our orbit model requires mean (averaged) orbital elements, not an osculating position and velocity. Moreover, the tracking is not perfect so the data will contain some uncertainty. We certainly would not want to start our prediction from data that contain errors.

Fortunately, Gauss published his method of least squares for calculating the orbits of celestial bodies in 1809. The method of least squares is a standard approach in regression analysis to approximate the solution of overdetermined systems (sets of equations in which there are more equations than unknowns) by minimizing the sum of the squares of the residual differences between the measurements and the model predictions.

Thanks to Gauss, we have a well-defined process to determine initial conditions for our model that will assure the closest possible match between the observations and the model over some time period for which we have data. We refer to this process as model initialization or orbit determination. After the initial conditions are determined, we speak of using the model for prediction beyond the observation period.

In this chapter, we follow Hoots (2007) to provide a theoretical development of the method of least squares and describe how it is applied to orbit determination. The Gauss least squares method forms the basis for numerous extensions and variations

© The Author(s), under exclusive license to Springer Nature Switzerland AG 2022
J. M. Longuski et al., *Introduction to Orbital Perturbations*, Space Technology
Library 40, https://doi.org/10.1007/978-3-030-89758-1_13

of the technique. Examples are Kalman filters, unscented filters, particle filters, and others. Regardless of the technique, the end objective is to determine an initial state for the orbit model, so that it will provide the most accurate prediction.

The Gauss least squares method is easy to describe, but it is difficult to understand how it actually works. As we will see, it involves summations of data components, matrix multiplication, and matrix inversion. Basically, it seems to be a "magic box" into which one pours observations and out of the bottom comes the optimum orbit state. With all those operations, it is virtually impossible to follow the trail of how a given piece of data may affect the final solution. Furthermore, the Gauss method is the least complicated of all the modern variations of orbit determination methods.

For this reason, we spend the remainder of the chapter developing closed-form analytical expressions for the least squares solution that provide a clear understanding of exactly how the least squares solution is affected by such factors as the type of measurement, the number of measurements, the quality of the measurements, and the time distribution of the measurements.

13.2   Gauss Least Squares Method

Satellites can be tracked by various devices, the most common of which are radars and telescopes. A radar measurement typically consists of a range, azimuth, and elevation at a given instant of time. Collectively this information specifies a unique point in space where the satellite was located at the measurement time. This set of data is called an observation. A telescope measurement consists of a pair of angles (e.g., right ascension and declination or azimuth and elevation) at a given instant, but it does not include range. So, a telescope observation provides a line of sight in the direction of the satellite but does not specify where along the line that the satellite lies.

We begin by letting m_{ob} denote a measurement at some time t. The measurement could be any of the observation components described. Let $\mathbf{X}$ denote the 6×1 orbital elements vector of the satellite at some epoch t_0. Generally, the observation location will not exactly match the predicted location of the satellite at the time of the observation. The purpose of the Gauss least squares method is to provide a way to adjust the initial conditions of the element set such that the prediction will come as close to the observation as possible. Let $m\,(\mathbf{X})$ denote the estimate of the measurement by the current orbital elements $\mathbf{X}$.

We assume that the measurement is sufficiently close to the predicted orbit to be approximated with the first term of a Taylor series as

$$m_{ob} = m\,(\mathbf{X}) + \left(\frac{\partial m}{\partial x_1}\right)\Delta x_1 + \left(\frac{\partial m}{\partial x_2}\right)\Delta x_2 + \left(\frac{\partial m}{\partial x_3}\right)\Delta x_3$$

$$+ \left(\frac{\partial m}{\partial x_4}\right)\Delta x_4 + \left(\frac{\partial m}{\partial x_5}\right)\Delta x_5 + \left(\frac{\partial m}{\partial x_6}\right)\Delta x_6. \tag{13.1}$$

We can rearrange Eq. 13.1 into a more convenient form given by

$$m_{ob} - m(\mathbf{X}) = \left[\frac{\partial m}{\partial x_1} \; \frac{\partial m}{\partial x_2} \; \frac{\partial m}{\partial x_3} \; \frac{\partial m}{\partial x_4} \; \frac{\partial m}{\partial x_5} \; \frac{\partial m}{\partial x_6}\right]\Delta\mathbf{X}. \tag{13.2}$$

For each measurement type, there will be a closed-form expression that relates that measurement type to the orbital elements through the partial derivatives. Also, for each measurement type from a given radar or telescope, we will have an expected measurement noise characterized by a standard deviation of the ensemble of measurements.

Thus, each measurement m_i contains some unknown error that we characterize by its standard deviation, σ_i. For a collection of k measurements, we introduce matrix notation defined by

$$A_{ij} = \frac{1}{\sigma_i} \frac{\partial m_i}{\partial x_j}\bigg|_{\mathbf{X}} \qquad i = 1, 2, \ldots, k \qquad j = 1, 2, \ldots, 6 \tag{13.3}$$

$$B_i = \frac{1}{\sigma_i}[m_i - m_i(\mathbf{X})]. \tag{13.4}$$

Using the assumption in Eq. 13.2 for each of the k measurements, the notation defined in Eqs. 13.3 and 13.4 allows us to write

$$B = A\Delta\mathbf{X}, \tag{13.5}$$

where $\Delta\mathbf{X}$ is the correction to the orbital elements vector that will bring the orbit into closer agreement with the measurements. In general, we will have $k \gg 6$ in our overdetermined system.

We multiply Eq. 13.5 by A^T, the transpose of the A matrix, to obtain

$$A^T B = \left(A^T A\right)\Delta\mathbf{X}. \tag{13.6}$$

If we multiply Eq. 13.6 by $\left(A^T A\right)^{-1}$, the inverse of the $A^T A$ matrix, we obtain

$$\left(A^T A\right)^{-1}\left(A^T B\right) = \left(A^T A\right)^{-1}\left(A^T A\right)\Delta\mathbf{X} = \Delta\mathbf{X}.$$

Thus, the correction to the orbital elements is

$$\Delta\mathbf{X} = \left(A^T A\right)^{-1}\left(A^T B\right). \tag{13.7}$$

This solution was based on a Taylor series representation in Eq. 13.1. Since our objective is to minimize the residuals (i.e., the difference between the measurements and the prediction), we introduce the quantity

$$B^T B, \tag{13.8}$$

which is the sum of the squares of the weighted residuals. This sum provides a measure of how well the orbit is fitting the measurements.

Since we only retained the linear term in the Taylor series, we expect that Eq. 13.7 will provide a first approximation of the required correction. If we apply the correction to the orbital elements as specified in Eq. 13.7, we can then repeat the process to obtain a further refinement of the orbital elements. We repeat this iterative process until $B^T B$ is only changing by a small percentage. We say that the iteration in which this occurs has converged to the best set of orbital elements that minimizes the sum of the squares of the residuals. The quantity

$$C = \left(A^T A\right)^{-1} \tag{13.9}$$

is called the covariance. This completes the description of the Gauss least squares process applied to the orbit determination of a satellite.

13.3 Unpacking the Least Squares Magic Box

One can treat an orbit determination tool as simply a magic process that provides orbits that fit the data. However, as we will see, the results depend significantly on the nature and mixture of data used. If we have gone to all the trouble to develop an analytical orbit model, why not also understand how the data will influence the orbit determination results?

To gain such insight, we employ a very simple model of the real world. In doing so, we must make sure that the model still retains all the geometrical features that control the behavior of the model while providing enough algebraic simplification that we can achieve an explicit analytical solution.

13.3.1 Simple Model of Our Problem

To develop a simple model for our problem, we restrict our analysis to a two-body circular orbit. Since the two-body motion is confined to a plane, we formulate our analysis in the plane of motion. The motion can be described by

$$x = a_0 \cos M$$

$$y = a_0 \sin M$$

$$M = M_0 + n_0 t, \tag{13.10}$$

where

$$a_0 = \text{epoch value of semimajor axis}$$

$$M_0 = \text{epoch value of mean anomaly}$$

$$n_0 = \text{epoch value of mean motion}$$

$$t = \text{time since epoch.}$$

Thus, our orbit model has only two orbital elements—the initial mean motion n_0 and the initial mean anomaly M_0. We make another simplifying assumption that the measurements are taken by a radar located at the center of the Earth. Then the measurements can be characterized by a range and a single angle since the radar is always in the plane of the satellite motion.

Figure 13.1 illustrates the coordinate frame where measurements of a satellite are taken. The coordinate system has been selected so that both the sensor angle, θ, and the mean anomaly, M, of the satellite are measured from a common line, the inertial x axis. The orange line represents the arc over which sensor S_1 tracks, and the blue line represents the arc over which sensor S_2 tracks.

The radar measurements will not be perfect, but rather can be characterized by their standard deviations, $\sigma_{1\rho}$ and $\sigma_{1\theta}$, respectively for sensor S_1. We will similarly characterize measurements from sensor S_2 by their standard deviations, $\sigma_{2\rho}$ and $\sigma_{2\theta}$, respectively. The development that follows assumes the errors are normally distributed with zero mean error.

The ith radar measurement from the first sensor can be represented as

$$\rho_{1i} = \rho_{1T_i} + \delta_{1\rho_i}$$

$$\theta_{1i} = \theta_{1T_i} + \delta_{1\theta_i},$$

Fig. 13.1 Two sensors located at the Earth center observe a satellite with tracks indicated by the arcs S_1 and S_2

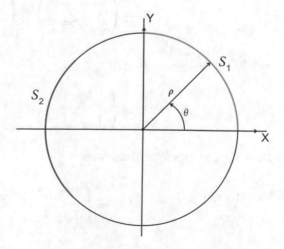

and the ith measurement from the second sensor can be represented as

$$\rho_{2i} = \rho_{2T_i} + \delta_{2\rho_i}$$
$$\theta_{2i} = \theta_{2T_i} + \delta_{2\theta_i},$$

where

$$\rho_{T_i} = \text{true range for } i\text{th observation}$$
$$\delta_{\rho_i} = \text{radar range noise for } i\text{th observation}$$
$$\delta_{\theta_i} = \text{radar angle noise for } i\text{th observation}$$
$$\theta_{T_i} = M_0 + n_0 t_i, \text{ true angle for } i\text{th observation}.$$

It is important to recognize the notation difference between the standard deviation σ expected for a measurement from a given radar versus the noise δ of a particular measurement from that radar.

The differences between the measurements and the estimate of the satellite orbit are described by

$$\Delta\rho = \left(\frac{\partial\rho}{\partial M}\right)\Delta M_0 + \left(\frac{\partial\rho}{\partial n}\right)\Delta n_0$$

$$\Delta\theta = \left(\frac{\partial\theta}{\partial M}\right)\Delta M_0 + \left(\frac{\partial\theta}{\partial n}\right)\Delta n_0. \tag{13.11}$$

For a single observation consisting of a range and an angle measurement, Eqs. 13.3 and 13.4 will give

$$A_i = \begin{pmatrix} \frac{1}{\sigma_\rho}\frac{\partial\rho}{\partial M} & \frac{1}{\sigma_\rho}\frac{\partial\rho}{\partial n} \\ \frac{1}{\sigma_\theta}\frac{\partial\theta}{\partial M} & \frac{1}{\sigma_\theta}\frac{\partial\theta}{\partial n} \end{pmatrix} = \begin{pmatrix} 0 & \frac{1}{\sigma_\rho}H \\ \frac{1}{\sigma_\theta} & \frac{1}{\sigma_\theta}t_i \end{pmatrix} \tag{13.12}$$

$$H = -\frac{2a_0}{3n_o} \tag{13.13}$$

$$B_i = \begin{pmatrix} \frac{1}{\sigma_\rho}\Delta\rho_i \\ \frac{1}{\sigma_\theta}\Delta\theta_i \end{pmatrix}. \tag{13.14}$$

From Eq. 13.6, we need to calculate

$$A_i^T B_i = \begin{pmatrix} 0 & \frac{1}{\sigma_\theta} \\ \frac{1}{\sigma_\rho}H & \frac{1}{\sigma_\theta}t_i \end{pmatrix}\begin{pmatrix} \frac{1}{\sigma_\rho}\Delta\rho_i \\ \frac{1}{\sigma_\theta}\Delta\theta_i \end{pmatrix}$$

$$A_i^T B_i = \begin{pmatrix} \frac{1}{\sigma_\theta^2} \Delta\theta_i \\ \frac{1}{\sigma_\rho^2} H \Delta\rho_i + \frac{1}{\sigma_\theta^2} t_i \Delta\theta_i \end{pmatrix}. \qquad (13.15)$$

And from Eq. 13.6, we also need to calculate

$$A_i^T A_i = \begin{pmatrix} 0 & \frac{1}{\sigma_\theta} \\ \frac{1}{\sigma_\rho} H & \frac{1}{\sigma_\theta} t_i \end{pmatrix} \begin{pmatrix} 0 & \frac{1}{\sigma_\rho} H \\ \frac{1}{\sigma_\theta} & \frac{1}{\sigma_\theta} t_i \end{pmatrix}$$

$$A_i^T A_i = \begin{pmatrix} \frac{1}{\sigma_\theta^2} & \frac{1}{\sigma_\theta^2} t_i \\ \frac{1}{\sigma_\theta^2} t_i & \frac{1}{\sigma_\rho^2} H^2 + \frac{1}{\sigma_\theta^2} t_i^2 \end{pmatrix}. \qquad (13.16)$$

We assume that a radar collects k observations. Then Eqs. 13.15 and 13.16 can be written as

$$A^T B = \begin{pmatrix} \frac{1}{\sigma_\theta^2} \sum_{i=1}^{k} \Delta\theta_i \\ \frac{1}{\sigma_\rho^2} H \sum_{i=1}^{k} \Delta\rho_i + \frac{1}{\sigma_\theta^2} \sum_{i=1}^{k} t_i \Delta\theta_i \end{pmatrix} \qquad (13.17)$$

and

$$A^T A = \begin{pmatrix} \frac{1}{\sigma_\theta^2} \sum_{i=1}^{k} 1 & \frac{1}{\sigma_\theta^2} \sum_{i=1}^{k} t_i \\ \frac{1}{\sigma_\theta^2} \sum_{i=1}^{k} t_i & \frac{1}{\sigma_{\rho_i}^2} H^2 \sum_{j=1}^{k} 1 + \frac{1}{\sigma_\theta^2} \sum_{j=1}^{k} t_i^2 \end{pmatrix}. \qquad (13.18)$$

Equations 13.17 and 13.18 can be simplified to

$$A^T B = \frac{1}{\sigma_\theta^2} \begin{pmatrix} \sum_{i=1}^{k} \Delta\theta_i \\ GH \sum_{i=1}^{k} \Delta\rho_i + \sum_{i=1}^{k} t_i \Delta\theta_i \end{pmatrix} \qquad (13.19)$$

$$A^T A = \frac{1}{\sigma_\theta^2} \begin{pmatrix} \sum_{i=1}^{k} 1 & \sum_{i=1}^{k} t_i \\ \sum_{i=1}^{k} t_i & GH^2 \sum_{j=1}^{k} 1 + \sum_{j=1}^{k} t_i^2 \end{pmatrix}, \qquad (13.20)$$

where

$$G = \frac{\sigma_\theta^2}{\sigma_\rho^2}.$$

We further assume that a radar collects k observations spread over $k - 1$ equal time intervals spanning T and beginning at time $-\tau$. The time of the ith observation is

$$t_i = -\tau + (i - 1)\,\Delta t \quad \text{(start time is prior to epoch)}, \tag{13.21}$$

where

$$\Delta t = \frac{T}{k - 1}. \tag{13.22}$$

We need the following sums in the subsequent development:

$$\sum_{i=1}^{k} 1 \quad \sum_{i=1}^{k} t_i \quad \sum_{i=1}^{k} t_i^2. \tag{13.23}$$

The first of these summations has the simple result

$$\sum_{i=1}^{k} 1 = k. \tag{13.24}$$

Using Eq. 13.21 in the second summation of Eq. 13.23, we obtain

$$\sum_{i=1}^{k} t_i = \sum_{i=1}^{k} [-\tau + (i - 1)\Delta t] = \sum_{i=1}^{k} -\tau + \sum_{i=1}^{k} (i - 1)\Delta t$$

$$= -\tau \sum_{i=1}^{k} 1 + \Delta t \sum_{i=1}^{k} i - \Delta t \sum_{i=1}^{k} 1. \tag{13.25}$$

Using Eq. 13.22 and the known formula for the sum of integers, Eq. 13.25 reduces to

$$\sum_{i=1}^{k} t_i = -\tau k + \Delta t \frac{k\,(k + 1)}{2} - \Delta t k = -\tau k + \Delta t \frac{k^2 + k - 2k}{2}$$

$$= -\tau k + \Delta t \frac{k\,(k - 1)}{2}. \tag{13.26}$$

Substituting Eq. 13.22 into Eq. 13.26 gives

$$\sum_{i=1}^{k} t_i = -k\tau + \frac{T}{(k-1)} \frac{k\,(k-1)}{2} = -k\tau + \frac{k}{2}T$$

$$= k\left(-\tau + \frac{T}{2}\right). \tag{13.27}$$

Using Eq. 13.21 in the third summation of Eq. 13.23, we obtain

$$\sum_{i=1}^{k} t_i^2 = \sum_{i=1}^{k} [-\tau + (i-1)\,\Delta t]^2 = \sum_{i=1}^{k} \tau^2 - \sum_{i=1}^{k} 2\tau\,(i-1)\,\Delta t + \sum_{i=1}^{k} (i-1)^2 \Delta t^2$$

$$= \tau^2 \sum_{i=1}^{k} 1 - 2\tau\,\Delta t \sum_{i=1}^{k} (i-1) + \Delta t^2 \sum_{i=1}^{k} (i-1)^2$$

$$= \tau^2 \sum_{i=1}^{k} 1 - 2\tau\,\Delta t \sum_{i=1}^{k} i + 2\tau\,\Delta t \sum_{i=1}^{k} 1 + \Delta t^2 \left[\sum_{i=1}^{k} i^2 - 2 \sum_{i=1}^{k} i + \sum_{i=1}^{k} 1 \right].$$
$$\tag{13.28}$$

Using the known formulas for the sums of the integers and the squares of the integers in Eq. 13.28, we obtain

$$\sum_{i=1}^{k} t_i^2 = \tau^2 k - 2\tau\,\Delta t \frac{k\,(k+1)}{2} + 2\tau\,\Delta t k$$

$$+ \Delta t^2 \left[\frac{k\,(k+1)\,(2k+1)}{6} - 2\frac{k\,(k+1)}{2} + k \right]$$

$$= \tau^2 k - 2\tau\,\Delta t \frac{k\,(k+1) - 2k}{2} + \Delta t^2 \frac{k}{6} [(k+1)\,(2k+1) - 6\,(k+1) + 6]$$

$$= \tau^2 k - \tau\,\Delta t\,(k)\,(k-1) + \Delta t^2 \frac{k}{6} \left[2k^2 - 3k + 1 \right]$$

$$= \tau^2 k - \tau\,\Delta t\,(k)\,(k-1) + \Delta t^2 \frac{k}{6}\,(k-1)\,(2k-1). \tag{13.29}$$

Substituting Eq. 13.22 into Eq. 13.29 gives

$$\sum_{i=1}^{k} t_i^2 = \tau^2 k - \tau \frac{T}{(k-1)} k\,(k-1) + \frac{T^2}{(k-1)^2} \frac{k}{6}\,(k-1)\,(2k-1)$$

$$= k\tau^2 - k\tau T + k \frac{T^2}{6} \frac{(2k-1)}{(k-1)}. \tag{13.30}$$

In summary, we have

$$\sum_{i=1}^{k} 1 = k \tag{13.31}$$

$$\sum_{i=1}^{k} t_i = k \left(-\tau + \frac{T}{2} \right) \tag{13.32}$$

$$\sum_{i=1}^{k} t_i^2 = k \left(\tau^2 - \tau T + \frac{T^2}{3} \right), \tag{13.33}$$

where we have assumed $k \gg 1$ in Eq. 13.33.

We assume that the measurement errors are normally distributed with zero mean so that

$$\sum \Delta \rho = \sum (\rho - \rho_{\text{truth}}) + \sum \delta_\rho = \sum \delta_\rho$$

$$\sum \Delta \theta = \sum (\theta - \theta_{\text{truth}}) + \sum \delta_\theta = \sum \delta_\theta.$$

Thus, Eq. 13.19 becomes

$$A^T B = \frac{1}{\sigma_\theta^2} \left(\begin{array}{c} \sum_{i=1}^{k} \delta_{\theta_i} \\ GH \sum_{i=1}^{k} \delta_{\rho_i} + \sum_{i=1}^{k} t_i \delta_{\theta_i} \end{array} \right). \tag{13.34}$$

We introduce the following shorthand notation:

$$Q = \sum_{i=1}^{k} \delta_{\theta_i}$$

$$Q_t = \sum_{i=1}^{k} t_i \delta_{\theta_i} \tag{13.35}$$

$$R = \sum_{i=1}^{k} \delta_{\rho_i}$$

into Eq. 13.34 to write

$$A^T B = \left(\begin{array}{c} Q \\ GHR + Q_t \end{array} \right). \tag{13.36}$$

Note that Q and R only depend on the total noise in the observations, whereas Q_t contains a correlation between the noise and the time at which the noise occurred.

Substituting Eqs. 13.31, 13.32, and 13.33 into Eq. 13.20 yields

$$A^T A = \frac{1}{\sigma_\theta^2} \begin{bmatrix} k & k\left(-\tau + \frac{T}{2}\right) \\ k\left(-\tau + \frac{T}{2}\right) & kGH^2 + k\left(\tau^2 - \tau T + \frac{T^2}{3}\right) \end{bmatrix}$$

$$A^T A = \frac{k}{\sigma_\theta^2} \begin{bmatrix} 1 & \left(-\tau + \frac{T}{2}\right) \\ \left(-\tau + \frac{T}{2}\right) & GH^2 + \left(\tau^2 - \tau T + \frac{T^2}{3}\right) \end{bmatrix}. \qquad (13.37)$$

Equations (13.36) and (13.37) describe the effects of k evenly spaced observations from a single radar.

13.3.2 Application to Multiple Tracks

Returning to Fig. 13.1, suppose we have two sensors, S_1 and S_2. Further, suppose that each tracks for a non-overlapping time span T_1 and T_2 while collecting k and m observations, respectively. We illustrate the track times in Fig. 13.2.
    Equation 13.37 then becomes

$$A^T A = A_1^T A_1 + A_2^T A_2 = \begin{bmatrix} m_{11} & m_{12} \\ m_{21} & m_{22} \end{bmatrix}, \qquad (13.38)$$

where

$$m_{11} = \frac{k}{\sigma_{1\theta}^2} + \frac{m}{\sigma_{2\theta}^2}$$

$$m_{12} = m_{21} = \frac{k}{\sigma_{1\theta}^2}\left(-\tau_1 + \frac{T_1}{2}\right) + \frac{m}{\sigma_{2\theta}^2}\left(-\tau_2 + \frac{T_2}{2}\right)$$

$$m_{22} = \frac{kH^2}{\sigma_{1\rho}^2} + \frac{mH^2}{\sigma_{2\rho}^2} + \frac{k}{\sigma_{1\theta}^2}\left(\tau_1^2 - \tau_1 T_1 + \frac{T_1^2}{3}\right) + \frac{m}{\sigma_{2\theta}^2}\left(\tau_2^2 - \tau_2 T_2 + \frac{T_2^2}{3}\right).$$

It is useful to recognize that

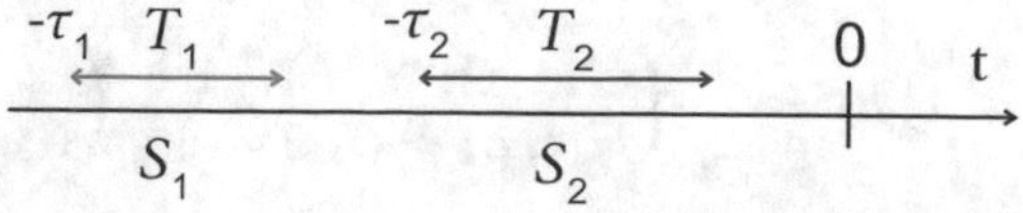

Fig. 13.2 Timeline for tracks of length T_1 and T_2 from sensors S_1 and S_2, respectively. Both tracks occur prior to the epoch time 0

$$\tau^2 - \tau T + \frac{T^2}{3} = \tau^2 - \tau T + \frac{T^2}{4} + \frac{T^2}{12} = \left(\tau - \frac{T}{2}\right)^2 + \frac{T^2}{12}. \tag{13.39}$$

Furthermore, we introduce the following notation and recognize that

$$\hat{T} = \frac{T}{2} - \tau \tag{13.40}$$

is the midpoint of the tracking span. Using Eqs. 13.39 and 13.40 in Eq. 13.38, we obtain

$$A^T A = \begin{pmatrix} \frac{k}{\sigma_{1\theta}^2} + \frac{m}{\sigma_{2\theta}^2} & \frac{k}{\sigma_{1\theta}^2}\hat{T}_1 + \frac{m}{\sigma_{2\theta}^2}\hat{T}_2 \\ \frac{k}{\sigma_{1\theta}^2}\hat{T}_1 + \frac{m}{\sigma_{2\theta}^2}\hat{T}_2 & \frac{kH^2}{\sigma_{1\rho}^2} + \frac{mH^2}{\sigma_{2\rho}^2} + \frac{k}{\sigma_{1\theta}^2}\hat{T}_1^2 + \frac{m}{\sigma_{2\theta}^2}\hat{T}_2^2 + \frac{k}{\sigma_{1\theta}^2}\left(\frac{T_1^2}{12}\right) + \frac{m}{\sigma_{2\theta}^2}\left(\frac{T_2^2}{12}\right) \end{pmatrix}. \tag{13.41}$$

We introduce the following symbols, all of which are constants:

$$F = \frac{k}{\sigma_{1\theta}^2} + \frac{m}{\sigma_{2\theta}^2}$$

$$J = \frac{k}{\sigma_{1\theta}^2}\hat{T}_1 + \frac{m}{\sigma_{2\theta}^2}\hat{T}_2$$

$$K = \frac{k}{\sigma_{1\theta}^2}\hat{T}_1^2 + \frac{m}{\sigma_{2\theta}^2}\hat{T}_2^2 \tag{13.42}$$

$$L = \frac{k}{\sigma_{1\rho}^2} + \frac{m}{\sigma_{2\rho}^2}$$

$$M = \frac{1}{12}\left(\frac{k}{\sigma_{1\theta}^2}T_1^2 + \frac{m}{\sigma_{2\theta}^2}T_2^2\right).$$

Then Eq. 13.41 takes the more compact form

$$A^T A = \begin{pmatrix} F & J \\ J & LH^2 + K + M \end{pmatrix}. \tag{13.43}$$

Similarly, Eq. 13.36 applied to the two radar tracks becomes

$$A^T B = A_1^T B_1 + A_2^T B_2 = \frac{1}{\sigma_{1\theta}^2}\begin{pmatrix} Q_1 \\ G_1 H R_1 + Q_{t1} \end{pmatrix} + \frac{1}{\sigma_{2\theta}^2}\begin{pmatrix} Q_2 \\ G_2 H R_2 + Q_{t2} \end{pmatrix}. \tag{13.44}$$

From Eq. 13.7, we need to solve the equation

$$\Delta \mathbf{X} = \left(A^T A\right)^{-1} \left(A^T B\right),$$ (13.45)

Thus, we need to compute the inverse of the matrix given by Eq. 13.43. Let

$$D = FLH^2 + FK + FM - J^2.$$ (13.46)

Using equations (13.42) we note that

$$
\begin{aligned}
FK - J^2 &= \left(\frac{k}{\sigma_{1\theta}^2} + \frac{m}{\sigma_{2\theta}^2}\right)\left(\frac{k}{\sigma_{1\theta}^2}\hat{T}_1^2 + \frac{m}{\sigma_{2\theta}^2}\hat{T}_2^2\right) - \left(\frac{k}{\sigma_{1\theta}^2}\hat{T}_1 + \frac{m}{\sigma_{2\theta}^2}\hat{T}_2\right)^2 \\
&= \frac{k^2}{\sigma_{1\theta}^4}\hat{T}_1^2 + \frac{km}{\sigma_{1\theta}^2\sigma_{2\theta}^2}\hat{T}_2^2 + \frac{km}{\sigma_{1\theta}^2\sigma_{2\theta}^2}\hat{T}_1^2 + \frac{m^2}{\sigma_{2\theta}^4}\hat{T}_2^2 \\
&\quad - \frac{k^2}{\sigma_{1\theta}^4}\hat{T}_1^2 - 2\frac{km}{\sigma_{1\theta}^2\sigma_{2\theta}^2}\hat{T}_1\hat{T}_2 - \frac{m^2}{\sigma_{2\theta}^4}\hat{T}_2^2 \\
&= \frac{km}{\sigma_{1\theta}^2\sigma_{2\theta}^2}\hat{T}_2^2 + \frac{km}{\sigma_{1\theta}^2\sigma_{2\theta}^2}\hat{T}_1^2 - 2\frac{km}{\sigma_{1\theta}^2\sigma_{2\theta}^2}\hat{T}_1\hat{T}_2 = \frac{km}{\sigma_{1\theta}^2\sigma_{2\theta}^2}\left(\hat{T}_2 - \hat{T}_1\right)^2 \\
&= \frac{km}{\sigma_{1\theta}^2\sigma_{2\theta}^2}\left(\hat{T}_2 - \hat{T}_1\right)^2,
\end{aligned}
$$

so that Eq. 13.46 becomes

$$D = FLH^2 + FM + \frac{km}{\sigma_{1\theta}^2\sigma_{2\theta}^2}\left(\hat{T}_2 - \hat{T}_1\right)^2.$$ (13.47)

Then the inverse of the matrix in Eq. 13.43 is

$$\left(A^T A\right)^{-1} = \frac{1}{D}\begin{pmatrix} LH^2 + K + M & -J \\ -J & F \end{pmatrix}.$$ (13.48)

Substituting Eqs. 13.48 and 13.44 into Eq. 13.45 gives

$$
\Delta \mathbf{X} = \frac{1}{D}\begin{pmatrix} LH^2 + K + M & -J \\ -J & F \end{pmatrix}\left[\frac{1}{\sigma_{1\theta}^2}\begin{pmatrix} Q_1 \\ G_1 H R_1 + Q_{t1} \end{pmatrix}\right.
$$

$$
\left. + \frac{1}{\sigma_{2\theta}^2}\begin{pmatrix} Q_2 \\ G_2 H R_2 + Q_{t2} \end{pmatrix}\right].
$$ (13.49)

Thus, the corrections to the orbital elements from the least squares process are

$$\Delta M_0 = \frac{1}{D} \frac{1}{\sigma_{1\theta}} \left[\left(LH^2 + K + M \right) Q_1 - JG_1 H R_1 - J Q_{t1} \right]$$
$$+ \frac{1}{D} \frac{1}{\sigma_{2\theta}} \left[\left(LH^2 + K + M \right) Q_2 - JG_2 H R_2 - J Q_{t2} \right] \tag{13.50}$$

$$\Delta n_0 = \frac{1}{D} \frac{1}{\sigma_{1\theta}} [-J Q_1 + F G_1 H R_1 + F Q_{t1}] + \frac{1}{D} \frac{1}{\sigma_{2\theta}} [-J Q_2 + F G_2 H R_2 + F Q_{t2}]. \tag{13.51}$$

13.3.3   Covariance

The prediction errors can be estimated using the covariance matrix. The covariance matrix at a time t is given by

$$C(t) = A_t \left(A^T A \right)^{-1} A_t^T, \tag{13.52}$$

where the subscript t indicates evaluation at the desired prediction time, and where

$$A_t = \begin{pmatrix} \frac{\partial \rho}{\partial M} & \frac{\partial \rho}{\partial n} \\ \frac{\partial \theta}{\partial M} & \frac{\partial \theta}{\partial n} \end{pmatrix} = \begin{pmatrix} 0 & H \\ 1 & t \end{pmatrix}. \tag{13.53}$$

The covariance is

$$\begin{aligned}
C(t) &= \begin{pmatrix} 0 & H \\ 1 & t \end{pmatrix} \frac{1}{D} \begin{pmatrix} LH^2 + K + M & -J \\ -J & F \end{pmatrix} \begin{pmatrix} 0 & 1 \\ H & t \end{pmatrix} \\
&= \frac{1}{D} \begin{pmatrix} -HJ & FH \\ LH^2 + K + M - Jt & -J + Ft \end{pmatrix} \begin{pmatrix} 0 & 1 \\ H & t \end{pmatrix} \\
&= \frac{1}{D} \begin{pmatrix} FH^2 & -HJ + FHt \\ -HJ + FHt & LH^2 + M + K - 2Jt + Ft^2 \end{pmatrix}. \tag{13.54}
\end{aligned}$$

Thus, at some predicted time t the variances and covariances of the coordinates are

$$\sigma_{\rho\rho}^2 = \frac{FH^2}{D} \tag{13.55}$$

$$\sigma_{\theta\theta}^2 = \frac{1}{D} \left[LH^2 + M + K - 2Jt + Ft^2 \right] \tag{13.56}$$

$$\sigma_{\rho\theta}^2 = \frac{H}{D} [-J + Ft]. \tag{13.57}$$

Using the definitions from Eqs. 13.42 in Eq. 13.57, we obtain

$$\frac{H}{D}(-J + Ft) = \frac{H}{D}\left(-\frac{k}{\sigma_{1\theta}^2}\hat{T}_1 - \frac{m}{\sigma_{2\theta}^2}\hat{T}_2 + \frac{k}{\sigma_{1\theta}^2}t + \frac{m}{\sigma_{2\theta}^2}t\right)$$

$$= \frac{H}{D}\left[\frac{k}{\sigma_{1\theta}^2}\left(t - \hat{T}_1\right) + \frac{m}{\sigma_{2\theta}^2}\left(t - \hat{T}\right)\right]. \tag{13.58}$$

We examine the last three terms in the bracket in Eq. 13.56 and use the definitions in Eqs. 13.42 to obtain

$$K - 2Jt + Ft^2 = \frac{k}{\sigma_{1\theta}^2}\hat{T}_1^2 + \frac{m}{\sigma_{2\theta}^2}\hat{T}_2^2 - 2\left(\frac{k}{\sigma_{1\theta}^2}\hat{T}_1 + \frac{m}{\sigma_{2\theta}^2}\hat{T}_2\right)t + \left(\frac{k}{\sigma_{1\theta}^2} + \frac{m}{\sigma_{2\theta}^2}\right)t^2$$

$$= \frac{k}{\sigma_{1\theta}^2}\left(t^2 - 2\hat{T}_1 t + \hat{T}_1^2\right) + \frac{m}{\sigma_{2\theta}^2}\left(t^2 - 2\hat{T}_2 t + \hat{T}_2^2\right)$$

$$= \frac{k}{\sigma_{1\theta}^2}\left(t - \hat{T}_1\right)^2 + \frac{m}{\sigma_{2\theta}^2}\left(t - \hat{T}_2\right)^2. \tag{13.59}$$

We substitute Eq. 13.58 into Eq. 13.57 to obtain

$$\sigma_{\rho\theta}^2 = \frac{H}{D}\left[\frac{k}{\sigma_{1\theta}^2}\left(t - \hat{T}_1\right) + \frac{m}{\sigma_{2\theta}^2}\left(t - \hat{T}_2\right)\right],$$

and substitute Eq. 13.59 into Eq. 13.56 to obtain

$$\sigma_{\theta\theta}^2 = \frac{1}{D}\left[LH^2 + M + \frac{k}{\sigma_{1\theta}^2}\left(t - \hat{T}_1\right)^2 + \frac{m}{\sigma_{2\theta}^2}\left(t - \hat{T}_2\right)^2\right].$$

Thus, at some predicted time t the variances and covariances of the coordinates are

$$\sigma_{\rho\rho}^2 = \frac{FH^2}{D} \tag{13.60}$$

$$\sigma_{\theta\theta}^2 = \frac{1}{D}\left[LH^2 + M + \frac{k}{\sigma_{1\theta}^2}\left(t - \hat{T}_1\right)^2 + \frac{m}{\sigma_{2\theta}^2}\left(t - \hat{T}_2\right)^2\right] \tag{13.61}$$

$$\sigma_{\rho\theta}^2 = \frac{H}{D}\left[\frac{k}{\sigma_{1\theta}^2}\left(t - \hat{T}_1\right) + \frac{m}{\sigma_{2\theta}^2}\left(t - \hat{T}_2\right)\right], \tag{13.62}$$

with the values of the constants being

$$F = \frac{k}{\sigma_{1\theta}^2} + \frac{m}{\sigma_{2\theta}^2}$$

$$J = \frac{k}{\sigma_{1\theta}^2}\hat{T}_1 + \frac{m}{\sigma_{2\theta}^2}\hat{T}_2$$

$$K = \frac{k}{\sigma_{1\theta}^2}\hat{T}_1^2 + \frac{m}{\sigma_{2\theta}^2}\hat{T}_2^2$$

$$L = \frac{k}{\sigma_{1\rho}^2} + \frac{m}{\sigma_{2\rho}^2}$$

$$M = \frac{1}{12}\left(\frac{k}{\sigma_{1\theta}^2}T_1^2 + \frac{m}{\sigma_{2\theta}^2}T_2^2\right)$$

$$H = -\frac{2a_0}{3n_o}$$

$$D = FLH^2 + FM + \frac{km}{\sigma_{1\theta}^2\sigma_{2\theta}^2}\left(\hat{T}_2 - \hat{T}_1\right)^2.$$

We can characterize the constant D as follows:

- Directly proportional to the square of the number of measurements—FL and FM.
- Inversely proportional to the fourth power of the measurements noise—every term.
- Directly proportional to the square of the length of the measurement spans—M.
- Directly proportional to the square of the time span separating the two tracks—last term.

Every component of the covariance has the following properties:

- Property 1—Inversely proportional to number of observations.
- Property 2—Directly proportional to square of the measurement noise.
- Property 3—Inversely proportional to square of the length of the measurement spans.
- Property 4—Inversely proportional to the square of the time span separating the two tracks.

Additionally, we see that $\sigma_{\rho\rho}^2$ is a constant, $\sigma_{\rho\theta}^2$ depends linearly on the time of prediction from the midpoint of the track intervals, and $\sigma_{\theta\theta}^2$ depends quadratically on the time of prediction from the midpoint of the track intervals. Each of these comments shows that the results of Eqs. 13.60–13.62 provide explicit proof of our intuition concerning how an orbit determination should work. We also note that the $\sigma_{\theta\theta}^2$ and $\sigma_{\rho\theta}^2$ terms will be smaller at the middle of the tracking spans.

We make one last comment about the corrections to the orbital elements provided by the least squares. In Eq. 13.35, we introduced the constants

$$Q = \sum_{i=1}^{k} \delta_{\theta_i}$$

$$Q_t = \sum_{i=1}^{k} t_i \delta_{\theta_i}$$

$$R = \sum_{i=1}^{k} \delta_{\rho_i},$$

so if we have a large number of observations, we can assume that

$$Q \approx 0$$

$$R \approx 0.$$

We cannot make the same assumption for Q_t because the observations in the summation are weighted according to their occurrence in time. Then Eqs. 13.50 and 13.51 reduce to

$$\Delta M_0 = -\frac{J}{D} \left(\frac{1}{\sigma_{1\theta}} Q_{t1} + \frac{1}{\sigma_{2\theta}} Q_{t2} \right)$$

$$\Delta n_0 = \frac{F}{D} \left(\frac{1}{\sigma_{1\theta}} Q_{t1} + \frac{1}{\sigma_{2\theta}} Q_{t2} \right).$$

Therefore, even if the orbit determination is started with the truth values of n_0 and M_0, the least squares correction may not be zero. It will attempt to remove the bias caused by the Q_t time-dependent noise term.

13.3.4   Examples

We now explore how the various parameters affect the covariance and hence the model prediction accuracy. Because of the nature of the dynamics, the largest effect occurs along the track of the satellite. So, our graphs focus on the metric of in-track position error. The following parameters affect the covariance:

- Time of track relative to epoch.
- Number of observations (evenly spaced within the track length).
- Track length.
- Angle noise.
- Range noise.

Since we are modeling two separate tracking systems, we are interested in how variations between the two systems affect the covariance. Thus, we provide a series of graphs where we vary a single parameter between the two systems. The nominal values of the parameters are

- Time of tracks relative to epoch $= -190$ min, -10 min.
- Number of observations $= 100$.
- Track length $= 20$ min.
- Angle noise $= 0.015°$.
- Range noise $= 10$ m.

For each example case we use the nominal values except for the parameter being studied. The epoch is at time 0.

Case 1: Placement in Time of the Tracks

This case explores the effect of placement in time of the tracks from the two systems. In Fig. 13.3, the legend identifies the time (in minutes) of the two tracks relative to the reference time 0.

Having both tracking events occur at the end of the fitting interval (times 0 and 0) gives the best short-term prediction illustrated by the gray curve. But having the two tracks span the entire fitting interval (times -200 and 0) gives the best long-term prediction illustrated by the green curve. This outcome is consistent with covariance Property 4.

Case 2: Number of Observations in Each of the Tracks

This case explores the effect of the number of observations in the tracks from the two systems. In Fig. 13.4, the legend identifies the observation count of the two tracks.

We see that the accuracy improves as the total number of observations increases. It matters little which of the trackers is producing the greater number of observations. A larger number of total observations has a slightly slower error growth.

Case 3: Length of Time for Each of the Tracks

This case explores the effect of the lengths of time of the tracks from the two systems. In Fig. 13.5, the legend identifies the time lengths (in minutes) of the two tracks.

Covariance Property 3 states that accuracy has an inversely proportional dependence on the length of the measurement spans. While that relationship can be seen by examining the constant M, we see that it is a factor of 12 smaller than other terms in Eq. 13.61. Therefore, the effect does not stand out in the plots.

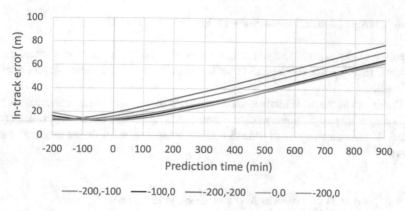

Fig. 13.3 Case 1—The minimum of each curve is at the midpoint of the two tracking events. This property is analytically predicted by Eq. 13.61

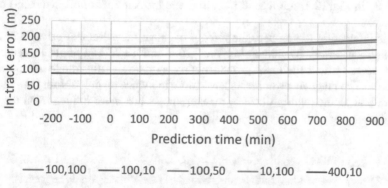

Fig. 13.4 Case 2—The blue curve has the largest number of total observations (410) and has the smallest error throughout the time span. This result is consistent with covariance Property 1

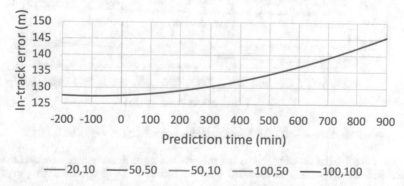

Fig. 13.5 Case 3—Each of the 5 samples in this case has the same timing of the two tracks and the same number of measurements in each track. The difference is the time length of each track, and no perceptible difference can be seen among the 5 samples

Case 4: Angle Measurement Noise in Each of the Tracks

This case explores the effect of angle measurement noise in the tracks from the two systems. In Fig. 13.6, the legend identifies the measurement noise (degrees) of the two tracks.

The errors in the orbit increase as the angular measurement noise increases. Note that a single low angle noise tracker dramatically improves the overall results (gray curve). It is interesting that reduced tracker noise results in faster prediction error growth.

Case 5: Range Measurement Noise in Each of the Tracks

This case explores the effect of range measurement noise in the tracks from the two systems. In Fig. 13.7, the legend identifies the measurement noise (meters) of the two tracks.

It is interesting to compare the effects of range measurement errors and angle measurement errors by comparing Figs. 13.6 and 13.7. The angle measurement errors do not affect the error growth nearly as much as the range measurement errors do. The range measurements have more influence on determining the satellite period, and so the quality of those measurements has more influence on the error growth.

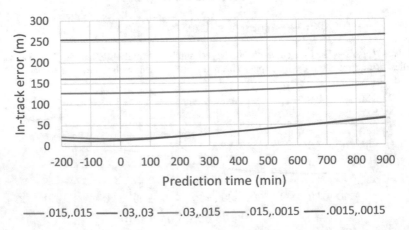

Fig. 13.6 Case 4—The dark blue curve has the smallest angle measurement error and has the smallest error throughout the time span. This behavior is consistent with covariance Property 2

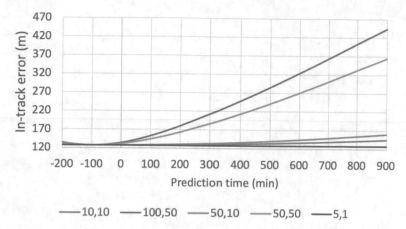

Fig. 13.7 Case 5—The dark blue curve has the smallest range measurement error and has the smallest error throughout the time span. This behavior is consistent with covariance Property 2

13.4 Conclusions

With suitable simplifying assumptions, we can formulate the orbital covariance into analytical expressions that reveal the functional dependence of covariance on the number of observations, accuracy of those observations, and timing of the data sample. Furthermore, we can observe the behavior of the covariance as a function of prediction time. These analytical formulas nicely correspond to both intuition and observed behavior of real-world cases. Since the formulas include individual modeling of two separate tracking systems, we can understand the effects of combining data from two disparate systems.

Reference

Hoots, F. R. (2007). Radar-optical observation mix. AAS/AIAA Astrodynamics Specialist Conference, Paper 07-287, Mackinac Island, MI.

Bibliography

Abramowitz, M., & Stegun, I. A. (1972). *Handbook of mathematical functions*. New York: Dover Publications.

Bate, R. R., Mueller, D. D., & White, J. E. (1971). *Fundamentals of astrodynamics*. New York: Dover Publications, Inc.

Battin, R. H. (1999). *An introduction to the mathematics and methods of astrodynamics* (Revised ed.). Reston: American Institute of Aeronautics and Astronautics, Inc.

Beutler, G. (2005). *Methods of celestial mechanics volume I: Physical, mathematical, and numerical principles*. Berlin: Springer.

Beutler, G. (2005). *Methods of celestial mechanics volume II: Application to planetary system, geodynamics and satellite geodesy*. Berlin: Springer.

Bomford, B. G. (1962). *Geodesy* (2nd ed.). Oxford: Oxford University Press.

Brouwer, D., & Clemence, G. M. (1961). *Methods of celestial mechanics*. New York: Academic Press.

Brown, E. W. (1896). *An introductory treatise on lunar theory*. New York: Cambridge University Press.

Burns, J. A. (Ed.) (1977). *Planetary satellites*. Tucson: University of Arizona Press.

Cefola, P. J., Folcik, Z., Di-Costanzo, R., Bernard, N., Setty, S., & San Juan, J. F. (2014). Revisiting the DSST standalone orbit propagator. *Advances in the Astronautical Sciences, 152*, 2891–2914..

Chao, C.-C., & Hoots, F. R. (2018). *Applied orbit perturbations and maintenance* (2nd ed.). El Segundo: The Aerospace Press.

Chobotov, V. A. (Ed.) (2002). *Orbital mechanics* (3rd ed.). Reston: American Institute of Aeronautics and Astronautics, Inc.

Ciufolini, I., & Wheeler, J. A. (1995). *Gravitation and inertia*. New Jersey: Princeton University Press.

Danby, J. M. A. (1992). *Fundamentals of celestial mechanics* (2nd ed.). Richmond: Willman-Bell, Inc.

Danielson, D. A., Neta, B., & Early, L. W. (1994). *Semianalytic satellite theory (SST): Mathematical algorithms*. Naval Postgraduate School. Report Number NPS-MA94-001.

Escobal, P. R. (1985). *Methods of orbit determination* (2nd ed.). Malabar: Krieger Publishing Co.

Fitzpatrick, P. M. (1970). *Principles of celestial mechanics*. New York: Academic Press.

Fitzpatrick, R. (2012). *An introduction to celestial mechanics*. New York: Cambridge University Press.

Greenwood, D. T. (1988). *Principles of dynamics* (2nd ed.). Englewood Cliffs: Prentice Hall.

© The Author(s), under exclusive license to Springer Nature Switzerland AG 2022
J. M. Longuski et al., *Introduction to Orbital Perturbations*, Space Technology
Library 40, https://doi.org/10.1007/978-3-030-89758-1

Hoots, F. R. (2007). Radar-optical observation mix. AAS/AIAA Astrodynamics Specialist Conference, Paper 07-287, Mackinac Island, MI.

Kaplan, M. H. (1976). *Modern spacecraft dynamics & control*. New York: Wiley.

Kaula, W. M. (1966). *Theory of satellite geodesy*. Waltham: Blaisdell Publishing Company.

King-Hele, D. (1964). *Theory of satellite orbits in an atmosphere*. London: Butterworths.

Kovalevsky, J. (1967). *Introduction to celestial mechanics*. Netherlands: Springer.

Kovalevsky, J., & Sagnier, J. (1977). Motions of natural satellites. In J. A. Burns (Ed.) *Planetary satellites*. Tucson: University of Arizona Press.

Kozai, Y. (1959). The motion of a close earth satellite. *The Astronomical Journal, 64*(1274), 367–377.

Kwok, J. H. (1986). *The long-term orbit predictor (LOP)*. JPL Technical Report EM 312/86-151.

Liu, J. F. .F. & Alford, R. L. (1980). Semianalytic theory for a close-earth artificial satellite. *Journal of Guidance and Control, 3*(4).

Longuski, J. M. (1979). Analytic theory of orbit contraction and ballistic entry into planetary atmospheres. Ph.D. Thesis, The University of Michigan, Ann Arbor.

Longuski, J. M., & Vinh, N. X. (1980). Analytic theory of orbit contraction and ballistic entry into planetary atmospheres. Jet Propulsion Laboratory, California Institute of Technology, JPL Publication 80-58, Pasadena, California.

Longuski, J. M., Todd, R. E., & König, W. W. (1992). Survey of nongravitational forces and space environmental torques: applied to the Galileo. *Journal of Guidance, Control, and Dynamics, 15*(3), 545–553.

Longuski, J. M., Fischbach, E., & Scheeres, D. J. (2001). Deflection of spacecraft trajectories as a new test of general relativity. *Physical Review Letters, 86*(14), 2942–2945.

Longuski, J. M., Fischbach, E., Scheeres, D. J., Giampieri, G., & Park, R. (2004). Deflection of spacecraft trajectories as a new test of general relativity: determining the parameterized post-Newtonian parameters β and γ. *Physical Review D, 69*, 42001-1–42001-15.

McCuskey, S. W. (1963). *Introduction to celestial mechanics*. Reading: Addison-Wesley Publishing Company, Inc.

Meirovitch, L. (1970). *Methods of analytical dynamics*. New York: McGraw-Hill Book Company.

Morrison, J. A. (1966). Generalized method of averaging and the von Zeipel method. In R. Duncombe & V. Szebehely (Eds.) *Progress in astronautics and aeronautics—methods in astrodynamics and celestial mechanics* (Vol. 17). New York and London: Academic Press.

Moulton, F. R. (1914). *An introduction to celestial mechanics* (2nd ed.). New York: The Macmillan Company.

Pars, L. A. (1979). *A treatise on analytical dynamics*. Woodbridge: Ox Bow Press.

Plummer, H. C. (1918). *An introductory treatise on dynamical astronomy*. Cambridge: Cambridge University Press.

Prussing, J. E., & Conway, B. A. (2013). *Orbital mechanics* (2nd ed.). New York: Oxford University Press.

Rindler, W. (1969). *Essential relativity: special, general, and cosmological*. New York: Springer-Verlag.

Roy, A. E. (2005). *Orbital motion* (4th ed.). New York: Taylor & Francis Group.

Schaub, H., & Junkins, J. L. (2018). *Analytical mechanics of space systems* (4th ed.). Reston: American Institute of Aeronautics and Astronautics, Inc.

Smart, W. M. (1953). *Celestial mechanics*. New York: John Wiley & Sons, Inc.

Spier, G. W. (1971). Design and implementation of models for the double precision trajectory program (DPTRAJ). Jet Propulsion Laboratory, California Institute of Technology, Technical Memorandum 33–451, Pasadena, California.

Steinberg, S. (1984). Lie series and nonlinear ordinary differential equations. *Journal of Mathematical Analysis and Applications, 101*(1), 39–63.

Szebehely, V. G. (1989). *Adventures in celestial mechanics, a first course in the theory of orbits*. Austin: University of Texas Press.

Taff, L. G. (1985). *Celestial mechanics, a computational guide for the practitioner*. New York: John Wiley & Sons.

Tapley, B. D., Schutz, B. E., & Born, G. H. (2004). *Statistical orbit determination*. Burlington: Elsevier Academic Press.

Tragesser, S. G., & Longuski, J. M. (1999). Modeling issues concerning motion of the Saturnian satellites. *Journal of the Astronautical Sciences, 47*(3 and 4), 275–294.

Vallado, D. A. (2013). *Fundamentals of astrodynamics and applications* (4th ed.). El Segundo: Microcosm Press.

Vinh, N. X., Busemann, A., & Culp, R. D. (1980). *Hypersonic and planetary flight mechanics*. Ann Arbor: The University of Michigan Press.

von Zeipel, H. (1916). Recherches sur le mouvement des petites planètes. *Arkiv för Matematik, Astronomi och Fysik, 11*(1).

Weinberg, S. (1972). *Gravitation and cosmology: principles and applications of the general theory of relativity*. New York: Wiley.

Wiesel, W. (2003). *Modern astrodynamics*. Beavercreek: Aphelion Press.

Suggested Student Projects in Orbital Perturbations

Project Guidelines

The final project should have significant mathematical content, but it should be fun also. The types of topics allowed and encouraged include biographies, tutorials, history, reviews of journal articles or book chapters, simulations, and mathematical analyses.

The length of the report may be (roughly) 20 double-spaced pages at about 12-point font; or approximately 5000 to 10,000 words (including equations and equivalent space for figures and tables).

This guideline is merely a suggestion, as there is no specific required length for the course final report. Shorter or longer reports may be perfectly acceptable.

To encourage and inspire the reader, we have compiled a list of topics and titles that students have enjoyed working on over the years. These topics and titles are not to be interpreted as specific assignments, but rather as ideas to stimulate creativity and imagination. Without further ado, here is our list:

Harmonics
- Derivation of Tesseral Harmonics in an Earth Gravity Model and Resonance Effects
- The Effect of Spherical Harmonics on Low Earth Orbit
- Effect of Zonal Harmonics during Aerobraking at Jupiter
- Study on Orbits as Perturbed by Zonal Harmonics of the Moon

n-Body Problem
- The Three-body Problem and the Lagrange Points
- Derivation of the Disturbing Function for the Restricted 4-Body Problem

General Relativity
- The Effects of General Relativity on the Advance of Mercury's Perihelion

© The Author(s), under exclusive license to Springer Nature Switzerland AG 2022
J. M. Longuski et al., *Introduction to Orbital Perturbations*, Space Technology
Library 40, https://doi.org/10.1007/978-3-030-89758-1

- The Motion of the Perihelion of Mercury due to General Relativity using Post-Newtonian Expansions
- Mercury, Vulcan, and the History of General Relativity
- Analysis of an Extended GPS Propagation Ephemerides

Solar Radiation
- The Effect of Solar Radiation Pressure on Orbit Perturbation
- Review of Orbital Perturbations Due to Radiation Pressure For Spacecraft of Complex Shape
- Solar Radiation Pressure Modeling Issues for High Altitude Satellites
- Sun-Synchronous Orbit Using Solar Sails for Geomagnetic Tail Exploration
- Non-Gravitational Perturbations Due to Atmospheric Drag and Solar and Earth Radiation

The Effect of Drag
- Numerical Simulations and Analytical Predictions of Satellite Orbit Contraction due to Atmospheric Drag
- Atmospheric Drag and Orbit Decay: A Comparison of Analytical and Numerical Solutions
- The Poynting–Robertson Effect

Lunar Theory
- A History of Lunar Theory: The Theories of Ptolemy, Brahe, and Brown
- Perturbing Function for Lunar Motion
- Lunar Frozen Orbits for South Pole Coverage
- The Historical Development of the Perturbing Effects of the Sun: A Lunar Theory
- Tidal Acceleration of the Moon
- Orbit Perturbations: The Earth–Moon System
- An Investigation of Sun, Earth, Moon L2 Perturbations

Tidal Effects
- Tidal Acceleration of the Moon
- Motion of an Orbiter under Tidal Perturbation

Equinoctial Orbit Elements
- Equinoctial Coordinates for General Perturbations
- Numerical Comparison of Lagrange's Planetary Equations to Equinoctial Orbit Elements

The Lorentz Force
- Development of Variational Equations for the Lorentz Force
- Inclination Change in LEO via the Geomagnetic Lorentz Force
- Electric and Magnetic Forces and Their Impact On a Satellite's Motion

Earth Orbit
- The Effects of the J_2
- Perturbation Incorporation of Perturbations into Equations of Motion for the Orbit Dynamics of Low Earth Orbiting Spacecraft in Formation

- Relative Perturbed Satellite Motion in Earth Orbit
- The Gravity Recovery and Climate Experiment (GRACE)

Uranus
- Inclination Change at Uranus with a Lifting Body
- Changes in the Inclination of Uranian Moons after Shifting in Uranus's Tilt

Gauss's Equations
- Derivation of the Alternate form of Gauss's and Lagrange's Equations
- Gauss's Form for Continuous Thrust Maneuvers

Perturbations
- General Perturbation Techniques
- A General Perturbations Approach to Relative Orbital Elements
- Perturbations on a Europa Orbiter
- Exploration of Orbital Perturbations using Mathematica
- Impact of First-Order Perturbations on the Probability of Collision
- Pan and Daphnis: Perturbations of Saturn's Inner Moons
- Comparison of General and Special Perturbations for an Oblate Spheroid

Miscellaneous
- Review of Orbital Motion by A.E. Roy
- A Review of the Works of Dr. Sosnitskii
- On the Discovery of Neptune
- Analytical Solutions for the Motion of the Galilean Satellites
- The Seasonal Yarkovsky Effect on Stony Asteroids
- Using Resonant Pumping to Increase the Altitude of a Tethered Satellite's Orbit Around an Oblate Planet
- The Detection of Exoplanets
- Gravimetry in Space
- Computational Modeling of the Dynamics of Circumbinary Planets
- The Giants: Copernicus, Brahe, Kepler, Newton, and Halley
- Brief Survey of MMR (Mean-Motion-Resonance)
- An Exploration of the Inertial Frame
- Investigation of the Definition and Application of the Sphere of Influence

Suggested Homework in Orbital Perturbations

In writing this text we had to address the problem of homework assignments. How should we do (or could we do) homework? The analysis of even the simplest problems in orbital perturbations can run several pages in length—in which terms in series expansions can rapidly pile up as terms of the first-, second-, third-, and fourth-order mushroom into larger and larger expressions. The derivation of Lagrange's planetary equations consumes over 50 pages of text!

Clearly the answer (to the homework problems) will not "be found at the back of the book." What we found in a classroom setting is that students should bravely march through at least some of these monstrous expressions. Lagrange's planetary equations must be proven.

Our solution is to have the students write a journal that verifies the results of the text (and lecture), while skipping some of the intermediate steps and truncating many of the expansions. Sometimes a second-order or even first-order solution may suffice.

It turns out that the prepared students (typically advanced graduate students) actually welcome the challenge and take pride in developing the analytical theories presented in the text. They find it to be fun!

Guidelines for Homework

Homework is due every one to two weeks. In the journal, students verify the results of the lecture and text without writing out all the details of the derivations—they are too long. In addition to the text material, students develop related material of their own interest. Outside topics may include: analytical work, numerical solutions, tutorials, book reviews, paper reviews, or historical studies (provided they contain significant mathematical work). A brief report on outside topics should suffice.

© The Author(s), under exclusive license to Springer Nature Switzerland AG 2022   333
J. M. Longuski et al., *Introduction to Orbital Perturbations*, Space Technology
Library 40, https://doi.org/10.1007/978-3-030-89758-1

Index

A

Abramowitz, M., 285, 288, 290, 295
Advance in perihelion of Mercury, 270
Alford, R.L., 181, 184, 201
Analysis of evection, 252–258
Angular momentum, 10, 19–20, 24–27, 229, 234, 235, 272
Anomaly, eccentric, 34–38, 68, 174, 175, 192, 279, 280
Anomaly, mean, 34–38, 61, 71, 72, 104–105, 110–111, 131, 132, 174, 175, 182, 188, 191, 210, 214, 215, 242, 246, 262, 268, 280, 282, 307
Anomaly, true, 25, 26, 32, 34, 36, 37, 61, 107, 154, 174, 175, 192, 217, 229, 232, 245, 256, 263, 274, 279
Aphelion, 27
Apoapsis, 27, 298, 299, 301
Apogee, 27, 214
Apse, 229, 232, 234, 242
Areal velocity, 23, 24
Area of the ellipse, 28
Area swept out, 28
Argument of periapsis, 25, 32, 33, 78, 107, 114, 119, 202–206, 210, 211, 280, 282
Argument of perigee, 213, 214, 217, 259
Argument of perihelion, 266, 268
Ascending node, 33, 182, 212, 214, 217
Ascending node angle, 33, 78, 113
Asteroid Perturbation by Jupiter, 47
Asymptotic true anomaly, 26, 274
Atmospheric drag, effects of, v, 121, 277–280, 282, 303
Averaged equations for orbit decay, 285–292
Averaging, generalized method, 173–215

B

Battin, R.H., 59, 96, 104
Bessel functions, 285, 287, 289, 290, 295
Brouwer, D., 258
Brown, E.W., 243, 252, 259
Busemann, A., 281

C

Cassini division in Saturn's rings, 47
Center of mass, 4–6, 8, 43, 143, 146–147, 152–154
Chao, C.-C., 165, 181, 243
Chobotov, V.A., 258, 259
Clemence, G.M., 258
Concept of averaging, 173–181
Conic equation, 26, 29, 232, 238, 241, 274
Conic section, 26, 271
Connecting models to the real world, 303–304
Conservation of energy, 28–33, 271
Conservative force, 28, 97, 181
Constants of integration, 8, 10, 12, 19, 24–26, 28, 31, 185, 190, 192, 196, 264
Conway, B.A., 33, 36
Covariance, 306, 316–323
Critical inclination, 214
Cross-sectional area, 277
Culp, R.D., 281

D

Defective method, 218–222
Deflection of light, 261, 271, 275, 276
Dependent and independent variables, 60–61, 65

© The Author(s), under exclusive license to Springer Nature Switzerland AG 2022
J. M. Longuski et al., *Introduction to Orbital Perturbations*, Space Technology
Library 40, https://doi.org/10.1007/978-3-030-89758-1

Printed in the United States
by Baker & Taylor Publisher Services